AF528556

Root Vegetables

THE AUTHORS

Dr. Sanyat Misra is presently Head, Department of Horticulture, Birsa Agricultural University, Kanke, Ranchi-834006. He is serving that University for the last 19 years, apart from serving NBPGR, New Delhi as Research Associate for three years. He did his Ph.D.in 2002 from Dr. Y.S. Parmar University of Horticulture & Forestry, Nauni, Solan (H.P.), M.Sc. from GPB University of Agriculture & Technology, Pantnagar in 1998, and B.Sc. Forestry from UAS, Bangalore in 1995. So far he has contributed about 45 research papers on various aspects of horticulture, guided 7 M.Sc.students and authored 10 books including the three voluminous books (Cut Flowers, Traditional & Loose Flowers and Commonly Used Ornamental Plants) published in 2017 from Kruger Brentt, Middlesex, England. In the University he is teaching almost all the horticultural courses and is involved in all the research and developmental activities of the Department.

Mr. Sanjay Mishra based in Lucknow is a **Computer Specialist** so it is his brain which has brought out this book to this stage. He has authored two books on Computer Science: (i) **Bits & Bytes of C++**, and (ii) **CCC-Course on Computer Concepts;** and (iii) a most popular book **Soilless Crop Production** which has gained its popularity throughout the country.

Dr. R.L. Misra served IARI, New Delhi for **40 years** as Scientist and retired in May 2008 as **Project Coordinator (Floriculture)**, and then further served IARI, New Delhi as **Adjunct Faculty** for 5 years in the Division of Floriculture & Landscaping,. He had been an active researcher throughout ~ serving **Shimla**, **Kullu** and **New Delhi** and to his credit he has more than 200 research papers, taught all the floricultural courses in the Division, **guided 6 M.Sc. students and 16 Ph.D. scholars**, apart from being Members of the Advisory Committee of many students. Single-handedly, he **bred 30 gladiolus varieties;** and brought out ***Journal of Ornamental Horticulture*** continuously for 12 years starting from 1993 to 2004. He is a **Life Member** in 12 professional societies. He was bestowed with **Urivi-Vikram Charitable Trust Award (2003)**; and is **Honorary Fellow of 4 Societies**, *viz.* **ISGPB**, New Delhi **(1980); ISOH**, New Delhi **(2009); HSI**, New Delhi **(2018);** and **Society for Promotion of Horticulturre**, IIHR, Bengaluru **(2022)**. He was bestowed with **HSI Gold Medal (2013)**, and **ISOH Lifetime Achievement Award in Floriculture (2018)**.

Dr. Pragya Ranjan is an agricultural graduate from BHU, Varanasi; and M. Sc and Ph.D in Horticulture from IARI, New Delhi. Dr Pragya joined ICAR through ARS in 2005 and presently working as a Principal Scientist in Germplasm Exchange & Policy Unit, National Bureau of Plant Genetic Resources, New Delhi. She has more than 100 publications to her credit including 40 research papers, 30 popular articles, 25 book chapters, 3 books and 12 technical articles.

Dr. J. K. Ranjan is a graduate {B. Sc. (Ag)} from BHU, Varanasi; and Postgraduate cum Doctorate in Horticulture from IARI, New Delhi. He joined ICAR through ARS in 2005 and presently working at ICAR-IARI, New Delhi in the Division of Vegetable Science as Principal Scientist. As a visiting scientist he had been to Clemson University, South Carolina (USA) during 2012. He has published 42 research papers in peer reviewed and international journals, 5 books and 105 other technical publications.

Root Vegetables

–Authors–

Sanyat Misra
Sanjay Mishra
R. L. Misra
J.K. Ranjan
Pragya Ranjan

2024

Daya Publishing House®
A Division of
Astral International Pvt. Ltd.
New Delhi – 110 002

ISBN 978-93-5461-986-1

Published by : **Daya Publishing House®**
A Division of
Astral International Pvt. Ltd.
– ISO 9001:2015 Certified Company –
4736/23, Ansari Road, Darya Ganj
New Delhi-110 002
Ph. 011-43549197, 23278134
e-mail: info@astralint.com
Website: www.astralint.com

Printed at : **Replika Press Pvt Ltd**

Preface

While writing the book 'Edible Tubers', the authors encountered certain hurdles as 'tubers' are entirely different when compared with 'corms', 'rhizomes' and 'tuberous roots' but even then various authoritative authors have taken these as one entity 'Tuber Crops' because out of all these categories if all are described separately only a few crops will fall in one category so they have put all these under one cover 'Tuber Crops' to have a quite comprehensive dealing about the subject. So is the case even with 'Root Crops' because most of the literature sources mention only about the major edible taproot crops such as carrot, beetroot, radish, turnip and rutabaga. Moreover, if one goes beyond these, he/she finds even horseradish, parsnip, salsify and yam beans but there are certain others too such as kudzu, winged bean, zombie pea, skirret, tatarican colewort, bitter root and chervil which though are lesser known but in certain parts of the continent these are used as famine or emergency foods in various forms though not only so and if popularized these may also become major root crops as to those of first types. Apart from these there may be a few more ones which also require to be explored.

One of the authors found a 'mobile vegetable vendor' in Ranchi as well as in Bhubaneswar selling one such roots as misrikand as he told which we found as 'jicama' (yam bean) roots. Moreover, there is one known locally in Hindi as 'gazarpunna ki jad'/'Kandmool'/'Ram Kandmool' which is an instant food being sold raw in pieces occasionally in certain 'Fairs' which is a giant-sized cylindrical root and is said to be found in Rajasthan, and also we concluded that this may be the root of a particular species of Agave. However, we found that it is *Agave sisalana.*

While watching 'National Geographic Channel' as well as 'Discovery' on T.V., we encounter that one quite large cylindrical root is dug up by explorers, a part is eaten but rest is again buried in the soil for further growth. Also we encounter the edible clustered roots of a plant which explorers collect as instant raw food. But somehow we could not manage finding their names even in very authentic literature sources. So we could not include these in this book. Otherwise the book on 'Root Vegetables' is complete and we expect that this will be liked all over the world by horticulturists, olericulturists, extension horticulturists, professors in horticultural education, the vegetable scientists, and the students of B.Sc. (Ag.), B.Sc. (Hort.) Hon., M.Sc. (Hort.), Ph.D. (Hort.) as well as farmers (vegetable growers). Since we have scanned the entire possible literature, so we expect it to be the first of its kind ~ a very comprehensive treatise dealing exclusively with 'root vegetables'.

Sanyat Misra
Sanjay Mishra
R.L. Misra
J.K. Ranjan
Pragya Ranjan

Contents

Introduction

Root vegetable is a type of edible plant where the parts of the plant that is consumed grow underground. These vegetables store energy and nutrients in their roots which make them valuable food source. Root vegetables come in various shape, size and flavours, and are used worldwide either row or after cooking.

Root vegetables

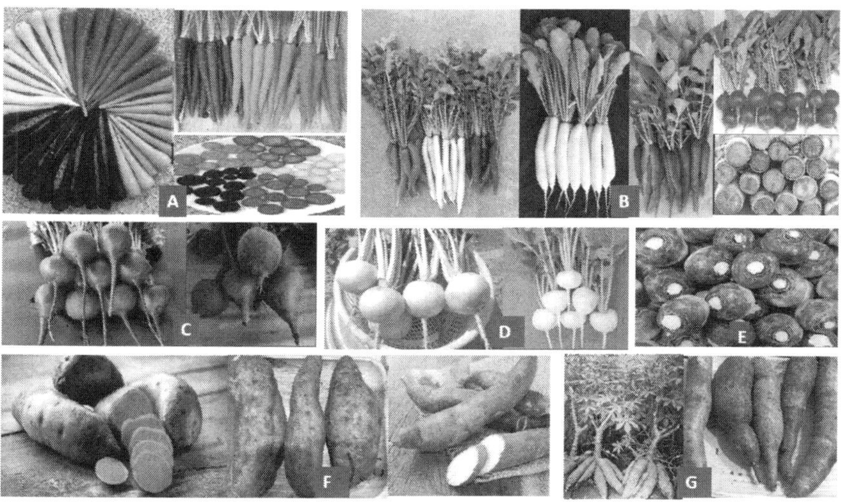

A: Carrots (*Daucas carota*); B: Radish (*Raphanus sativus*); C: Beetroot (*Beta vulgaris*); D: Turnip (*Brassica rapa*), E: Rutabaga (*Brassica napus*); F: Sweet potato (*Ipomoea batatas*); G: Cassava (*Manihot esculenta*)

Sources: ICAR-IIVR, Varanasi, India, ICAR-IARI, New Delhi, India, North Carolina State University, USA

Though in agricultural and culinary usage the term 'root vegetable' is generally applied for the geophytes eaten by humans as food. Botany distinguishes true roots such as taproots (*e.g.* radishes and carrots) and tuberous roots (*e.g.* sweet potatoes) from non-roots such as bulbs (*e.g.* onion and garlic), corms (*e.g.* *Sagittaria* spp. or arrowhead), rhizomes (*e.g.* arrowroot), and tubers (*e.g.* potato); however, some contain both hypocotyl and taproot tissue (Camelo and Andres, 2004). This book focuses mainly on taproot vegetables. For other type of root vegetables, refer to 'Tuber Crops' and 'Edible Alliums' (Misra *et al.*, 2024).

Taproots can be swollen and colourful or can remain quite drab but tasty. They are rich in nutrients, low in fat and calories, and have wildly varying characteristics such as radishes are pungent, carrots are sweet, beetroots have earthy flavour and others like parsnips and rutabaga have more subtle flavour. Moreover, they are inexpensive and usually available throughout the year. Examples of edible taproots include radishes, carrots, turnips, beets, *etc.*

Bioactive Components of Root Vegetables

Root vegetables such as carrot, sweet potato, turnip, radish, rutabaga, beetroot, *etc.* are a good source of health-promoting phytoconstituents including phenolic acids, flavonoids, essential oils, proteins and bioactive pigments. These bioactive compounds have wide range of pharmacological activities, including anti-hepatotoxicity, anti-hyperlipidemia, anti-inflammatory, anti-hypertension, anti-depressant and anti-hypoglycemia (Bashir *et al.*, 2023).

Phenolic Compounds

Root vegetables are a major source of phenolic compounds, and the phenolic content of some root vegetables is reported to be higher than non-root vegetables (Farahnaky *et al.*, 2018).

Vegetable	*Bioactive compound*	*Bioactivities*	*Reference*
Radish	Quercetin, ferulic acid, caffeic acid	Antioxidation	Li *et al.*, 2018; Chihoub *et al.*, 2019
Turnip	Isorhamnetin-3,7-di-O-glucoside, kaempferol-3-O-(feruloyl) sophoroside-7-O-glycoside, sinapic acid	Antioxidation	Chihoub *et al.*, 2019; Francisco *et al.*, 2009
Rutabaga	3-caffeoylquinic acid, 5-caffeoylquinic acid, 5-feruloylquinic acid, 4-*p*-coumaroylquinic acid	Antioxidant	Jakopic *et al.*, 2021; Stefanucci *et al.*, 2020

Vegetable	Bioactive compound	Bioactivities	Reference
Carrot	5′-caffeoylquinic acid, cis-5′-caffeoylquinic acid	Antioxidation, anti-inflammatory	Zhang and Hamauzu, 2004; Metzger and Barnes, 2009
Beetroot	Gallic acid, naringenin, myricertin, catechol	Antioxidation, anti-inflammatory, anti-hypertensive	Tumbas Saponjac *et al.*, 2020; El-Beltagi *et al.*, 2018; Shepherd *et al.*, 2019; Siervo *et al.*, 2020
Parsnip	Qucercetin 3,7-O-diglucoside, quercetin 3-O-rutinoside, 5-O-caffeoylshikimic acid	Antioxidation	Nikolic *et al.*, 2014

Radish has a high phenolic content among root vegetables having 1315.83μg gallic acid equivalent/g of extract, even higher than potatoes and kohlrabi (Farahnaky *et al.*, 2018). The phenolic compounds may vary in different parts of the plants (Goyeneche *et al.*, 2015). Carrot peels are reported to have more phenolic acids than vascular tissues. 5′-Caffeoylquinic acid is a major compound in different carrot varieties with the highest content value in peels (15.04mg/100g FW), followed by cis-5′-chlorogenic isomer (3.51mg/100g FW) of Caffeoylquinic acid (Zhang and Hamauzu, 2004). In another study the rutabaga sprouts were found with higher phenolic content (reaching up to 125.7mg GAE/g DW) as compared to seeds (6.9mg GAE/g DW) and roots (5.1mg GA/g DW) (Pasko *et al.*, 2013). Phenolic accumulation also differs depending upon the varieties as a study has indicated higher phenolic content in black radish (13.7mg GAE/g DW) compared to white and red varieties (Bors *et al.*, 2015). Higher phenolic content of plants correlates with antioxidant activity such as parsnip ~ a carrot-related root vegetable rich with phenolic glycosides that positively correlate with antioxidant activity (Nikolic *et al.*, 2014). Beetroot with high phenolic content of 56.65mg GAE/100ml of juice, encapsulated as soybean proteins reaching up to 150mg GAE/100ml (Tumbas Saponjac *et al.*, 2020). Turnip is reported to contain flavonoids and hydroxycinnamic derivatives. HPLC analysis of turnip tops has identified flavonoid glycosides and total phenolic content reaching up to 191.39mg/100g fresh weight (Romani *et al.*, 2006). Because turnip is usually consumed in processed form, phenolic compounds are negatively impacted upon applying heating procedures for processing. Franciso *et al.* (2010) have studied the impact of different heat treatments on the turnip. Flavonoid content decreased to 4.58 μmol/g dry weight upon conventional boil initially at 13.85μmol/g DW in raw samples. Likewise, high-pressure cooking also deteriorated flavonoids (5.00μmol/g DW).

A study has compared different techniques for black carrot extraction, reporting optimized microwave-assisted extraction (power = 348.06, time = 9.86 min, solvent: solid ratio = 19.3 g/mL, solvent = 19.8% ethanol) to be best suited for phenolic recovery with the content value of 264.9mg GAE/100ml while conventional extraction to be least suited with approximately 2.5 times less phenolic yield than former technique (Kumar *et al.*, 2019).

Vegetables stored at room temperature lose sensory and composition quality due to the action of polyphenol oxidase and peroxidase enzymes. A study has recommended 'refrigerate storing' to prolong the shelf life of vegetables (de Brito, *et al.*, 2017). Peels of vegetables are often discarded during food processing; however, incorporation of peels in food can be beneficial, as a study carried out on rutabaga has indicated higher phenolic content in peels (18.14mg/GAE/g) as compared to the pulp (11.57mg GAE/g) using ultrasound-assisted extraction (Stefanucci *et al.*, 2020).

Glucosinolates

Glucosinolates are sulphur and nitrogen-containing compounds that predominantly occur in brassica vegetables. Their presence in some of the root vegetables of the Brassicaceae family are shown in the following table.

Plant		Glucosinolate	Bioactivity	Reference
Radish	White	Glucorasphatin, glucoraphenin, glucobrassicanapin	Detoxification enzyme induction	Hanlon *et al.*, 2007; Ediage *et al.*, 2011; Yi *et al.*, 2016
	Black	Glucoraphenin, glucosisymbrin, glucosisaustricin		
Turnip		Progoitrin, glucoalyssin, gluconasturtiin,4-hydroxyglucobrassicin, glyconapin	Bone formation	Jeong *et al.*, 2015
Rutabaga		Progoitrin, sinigrin, glucoerrucin	Antimicrobial	Pasko *et al.*, 2018, Dastan *et al.*, 2011
Mustard roots		2-propenyl glucosinolate (Sinigrin), 2-hydroxy-2-phenylethyl glucosinolate (glucobarbarin),	Nematicidal	Gimsing and Kirkegaard, 2006; Oliveira *et al.*, 2011

Depending on their structure, these compounds are further divided into three subclasses, aliphatic, aromatic, and indole glucosinolates (Li *et al.*, 2007). Glucosinolates, well-known cancer-preventing bioactive compounds, and isothiocyanates provide characteristic flavours to brassica vegetables (Zhang *et al.*, 2008). Turnip is an excellent source of glucosinolates with 100–130 mg /100 g FW total glucosinolate content (Chen *et al.*, 2008). When compared with other Chinese leafy vegetables (cabbage, pakchoi, cai-Tai, choy-sum) of the brassica family, the turnip has shown the highest glucosinolate content, with gluconapin being the highest accumulated glucosinolate (65.84 mg/100 g FW) (Chen *et al.*, 2008). UPLC/MS analysis has indicated progoitrin as another high accumulated glucosinolate in turnip (Jeong *et al.*, 2015). Rutabaga is a hybrid plant of cabbage and turnip with a considerable amount of glucosinolates (7.34 μmol/g DW), with progoitrin as major glucosinolate; however, total glucosinolate content is reduced upon cooking(6.86 μmol/g DW) (Kapusta-Duch *et al.*, 2016). Boiling and high-pressure cooking can reduce glucosinolate content by three times (Francisco *et al.*, 2010).

Glucosinolate content of radish varies depending upon the radish cultivar; however, glucoraphanin is a major constituent of the glucosinolate profile of all radish cultivars. Red radish has reported glucosinolate content of 163.91 mg sinigrin equivalent/100 g FW (Jing *et al.*, 2012). 4-Methylthio-3-butenyl glucosinolate is prominent in white radish which is converted into 4-methylthio-3-butenyl isothiocyanate by the action of myrosinase enzyme. The resulting metabolite is responsible for the specific pungent flavour of radish. 4-Methylthio-3-butenyl glucosinolate comprises 90% of the total glucosinolate profile in Japanese varieties (Ishida *et al.*, 2015). Spanish black radish also has identified radish-specified glucosinolates, which induce detoxification enzymes in HepG2 cell lines. However, the glucoraphanin (4-methylthio-3-butenyl glucosinolate) ~ a major glucosinolate of radish is not preserved in black radish-based dietary supplements (Hanlon *et al.*, 2007; Ediage *et al.*, 2011). Tissue damage during cutting, chewing, cooking and fermentation leads to glucosinolate degradation into isothiocyanates, thiocyanates and cyanides. The major degradation product of radish glucosinolates other than erucin is 5-(methylthio)-4-pentenenitrile (Blazevic and Mastelic, 2009). Mustard is also reported for high glucosinolate levels and isothiocyanate byproducts. A study on mustard seeds has indicated higher levels of aliphatic glucosidase in all mustard varieties with total glucosidase content ranging from 51.96 to 64.36μmol/g FW in root mustard (Zhang *et al.*, 2022). Glucosinolates are highly sensitive to extraction techniques and pre-extraction treatments. A study conducted on brassica vegetables has indicated cold methanolic extraction with conventional wet tissue freeing is suitable for glucosinolate recovery compared to hot methanolic extraction. Also, freeze-drying was found to be avoidable for short time storage (Doheny-Adams *et al.*, 2017).

Natural Colour Pigments

Root vegetables possess natural-coloured pigments that give them their characteristic colours. These pigments cover a wide spectrum of colours from red to purple depending on their structure (Wiczkowski *et al.*, 2013).

Anthocyanins

Plants	Compounds	Colour	Ref.
Radish	Pelargonidin-3-caffeoyl-duglycoside-5-glucoside, Pelargonidin-3-feruloyl-duglycoside-5-glucoside, Pelargonidin-3-*p*-coumaroyl-duglycoside-5-glucoside	Red	Jing *et al.*, 2012
	Cyanidin 3- caffeoyl sophoroside-5-glucoside, cyanidin 3-feruloyl sophoroside-5-diglucoside, cyanidin 3-(glucosyl-p coumaroyl)sinapolysophoroside-5-Malonylglucoside etc.	Purple	Lin *et al.*, 2011
Turnip	Peonidin O-hexoside, Rosindin-O-hexoside, Delphinidin-O-malonylhexoside, cyaniding-3-O-glucoside, malvidin-3,5-diglucoside	Purple	Zhuang *et al.*, 2019
Carrot	Cyanidin 3-xylosyl-galactoside, cyaniding 3-xylosyl-feruloylglucosylgalactoside, cyaniding 3-xylosyl-p-coumaroylglucosylgalactoside	Black	Polat *et al.*, 2022
	Ferulic acid cyanidin 3-xylosyl-glucosylgalactoside, *p*-coumaric acid cyanidin 3-xylosyl-glusosylgalactose, cyanidin 3-xylosylglucosyl-galactoside	Purple	Macura *et al.*, 2019

Red radish contains anthocyanins that have similar properties as synthetic Red No. 40 and is stable to be used commercially (Otsuki *et al.*, 2002). The extraction and post-extraction concentration procedure affects the recovery of these natural colorants.

Hydroethanolic and acidified water extraction system with membrane perstraction concentration is suitable for high anthocyanin recovery which led to 62.58 mg/100 ml anthocyanin recovery from red radish (Patil *et al.*, 2009). Hydrogen-rich water is reported to influence anthocyanin accumulation in radish cv. 'Yanghua' (Su *et al.*, 2014).

Calcium chloride treatment promotes the accumulation of anthocyanins (reaching a relative anthocyanin content value of 42.06, which was at 9.72 in control plants) (Zhang *et al.*, 2018). Purple radish also witnessed the presence of acylated cyaniding sophoroside glycosides and diglycosides (Lin *et al.*, 2011). Within the biological system, anthocyanins are generated under the influence of light in the presence of chalcone synthase enzymes. Light-responsive genes are expressed in turnip at different levels in anthocyanin biosynthesis (Zhou *et al.*, 2013).

Purple carrots are found to be a good source of anthocyanin with content value reaching up to 33,876mg/kg DM; however, air or freeze-drying of fresh carrots leads to anthocyanins loss (Macura *et al.*, 2019). A study on black carrots also showed a considerable amount of anthocyanins (Ersus and Yurdagel, 2007).

Carotenoids

Plants	Compounds	Colour	Ref.
Carrot	β-Carotene, α-carotene, lutein	Orange	Wang *et al.*, 2020
		purple	Wang *et al.*, 2020
	Lycopene, β-carotene, α-carotene, lutein	Red	Wang *et al.*, 2020
Radish	Lutein, β-carotene	White	Park *et al.*, 2020
	Lutein, β-carotene, neoxanthin, zeaxanthin	Purple	Mamatha *et al.*, 2011

Carotenoids represent a class of natural products with lipid-soluble vibrant colour pigments. These pigments are responsible for yellow to red spectrum of colours in plants and other organisms (Elvira-Torales *et al.*, 2019). Unlike flavonoid-based anthocyanins, carotenoids are tetraterpenoid molecules containing eight isoprenoid units that generate a C40 carbon skeleton (Jaber *et al.*, 2022). In green tissues, carotenoids are located in chloroplasts while in coloured vegetables, accumulation occurs in chromoplasts imparting different colours to the plant (Clotault *et al.*, 2008). Among root vegetables, carrot is the most prominent source of carotenoids but with variance of the content depending upon the cultivars. It is reported to be in the range of 58.15–102.02mg/kg FW with β-carotene constituting the major portion of total carotenoid content (Koca Bozalan and Karadeniz, 2011). Studies have indicated the immense impact of light on carotenoids (Antonious and Kasperbauer, 2002). In carrots, the biosynthetic pathway of carotenoids is influenced by light as it can induce repression of gene expression of β-carotene and α-carotene biosynthesis (Fuentes *et al.*, 2012). Parsnip, on the other hand, despite being a carrot-like root vegetable, only has minor amounts of carotenoids (Fratianni *et al.*, 2019). β-Carotene is also dominant in sweet potato varieties with orange flesh. The carbon chain of the majority of carotenoids contains trans double bonds. Turnip leaves are also reported to possess carotenoids with content values reaching up to 250µg/g (Iqbal *et al.*, 2013); however, a study on white radish leaves has shown higher carotenoid content (486.95µg/g DW) as compared to turnip leaves. Moreover, despite being a white-fleshed variety, a minute amount of carotenoids is also identified in radish roots (Park *et al.*, 2020).

Betalains

Plants	Compounds	Colour	Ref.
Beetroot	Betanin, isobetanin, 2,17′-bidecarboxy-neobetanin, miraxanthin II, vulgaxanthin I	Deep red	Sawicki *et al.*, 2016
	Vulgaxanthin I, indicaxanthin, miraxanthin	Yellow	Stintzing *et al.*, 2002

Betalains: Betalain pigments are betalamic acid derivatives divided into two subclasses: betacyanin (red-violet pigments) and betaxanthin (yellow pigments), depending upon cyclo DOPA and amine condensation on betalamic acid, respectively (Sawicki *et al.*, 2016). Black radish peel extract appears yellowish and is reported to contain 22.5mg/100g DW of betaxanthin while only 7.7mg/100g DW of betacyanin pigments (Yucetepe *et al.*, 2021). Beetroot is the richest vegetable source of betalains (17.24mg/g DW), most prominently betacyanins responsible for its bright red colour. A study has indicated approximately three times higher betacyanin content than betaxanthins in peels and other root sections of beetroot (Sawicki *et al.*, 2016). Beetroot peels are also betalain rich. An optimized study on betalain extraction from beetroot peels has identified 1.5% citric acid, 50% ethanol, and 52.52°C temperature and 49.9 min extraction time to be best suited for betalain recovery with the content value of 1.20mg/g DW (Lazar *et al.*, 2021).

Vitamins

Vitamins are necessary for normal growth, and their deficiency can cause health complications. Following table lists major root vegetable sources of vitamins and health conditions related to deficiency.

Vitamin/ provitamins	*Plant source*	*Recommended intake/day*	*Deficiency*	*Refs.*
Vitamin A/ Provitamin A (β-carotene)	Carrot	700 µg	Night blindness, xerophthalmia	Coe *et al.*, 2021; Tang *et al.*, 2005; Cordeiro *et al.*, 2018
Vitamin B (folates)	Beetroot	400 µg dietary folate equivalent	Anaemia	Jastrebova *et al.*, 2003; Bailey, 2004; Odewole *et al.*, 2013
Vitamin C (ascorbic acid)	Turnip, Radish	75–90 mg	Skin-related complications, low wound healing	Azam *et al.*, 2013; Levine *et al.*, 2001; Dattola *et al.*, 2020
Vitamin E (tocopherols)	Radish, carrot	0.67–15 mg Tocopherol equivalent	Weak immune system	Liu *et al.*, 2021; Nicolle *et al.*, 2004; Lewis *et al.*, 2019; Troesch *et al.*, 2012

Radish is a major source of vitamin C as compared to turnips and carrots; however slight change in vitamin C accumulation is seen in plants grown in high CO2 conditions (Azam *et al.*, 2013). Turnip greens and turnip tops are good sources of vitamin C; however, the heat processing leads to a significant loss of vitamin C. Steaming can result in approximately 60% vitamin C loss, while boiling and high-pressure cooking can completely diminish vitamin C content from turnip (Francisco *et al.*, 2010). Other than heat treatment, the vitamin C content is also sensitive to long-term storage. A study on carrot cultivars reported 88–132mg/kg vitamin C content in fresh carrots which decreased by 58% (on average) upon 30 days of storage (Matejkova and Petríkova, 2010). Vitamin E refers to a group of lipid-soluble antioxidants that play an essential role in cell signal regulation and proliferation (Luby *et al.*, 2014). Among different types of vitamin K, α and γ tocopherol are the most prominent types (Liu *et al.*, 2021). Carrots serve as an excellent source of vitamin E; however, orange and purple carrot varieties have higher vitamin E content than white carrots (Nicolle *et al.*, 2004). Green leafy plant sections contain higher vitamin E content than roots as determined in radish leaves containing up to 48.5μg/g DW of vitamin E with α-tocopherol as major vitamin E source. At the same time, roots contain only up to 0.17μg/g DW with γ-tocopherol in the highest quantity; however, vitamin C and E are also sensitive to post-harvest storage. Vitamin K is also found in root vegetables but is predominant in green leafy portions. Phylloquinone, commonly known as vitamin K1, is the predominant form of vitamin K and is reported in considerable amounts in radish leaves (Liu *et al.*, 2021). As discussed in the previous section, carrots are rich in carotenoids among which α-carotene and β-carotene are major pro-vitamin A compounds that are converted into vitamin A (retinol) in a biological system. According to the US National academy of science, 24μg of pro-vitamin, A carotenoid is equivalent to 1μg of vitamin A (retinol) (Coe *et al.*, 2021; Tang *et al.*, 2005).

Other Bioactive Compounds

Phytosterols are plant-based steroid molecules with cardioprotective and anti-tumour activities. Seeds are a rich source of bioactive phytosterols such as radish seeds with good phytosterol profiles mainly consisting of brassicasterol, campesterol and sitosterol (Bellani *et al.*, 2012); however, considerable quantities of phytosterols are also reported in edible roots. A study has shown 366.16mg/100g phytosterol content in carrots with campesterol and sitosterol as major constituents (Aremu *et al.*, 2021). Radish sprouts also have shown campesterol (947μg/g) and β-sitosterol (899μg/g) in considerable amounts (Lee *et al.*, 2018). Beetroot is a rich source of triterpenoid saponins. Triterpenoid aglycones of saponins include oleanolic acid, hederagenin, akebonoic acid and glycogenin which were linked with hexose, uronic, deoxyhexose and pentose sugar to generate terpenoid saponins (Mikołajczyk-Bator *et al.*, 2016). Volatile fractions of plants contain useful alcohols, terpenes and hydrocarbons among which terpenes are most important due to their significant role in providing aroma and flavour to vegetables. The accumulation of terpenes in root vegetables varies depending upon the colour variation. Orange carrots have a high accumulation of β-caryophyllene, α-humulene and bornyl acetate while the yellow variety contains β-bisabolene and γ-bisabolene in higher amounts (Muchlinski *et al.*, 2020). Likewise, hydrodistillation of radish has also identified phytol as a major terpene with the highest abundance (69.7% of total compounds identified) in the white variety, followed by neophytadiene (1.5%) in the black variety and β-damascone (1.4%) in the red variety (Blazevic and Mastelic, 2009).

Abbreviations

DW = Dry weight, FW = Fresh weight, GAE = Gallic acid equivalent, HPLC-DAD = High performance liquid chromatography-Diode array detection

References

Antonious, G.F. and M.J. Kasperbauer, 2002. Color of light reflected to leaves modifies nutrient content of carrot roots. *Crop Sci.*, **42**(4):1211-6.

Aremu, M., P.L. Ajine, M.O. Omosebi, N.M. Baba, J.C. Onwuka and S.S. Audu, 2021. Lipid profiles and health promoting uses of carrot (*Daucus carota* L.) and cucumber (*Cucumis sativus* L.). *Intern. J. Sci.*, **10**(07): 22-9.

Azam, A., I. Khan, A. Mahmood and A. Hameed, 2013. Yield, chemical composition and nutritional quality responses of carrot, radish and turnip to elevated atmospheric carbon dioxide. *J. Sci. Food & Agric.*, **93**(13): 3237-44.

Bailey, L.B. 2004. Folate and vitamin B12 recommended intakes and status in the United States. *Nutr. Rev.*, 62(suppl_1):S14-S20.

Bashir, R., S. Tabassum, A. Rashid, S. Rehman, A. Adnan and R. Ghaffar, 2023. Bioactive Components of Root Vegetables. IntechOpen. doi: 10.5772/intechopen.105961.

Bellani, L.M., L. Salvini, A. Dell'Aquila and A. Scialabba, 2012. Reactive oxygen species release, vitamin E, fatty acid and phytosterol contents of artificially aged radish (*Raphanus sativus* L.) seeds during germination. *Acta Physiologiae Plant.*, **34**(5):1789-99.

Blazevic, I. and J. Mastelic, 2009. Glucosinolate degradation products and other bound and free volatiles in the leaves and roots of radish (*Raphanus sativus* L.). *Food Chem.*, **13**(1):96-102.

Bors, M.D., C.A. Semeniuc, S. Socaci, L. Varva, O. Moldovan and R. Vlaic, 2015. Total phenolic content and antioxidant capacity of radish as influenced by the variety and vegetative stage. Bull. *UASVM Food Sci. Techn.*, **72**(1):77-81.

Camelo, L. and F. Andres, 2004. Manual for the Preparation and Sale of Fruits and Vegetables. Food and Agriculture Organization of the United Nations, 6p. ISBN 92-5-104991-2. Retrieved 2009-07-31.

Chen, X., Z. Zhu, J. Gerendas and N. Zimmermann, 2008. Glucosinolates in Chinese Brassica campestris vegetables: Chinese cabbage, purple cai-tai, choysum, pakchoi, and turnip. *HortSci.*, **43**(2):571-4.

Chihoub, W., M.I. Dias, L. Barros, R.C. Calhelha, M.J. Alves and F. Harzallah-Skhiri, 2019. Valorisation of the green waste parts from turnip, radish and wild cardoon: Nutritional value, phenolic profile and bioactivity evaluation. *Food Res. Intern.*, **126**:108651.

Clotault, J., D. Peltier, R. Berruyer, M. Thomas, M. Briard and E. Geoffriau, 2008. Expression of carotenoid biosynthesis genes during carrot root development. *J. Exper. Bot.*, **59**(13):3563-73.

Coe, K.M., S. Ellison, D. Senalik, J. Dawson and P. Simon, 2021. The influence of the Or and carotene hydroxylase genes on carotenoid accumulation in orange carrots [Daucus carota (L.)]. *Theor. & Appl. Gen.*, **134**(10): 3351-62.

Cordeiro, A., C. Bento, A. Matos and A. Ramalho, 2018. Vitamin A deficiency is associated with body mass index and body adiposity in women with recommended intake of vitamin A. *Nutricion Hospitalaria*, **35**:1072.

Dastan, K., A. Tehranifard and M. Vazirian, 2011. Investigation of anti-pseudomanal activity of *Brassica napus* L. *Inter. Conf. on Environ. and Computer Sci.* (IPCBEE), held on Sept. 16, 2011 in Singapore. IACSIT Press, Singapore, pp. 144-6.

Dattola, A., M. Silvestri, L. Bennardo, M. Passante, E. Scali and C. Patruno, 2020. Role of vitamins in skin health: A systematic review. *Curr. Nutrit. Rep.*, **9**(3):226-35.

de Brito, F.A.L., K. Rabia, L.F.C. Janior, D.G. Coelho, M.R. da Silva Vieira and A. do Nascimento Simaµes, 2017. Changes in phenol metabolism of minimally processed baby cassava under different temperatures: An alternative to commercialization. *Afric. J. Biotechn.*, **16**(17):928-36.

Doheny-Adams, T., K. Redeker, V. Kittipol, I. Bancroft and S.E. Hartley, 2017. Development of an efficient glucosinolate extraction method. *Pl. Methods*, **13**(1):1-14.

Ediage, E.N., J.D. Di Mavungu, M.-L. Scippo, Y.-J. Schneider, Y. Larondelle and A. Callebaut, 2011. Screening, identification and quantification of glucosinolates in black radish (*Raphanus sativus* L. *niger*) based dietary supplements using liquid chromatography coupled with a photodiode array and liquid chromatography-mass spectrometry. *J. Chromat. A*, **1218**(28):4395-4405.

El-Beltagi, H.S., H.I. Mohamed, B.M.H. Megahed, M. Gamal and G. Safwat, 2018. Evaluation of some chemical constituents, antioxidant, antibacterial and anticancer activities of *Beta vulgaris* L. root. *Fresenius environ. Bull.*, **27**(9):6369-78.

Elvira-Torales, L.I., J. García-Alonso and M.J. Periago-Caston, 2019. Nutritional importance of carotenoids and their effect on liver health: A review. *Antioxidants*, **8**(7):229.

Ersus, S. and U. Yurdagel, 2007. Microencapsulation of anthocyanin pigments of black carrot (*Daucus carota* L.) by spray drier. *J. Food Eng.*, **80**(3):805-12.

Farahnaky, A., E. Kamali, M.T. Golmakani, M. Gavahian, G. Mesbahi and M. Majzoobi, 2018. Effect of ohmic and microwave cooking on some bioactive compounds of kohlrabi, turnip, potato and radish. *J. Food Meas. & Character.*, **12**(4):2561-9.

Francisco, M., D.A. Moreno, M.E. Cartea, F. Ferreres, C. García-Viguera and P. Velasco, 2009. Simultaneous identification of glucosinolates and phenolic compounds in a representative collection of vegetable *Brassica rapa. J. Chromat. A*, **1216**(38):6611-9.

Francisco, M., P. Velasco, D.A. Moreno, C. García-Viguera and M.E. Cartea, 2010. Cooking methods of Brassica rapa affect the preservation of glucosinolates, phenolics and vitamin *C. Food Res. Intern.*, **43**(5):1455-63.

Fratianni, A., S. Niro, M.C. Messia, G. Panfili, F. Marra and L. Cinquanta, 2019. Evaluation of carotenoids and furosine content in air dried carrots and parsnips pre-treated with pulsed electric field (PEF). *Eur. Food Res. & Techn.*, **245**(11):2529-37.

Fuentes, P., L. Pizarro, J.C. Moreno, M. Handford, M. Rodriguez-Concepcion and C. Stange, 2012. Light-dependent changes in plastid differentiation influence carotenoid gene expression and accumulation in carrot roots. *Pl. Molec. Biol.*, **79**(1):47-59.

Gimsing, A. And J. Kirkegaard, 2006. Glucosinolate and isothiocyanate concentration in soil following incorporation of Brassica biofumigants. *Soil Biol. & Biochem.*, **38**(8):2255-64.

Goyeneche, R., S. Roura, A. Ponce, A. Vega-Galvez, I. Quispe-Fuentes and E. Uribe, 2015. Chemical characterization and antioxidant capacity of red radish (*Raphanus sativus* L.) leaves and roots. *J. Funct. Foods,* **16**:256-64.

Hanlon, P.R., D.M. Webber and D.M. Barnes, 2007. Aqueous extract from Spanish black radish (*Raphanus sativus* L. *var. niger*) induces detoxification enzymes in the HepG2 human hepatoma cell line. *J. Agric. & Food Chem.*, **55**(16):6439-46.

Iqbal, S., U. Younas, K.W. Chan, Z. Saeed, M.A. Shaheen and N. Akhtar, 2013. Growth and antioxidant response of *Brassica rapa var. rapa* L. (turnip) irrigated with different compositions of paper and board mill (PBM) effluent. *Chemosphere*, **91**(8):1196-1202.

Ishida, M., T. Kakizaki, Y. Morimitsu, T. Ohara, K. Hatakeyama and H. Yoshiaki, 2015. Novel glucosinolate composition lacking 4-methylthio-3-butenyl glucosinolate in 'Japanese White' radish (*Raphanus sativus* L.). *Theor. & Appl. Gen.*, **128**(10): 2037-46.

Jaber, B.A., K.R. Majeed and A.G. Al-Hashimi, 2022. Isolation and identification of new strains of bacteria producing carotenoid pigments which were isolated from a different sources in Basrah, Iraq. *Basrah J. Agric. Sci.*, **35**(1): 81-93.

Jakopic, J., R. Veberic and A. Slatnar, 2021. Changes in quality parameters in rutabaga (*Brassica napus* var. *napobrassica*) roots during long term storage. *LWT.*, **147**:111587.

Jastrebova, J., C. WitthOft, A. Grahn, U. Svensson and M. Jägerstad, 2003. HPLC determination of folates in raw and processed beetroots. *Food Chem.*, **80**(4): 579-88.

Jeong, J., H. Park, H. Hyun, J. Kim, H. Kim and H.I. Oh, 2015. Effects of glucosinolates from turnip (*Brassica rapa* L.) root on bone formation by human osteoblast-like MG-63 cells and in normal young rats. *Phytother. Res.*, **29**(6):902-9.

Jing, P., S.-J. Zhao, S.-Y. Ruan, Z.-H. Xie, Y. Dong and L. Yu, 2012. Anthocyanin and glucosinolate occurrences in the roots of Chinese red radish (*Raphanus sativus* L.), and their stability to heat and pH. *Food Chem.*, **133**(4): 1569-76.

Kapusta-Duch, J., B. Kusznierewicz, T. Leszczynska and B. Borczak, 2016. Effect of cooking on the contents of glucosinolates and their degradation products in selected *Brassica vegetables. J. Funct. Foods*, **23**: 412-22.

Koca Bozalan, N. and F. Karadeniz, 2011. Carotenoid profile, total phenolic content, and antioxidant activity of carrots. *Intern. J. Food Prop.*, **14**(5): 1060-68.

Kumar, M., A. Dahuja, A. Sachdev, C. Kaur, E. Varghese and S. Saha, 2019. Valorisation of black carrot pomace: Microwave assisted extraction of bioactive phytoceuticals and antioxidant activity using Box-Behnken design. *J. Food Sci. & Techn.*, **56**(2): 995-1007.

Lazar, S., O.E. Constantin, N. Stanciuc, I. Aprodu, C. Croitoru and G. Rapeanu, 2021. Optimization of betalain pigment extraction using beetroot by-products as a valuable source. *Inventions*, **6**(3): 50.

Lee, K.B., Y.J. Kim, H.J. Kim, J. Choi and J.K. Kim, 2018. Phytochemical profiles of Brassicaceae vegetables and their multivariate characterization using chemometrics. *Appl. Biolog. Chem.*, **61**(2): 131-44.

Levine, M., Y. Wang, S.J. Padayatty and J. Morrow, 2001. A new recommended dietary allowance of vitamin C for healthy young women. *Proc. Nat. Acad. Sci.*, **98**(17): 9842-6.

Lewis, E.D., S.N. Meydani and D. Wu, 2019. Regulatory role of vitamin E in the immune system and inflammation. *IUBMB Life*, **71**(4): 487-94.

Li, S., I. Schonhof, A. Krumbein, L. Li, H. Stutzel and M. Schreiner, 2007. Glucosinolate concentration in turnip (*Brassica rapa* ssp. *rapifera* L.) roots as affected by nitrogen and sulfur supply. *J. Agric. & Food Chem.*, **55**(21): 8452-7.

Li, Z., H.W. Lee, X. Liang, D. Liang, Q. Wang and D. Huang, 2018. Profiling of phenolic compounds and antioxidant activity of 12 cruciferous vegetables. *Molecules*, **23**(5): 1139.

Lin, L-Z, J. Sun, P. Chen and J.A. Harnly, 2011. LC-PDA-ESI/MSn identification of new anthocyanins in purple bordeaux radish (*Raphanus sativus* L. *variety*). *J. Agric. & Food Chem.*, **59**(12): 6616-27.

Liu, M., H.Y. Seo, S. Min and K.-M. Ku, 2921. Phytonutrients and metabolism changes in topped radish root and its detached leaves during 1°C cold postharvest storage. *Horticulturae*, **8**(1): 42.

Luby, C.H., H.A. Maeda and I.L. Goldman, 2014. Genetic and phenological variation of tocochromanol (vitamin E) content in wild (*Daucus carota* L. var. *carota*) and domesticated carrot (*D. carota* L. var. *sativa*). *Hort. Res.*, **1**:14015.

Macura, R., M. Michalczyk, G. Fiutak and I. Maciejaszek, 2019. Effect of freeze-drying and air-drying on the content of carotenoids and anthocyanins in stored purple carrot. Acta Scientiarum Polonorum Technologia Alimentaria, **18**(2): 135-42.

Mamatha, B.S., R.K. Sangeetha and V. Baskaran, 2011. Provitamin-A and xanthophyll carotenoids in vegetables and food grains of nutritional and medicinal importance. *Intern. J. Food Sci. & Techn.*, **46**(2): 315-23.

Matejkova, J. and K. Petríkova, 2010. Variation in content of carotenoids and vitamin C in carrots. *Notulae Scientia Biologicae*, **2**(4): 88-91.

Metzger, B.T. and D.M. Barnes, 2009. Polyacetylene diversity and bioactivity in orange market and locally grown colored carrots (*Daucus carota* L.). *J. agric. & Food Chem.*, **57**(23):11134-9.

Mikołajczyk-Bator, K., A. Błaszczyk, M. CzyZniejewski and P. Kachlicki, 2016. Characterisation and identification of triterpene saponins in the roots of red beets (*Beta vulgaris* L.) using two HPLC–MS systems. *Food Chem.*, **192**: 979-90.

Misra, S., Sanjay Mishra and R.L. Misra, 2024. Edible Alliums. Pub. from Astral International Pvt. Ltd., New Delhi.

Misra, S., Sanjay Mishra and R.L. Misra, 2023. Tuber Crops. Pub. from Astral International Pvt. Ltd., New Delhi.

Muchlinski, A., M. Ibdah, S. Ellison, M. Yahyaa, B. Nawade and S. Laliberte, 2020. Diversity and function of terpene synthases in the production of carrot aroma and flavor compounds. *Scientific Reports*, **10**(1):9989.

Nicolle, C., S. Gr, E. Rock, P. Amouroux and C. Remesy, 2004. Genetic variability influences carotenoid, vitamin, phenolic, and mineral content in white, yellow, purple, orange, and dark-orange carrot cultivars. *J. Amer. Soc. Hort. Sci.*, **129**(4): 523-9.

Nikolic, N.C., M.M. Lazic, I.T. Karabegovic, G.S. Stojanovic and Z.B. Todorovic, 2014. A characterization of content, composition and scavenging capacity of phenolic compounds in parsnip roots of various weights. *Natural Prod. Commun.*, **9**(6):1934578X1400900620.

Odewole, O.A., R.S. Williamson, N.A. Zakai, R.J. Berry, S.E. Judd and Y.P. Qi, 2013. Near-elimination of folate-deficiency anemia by mandatory folic acid fortification in older US adults: Reasons for geographic and racial differences in stroke study 2003-2007. *Am. J. Clinical Nutrit.*, **98**(4): 1042-7.

Oliveira, R.D., O.D. Dhingra, A.O. Lima, G.N. Jham, M.A. Berhow and R.K. Holloway, 2011. Glucosinolate content and nematicidal activity of Brazilian wild mustard tissues against *Meloidogyne incognita* in tomato. *Plant & Soil,* **341**(1): 155-64.

Otsuki, T., H. Matsufuji, M. Takeda, M. Toyoda and Y. Goda, 2002. Acylated anthocyanins from red radish (*Raphanus sativus* L.). *Phytochem.*, **60**(1): 79-87.

Park, C.H., S.-Y. Park, Y.J. Park, J.K. Kim and S.U. Park, 2020. Metabolite profiling and comparative analysis of secondary metabolites in Chinese cabbage, radish, and hybrid *xBrassicoraphanus. J. Agric. & Food Chem.*, **68**(47): 13711-9.

Pasko, P., K. Bukowska-Strakova, J. Gdula-Argasinska and M. Tyszka-Czochara, 2013. Rutabaga (*Brassica napus* L. var. *napobrassica*) seeds, roots and sprouts: A novel kind of food with antioxidant properties and proapoptotic potential in Hep G2 hepatoma cell line. *J. Medic. Food*, **16**(8): 749-59.

Pasko, P., K. Okon, M. Krosniak, E. Prochownik, P. Zmudzki and J. Kryczyk-Kozioł, 2018. Interaction between iodine and glucosinolates in rutabaga sprouts and selected biomarkers of thyroid function in male rats. *J. Trace Elem. in Medic. & Biol.*, **46**: 110-16.

Patil, G., M.C. Madhusudhan, B. Ravindra Babu and K.S.M.S. Raghavarao, 2009. Extraction, dealcoholization and concentration of anthocyanin from red radish. *Chem. Engin. & Process: Process Intensific.*, **48**(1): 364-9.

Polat, S., G. Guclu, H. Kelebek, M. Keskin and S. Selli, 2022. Comparative elucidation of colour, volatile and phenolic profiles of black carrot (*Daucus carota* L.) pomace and powders prepared by five different drying methods. *Food Chem.*, **369**:130941.

Romani, A., P. Vignolini, L. Isolani, F. Ieri and D. Heimler, 2006. HPLC-DAD/MS characterization of flavonoids and hydroxycinnamic derivatives in turnip tops (*Brassica rapa* L. *subsp. sylvestris* L.). *J. Agric. & Food Chem.*, **54**(4): 1342-6.

Sawicki, T., N. Bączek and W. Wiczkowski, 2016. Betalain profile, content and antioxidant capacity of red beetroot dependent on the genotype and root part. *J. Funct. Foods*, **27**: 249-61.

Shepherd, A.I., J.T. Costello, S.J. Bailey, N. Bishop, A.J. Wadley and S. Young-Min, 2019. Beet the cold: Beetroot juice supplementation improves peripheral blood flow, endothelial function and anti-inflammatory status in individuals with Raynaud's phenomenon. J. Appl. Physiol., **127**(5): 1478-90.

Siervo, M., O. Shannon, N. Kandhari, M. Prabhakar, W. Fostier and C. Kochl, 2020. Nitrate-rich beetroot juice reduces blood pressure in tanzanian adults with elevated blood pressure: A double-blind randomized controlled feasibility trial. *J. Nutrit.*, **150**(9): 2460-68.

Stefanucci, A., G. Zengin, E.J. Llorent-Martinez, M.P. Dimmito, A. Della Valle and S. Pieretti, 2020. Chemical characterization, antioxidant properties and enzyme inhibition of Rutabaga root's pulp and peel (Brassica napus L.). *Arabian J. Chem.,* **13**(9): 7078-86.

Stintzing, F.C., A. Schieber and R. Carle, 2002. Identification of betalains from yellow beet (*Beta vulgaris* L.) and cactus pear [*Opuntia ficus-indica* (L.) Mill.] by high-performance liquid chromatography–electrospray ionization mass spectrometry. *J. agric. & Food Chem.*, **50**(8): 2302-7.

Su, N., Q. Wu, Y. Liu, J. Cai, W. Shen and K. Xia, 2014. Hydrogen-rich water reestablishes ROS homeostasis but exerts differential effects on anthocyanin synthesis in two varieties of radish sprouts under UV-A irradiation. *J. agric. & Food Chem.*, **62**(27): 6454-62.

Tang, G., J. Qin, G.G. Dolnikowski, R.M. Russell and M.A. Grusak, 2005. Spinach or carrots can supply significant amounts of vitamin A as assessed by feeding with intrinsically deuterated vegetables. *Amer. J. Clinical Nutr.*, **82**(4): 821-8.

Troesch, B., B. Hoeft, M. McBurney, M. Eggersdorfer and P. Weber, 2012. Dietary surveys indicate vitamin intakes below recommendations are common in representative western countries. *Brit. J. Nutrit.*, **108**(4): 692-8.

Tumbas Saponjac, V., J. Canadanovic-Brunet, G. Cetkovic, M. JakiSic, J. Vulic and S. StajCic, 2020. Optimisation of beetroot juice encapsulation by freeze-drying. *Polish J. Food & Nutrit. Sci.*, **70**(1): 25-34.

Wang, Y-H., T. Li, R.-R. Zhang, A. Khadr, Y.S. Tian and Z.-S. Xu, 2020. Transcript profiling of genes involved in carotenoid biosynthesis among three carrot cultivars with various taproot colors. *Protoplasma*, **257**(3): 949-63.

Wiczkowski, W., D. Szawara-Nowak and J. Topolska, 2013. Red cabbage anthocyanins: profile, isolation, identification, and antioxidant activity. *Food Res. Intern.*, **51**(1): 303-9.

Yi, G., S. Lim, W.B. Chae, J.E. Park, H.R. Park and E.J. Lee, 2016. Root glucosinolate profiles for screening of radish (*Raphanus sativus* L.) genetic resources. *J. agric. & Food Chem.*, **64**(1): 61-70.

Yucetepe, A., G. Altin and B. Ozçelik, 2021. A novel antioxidant source: Evaluation of in vitro bioaccessibility, antioxidant activity and polyphenol profile of phenolic extract from black radish peel wastes (*Raphanus sativus* L. var. *niger*) during simulated gastrointestinal digestion. Intern. *J. Food Sci. & Techn.*, **56**(3):1376-84.

Zhang, C., H. Di, P. Lin, Y. Wang, Z. Li and Y. Lai, 2022. Genotypic variation of glucosinolates and their breakdown products in mustard (*Brassica juncea*) seeds. *Scient. Hort.*, **294**: 110765.

Zhang, D. and Y. Hamauzu, 2004. Phenolic compounds and their antioxidant properties in different tissues of carrots (*Daucus carota* L.). *J. Food, Agric. & Environ.*, **2**: 95-100.

Zhang, H., I. Schonhof, A. Krumbein, B. Gutezeit, L. Li and H. Stutzel, 2008. Water supply and growing season influence glucosinolate concentration and composition in turnip root (*Brassica rapa* ssp. *rapifera* L.). *J. Pl. Nutrit. & Soil Sci.*, **171**(2): 255-65.

Zhang, X., J. Wei, Y. Huang, W. Shen, X. Chen and C. Lu, 2018. Increased cytosolic calcium contributes to hydrogen-rich water-promoted anthocyanin biosynthesis under UV-A irradiation in radish sprouts hypocotyls. *Frontiers in Pl. Sci.*, **9**: 1020.

Zhou, B., Y. Wang, Y. Zhan, Y. L and S. Kawabata, 2013. Chalcone synthase family genes have redundant roles in anthocyanin biosynthesis and in response to blue/UV-A light in turnip (*Brassica rapa*; Brassicaceae). *Amer. J. Bot.*, **100**(12): 2458-67.

Zhuang, H., Q. Lou, H. Liu, H. Han, Q. Wang and Z. Tang, 2019. Differential regulation of anthocyanins in green and purple turnips revealed by combined *de novo* transcriptome and metabolome analysis. *Intern J. Mol. Sci.*, **20**(18): 4387.

1

Carrot (*Daucus carota* L.)

Family: Apiaceae

Synonyms: *Ballimon maritimum* (Lam.) Raf.; *Carota sativa* (Hoffm.) Rupr.; *Carota sylvestris* (Mill.) Lobel ex Rupr.; *Caucalis carnosa* Roth; *C. carota* (L.) Crantz; *C. daucus* Crantz; *C. gingidium* (L.) Crantz; *C hispanica* Crantz; *C. pumila* Willd.; *Daucus abyssinicus* C.A.Mey.; *D. agrestis* Raf.; *D. alatus* Poir.; *D. allionii* Link; *D. annuus* (Beg.) Wojew., Reduron, Banasiak & Spalik; *D. asturiarum* Barnadez ex Cutanda; *D. australis* Guss.; *D. australis* Kotov; *D. blanchei* Reut.; *D. bocconei* Guss.; *D. brevicaulis* Raf.; *D. capillifolius* Gilli; *D. carnosus* Moench; *D. carota* f. *epurpuratus* Farw.; *D. carota* f. *fischeri* Moldenke; *D. carota* f. *goodmanii* Moldenke; *D. carota* f. *roseus* Farw.; *D. carota* f. *roseus* Millsp.; *D. carota* subsp. *boissieri* (Schweinf.) Hosni; *D. carota* var. *acaulis* (Breb.) P.D.Sell; *D. carota* var. *brachycaulos* Reduron; *D. carota* var. *fontanesii* (Thell.) Reduron; *D. carota* var. *gummifer* Syme; *D. carota* var. *linearis* Reduron; *D. carota* var. *pseudocarota* (Rouy & E.G.Camus) Reduron; *D. carota* var. *tenuisectus* (Degen ex Palyi) Reduron; *D. communis* Rouy & E.G.Camus; *D. communis* var. *pseudocarota* Rouy & E.G.Camus; *D. commutatus* (Paol.) Thell.; *D. dentatus* Bertol.; *D. drepanensis* (Arcang.) Tod. ex Lojac.; *D. esculentus* Salisb.; *D. exiguus* Steud.; *D. foliosus* Guss.; *D. gadecaei* (Rouy & E.G.Camus) Rouy & E.G.Camus; *D. gibbosus* Bertol.; *D. gingidium* Georgi; *D. gingidium* L.; *D. gingidium* subsp. *atlanticus* Rivas Mart.; *D. gingidium* subsp. *commutatus* (Prolongo) O.Bolòs & Vigo; *D. gingidium* subsp. *hispanicus* (Gouan) O.Bolòs & Vigo; *D. gingidium* subsp. *majoricus* (A.Pujadas) Mart.Flores, Juan, M.A.Alonso, A.Pujadas & M.B.Crespo; *D. gouanii* Nyman; *D. gummifer* All.; *D. gummifer* Lam.; *D. gummifer* var. *acaulis* Breb.; *D. gummifer* var. *intermedius* Corb.;

D. halophilus Brot.; *D. herculeus* Pau; *D. heterophylus* Raf.; *D. hispanicus* Gouan; *D. hispidus* Desf.; *D. hispidus* Gilib.; *D. hispidus* Mill.; *D. hispidus* var. *tenuisectus* Degen ex Palyi; *D. jolensis* Pomel; *D. kotovii* M.Hiroe; *D. levis* Raf.; *D. lucidus* L.f.; *D. marcidus* Timb.-Lagr.; *D. maritimus* Lam.; *D. maritimus* With.; *D. martellii* Gand. ex Calest.; *D. masclefii* Corb.; *D. matthiolii* Bubani; *D. mauritanicus* L.; *D. mauritanicus* Salzm. ex DC.; *D. maximus* Desf.; *D. micranthus* Pomel; *D. montanus* Schmidt ex Nyman; *D. nebrodensis* Strobl; *D. neglectus* Lowe; *D. nitidus* Gasp.; *D. nudicaulis* Raf.; *D. officinalis* Gueldenst. ex Ledeb.; *D. paralias* Pomel; *D. parviflorus* Desf.; *D. polygamus* Gouan; *D. polygamus* Jacq. ex Nyman; *D. rupestris* Guss.; *D. russeus* Heldr.; *D. sativus* (Hoffm.) Rohl. ex Pass.; *D. scariosus* Raf.; *D. sciadophylus* Raf.; *D. serotinus* Pomel; *D. serratus* Moris; *D. siculus* Tineo; *D. strigosus* Raf.; *D. sylvestris* Mill.; *D. tenuissimus* (A.Chev.) Spalik, Wojew., Banasiak & Reduron; *D. vulgaris* Garsault; *D. vulgaris* Lam.; *D. vulgaris* Neck.; *Melanoselinum annuum* (Beg.) A.Chev.; *M. tenuissimum* A.Chev.; *Peltactila hispida* Raf.; *P. parviflora* (Desf.) Raf.; *Platyspermum alatum* (Poir.) Schult.; *Thapsia annua* (A.Chev.) M.Hiroe; *T. tenuissima* (A.Chev.) M.Hiroe; *Tiricta daucoides* Raf.; *Tornabenea annua* Beg.; *T. tenuissima* (A.Chev.) A.Hansen & Sunding (powo.science.kew.org/taxon/841063-1; en.wikipedia.org/wiki/Daucus_carota)

Introduction and Origin

Daucus is an ancient Greek name for wild carrot, and the specific epithet *carota* simply means carrot (www.first-nature.com). The word carrot was borrowed from the Middle French carotte, itself from the Late Latin carota, from the ancient Greek karoton), originally from the Proto-Indo-European root ker- ('horn'), due to its horn-like shape of roots (www.etymonline.com/word/carrot).

Daucus has many wild forms growing mostly in the Mediterranean region and SW Asia. Afghanistan is believed to be the primary centre of genetic diversity. There are evidences that purple carrot together with a yellow variant spread from Afghanistan to the Mediterranean region as early as the 10th or 11th century (www.biologydiscussion.com).

Daucus carota L., native to temperate regions of the Old World and naturalized in the New World, has various common names such as wild carrot, bee's nest-plant, bird's-nest, bishop's lace, devil's plague, European wild carrot and Queen Anne's lace (North America) (Chapman, 2008; plants.ces.ncsu.edu; en.wikipedia.org/wiki/Daucus_carota). The native range of this species is Macaronesia to NW Africa and Europe to S. China (powo.science.kew.org/taxon/841063-1).

Botany of Wild Carrot

The wild carrot is somewhat variable biennial herb growing to a height between 30 and 120cm with a roughly hairy, stiff, solid stem, and finely divided *cum* lacy tripinnate leaves that are overall triangular in shape, 5–15cm long, bristly and alternate in a pinnate pattern that separate into thin segments. The small, dull white, umbellate and densely flat-clustered terminal flowers in 8-15cm wide clusters appear from May to September (plants.ces.ncsu.edu) and may be pink in bud though may have a few reddish or purple (Peterson *et al*,, 1968) flower in the centre of the umbel. Three-forked or pinnate lower bracts distinguish it from other white-flowered umbellifers. As the seeds develop, the umbel curls up at the edges becoming more congested, and develops a concave surface. The fruits are small, dry, bumpy, oval and flattened with short styles and hooked spines as well as protective hairs surrounding it (McClintock *et al.,* 1956). The fruit has two mericarps or bicarpellate. The endosperm of the fruit grows before the embryo (Wurtele *et al.,* 1993). The dried umbels detach from the plant becoming tumbleweeds (Faulkner, 1917).

Subspecies

Several subspecies of *D. carota* have evolved to different climates and atmospheres, such as two subspecies that are specifically from the Netherlands are *D.c.* ssp. *sativus* and *D.c.* ssp. *carota* (Baranski *et al.,* 2012). *D. carota* is the only parent of cultivated carrots (Banga, 1957) that are cultivars of *D.c.* ssp. *sativus* (en. wikipedia.org/wiki/Daucus_carota). The edible taproots of these cultivars have a thicker and sweeter root with wide range of colours. The bitter and inedible roots of *D.c.* ssp. *carota* are thin and white in colour. The middle umbellet of *D.c.* ssp. *carota* is not well developed (unlike in *D.c.* ssp. *sativus*) and the colour of the flower can vary from red to deep purple (Baranski *et al.,* 2012).

Uses of Wild Carrot

All tender parts of wild carrot are edible; like the cultivated carrot, the *D. carota* root is edible while young, though quickly becomes too woody to consume; flowers are sometimes battered and fried; leaves and seeds are also edible (plants.ces.ncsu.edu). When freshly cut, *D. carota* will draw or change colour depending on the colour of the water in which it is held. This effect is visible only on the 'head' or flower of the plant. Carnations also exhibit this effect. This occurrence is a popular science demonstration in grade school (en.wikipedia. org/wiki/Daucus_carota). The seeds and flowers have been used as a method of contraception and an abortifacient for centuries (Jansen and Wohlmuth, 2014; John, 2023; Peters, 2014). If used as a dyestuff, the flowers give a creamy off-white colour (en.wikipedia.org/wiki/Daucus_carota).

Toxicity of wild carrot

Although normal consumption of carrots has no toxic effect in humans, lab tests show that the compound falcarinol found in *D. carota* for protection against fungal diseases is toxic to mice and the water flea, *Daphnia magna* (Crosby and Aharonson, 1967).

Cultivated carrot

[*Daucus carota* ssp. *sativus* (Hoffm.) Schubl. & G. Martens]

Synonyms: *Carota sativa* (Hoffm.) Rupr., *Daucus sativus* (Hoffm.).

Introduction

Present day carrots are domesticated forms of *Daucus carota* ssp. *sativus* of wild carrot native to Europe and SW Asia (en.wikipedia.org/wiki/Carrot). The white and orange carrots are probably mutations of the yellow form. The domestic carrot readily crosses with widely adapted wild carrot known as Queen Anne's Lace.

Carrots have long been an important fresh market crop around the world. The domestic carrot has been selectively bred for its enlarged, more palatable and less woody-textured taproot that is the most commonly eaten part of the plant, however, the stems and leaves are also eaten (encyclopedia.pub/entry/11770). The carrot is one of the most important vegetables in the world because of its high yield potential and use as fresh or processed product. Carrots rank among the top 10 vegetable crops in the world (Dawid *et al.*, 2015). It plays a major role in human nutrition because of its high dietary value and good storage attributes (Leja *et al.*, 2013; Umar *et al.*, 2015).

It consists mainly of two parts ~ the stem and the root. Most of the root consists of the peel (periderm), a pulpy outer cortex (phloem) and an inner core (xylem). Cultivated carrots have orange, reddish, purple, black or yellow roots (encyclopedia.pub/entry/11770).

In 2021, world production of carrots and turnips was 41 million tonnes, with China producing 18.1 million tonnes, which is 59.84% of the world total, followed by Uzbekistan (3.1 million tonnes) and the United States (1.43 million tonnes) (FAOSTAT2021accessed on 8thSeptember2023)).

Genetics

Carrot is a diploid species (2n=18) with nine chromosomes having relatively short but uniform-length chromosomes (Iorizzo *et al.*, 2013, 2020). The genome size is estimated to be 473 million base pairs, which is four times larger than *Arabidopsis thaliana*, 1/5th the size of the maize genome, and about the same size as the rice genome (Bradeen *et al.*, 2007).

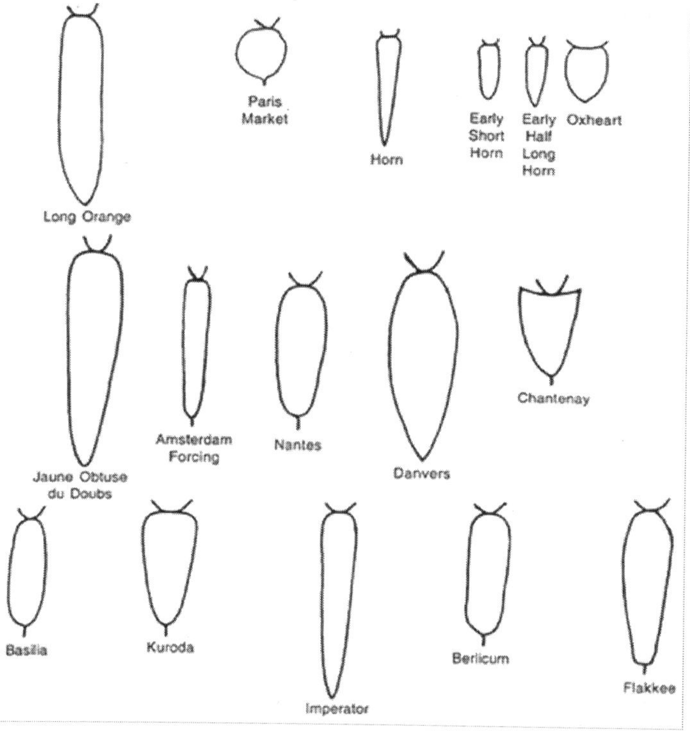

Taproot shapes in carrot
Source: *www.biologydiscussion.com*

Botany

Carrots are a biennial and entomophilous cross-pollinated species having a life cycle of 12-24 months. The stems of the carrot plants are so short that they are barely visible. The dark green trunk is slightly stiff, not woody, round and small in diameter. It has no branches but is overgrown by long leaf stalks so that it looks branched (Kartika *et al.*, 2001). The plant grows a rosette of leaves in the first year while building up the enlarged taproot that stores carbohydrates to provide energy for the plant to flower in the second year. A distinct demarcation between taproot and stem can be seen soon after germination; the stem lacks lateral roots and is thicker. The stem, located just above the ground is compressed with indistinct internodes. When the seed stalk elongates for flowering, tip of the stem narrows and becomes pointed and the stem extends upward 60-200cm tall to become a highly branched inflorescence (Rubatsky *et al.*, 1999).

The seed leaf is at the upper end of the stem, and as the plant grows, the bases of the seed leaves near the taproot are pushed apart. The first true leaf appears about 10–15 days after germination. The pinnately compound subsequent spirally arranged leaves are alternate (with a single leaf attached to a node) with leaf bases sheathing the stem (Rubatsky *et al.*, 1999).

Flowers consist of five petals, five stamens, and an entire calyx	**Top view of *Daucus carota* inflorescence, showing umbellets; the central flower is dark red**	**Carrots in a range of colours**
By Rasbak - Own work, CC BY-SA 3.0, https://commons.wikimedia.org/w/index.php?curid=10757932	**By Christian Fischer, CC BY-SA 3.0, https://commons.wikimedia.org/w/index.php?curid=15779192**	**By Stephen Ausmus – Public Domain, https://commons.wikimedia.org/w/index.php?curid=545126**

Taproots of carrots are generally orange in colour, though purple, black, red, white and yellow cultivars exist (Iorizzo *et al.*, 2013, 2020). Taproots are generally long and conical though cylindrical and nearly-spherical cultivars are also available. Depending on cultivars and growing conditions, the taproot diameter at the widest part can be between 1-10cm whereas the length may range from 5 to 50cm, however, in most cases the length is between 10 and 25cm. The taproot consists of a pulpy outer cortex (phloem) and an inner core (xylem). Although a completely xylem-free carrot is not possible, the taproot can appear to lack a core when the colour of the cortex and core are similar in intensity. A good quality carrot comprises a large proportion of cortex compared to core. Some cultivars have small and deeply pigmented cores (Rubatsky *et al.*, 1999).

Flowers start developing when its flat meristem stops producing leaves, becomes uplifted, conical and capable of producing stem elongation and then clustered flowers. The flowering period of individual umbel is usually 7 to 10 days, so a plant can be in the process of flowering for 30–50 days (Rubatsky *et al.*, 1999). Flowers are borne on a series of compound umbels, starting with a single central inflorescence known as the primary or king umbel, followed by a few secondary umbels, and then many more tertiary umbels. The multiple-flowered umbelets on each compound umbel usually have at least a dozen small epigynous and pentamerous flowers (sepals, petals, stamens and pistils) but with 2 carpels. A primary umbel can have over 1000 flowers at maturity whereas secondary, tertiary and quaternary umbels bear fewer flowers (seedalliance.org; www.biologydiscussion.com).

Floral development is centripetal, *i.e.* the flowers to dehisce first are on the outer edges of the outer umbellets. After straightening of filament, the pollen is shed and stamens quickly fall. After this, the petals open fully and the style elongates. The style is divided into two parts. The petals of petaloid plants are persistent unlike those of brown-anthered male-sterile plants. The ovary of each blossom has two locules, each producing a single seed when successful fertilization has occurred (seedalliance.org; www.biologydiscussion.com).

Its schizocarpic fruit comprises two mericarps that are true seeds which can be separated easily when they are dry. Premature separation (shattering) before harvest can result in seed loss. Mature seeds are flattened on the commissural side that faces the septum of the ovary. The flattened side has five longitudinal ribs. Bristly hairs protruding from some ribs are usually removed by abrasion during milling and cleaning. Seeds also contain oil ducts and canals. Seeds vary somewhat in size, ranging from less than 500 to more than 1000 seeds per gram (Rubatsky *et al.*, 1999).

Breeding and Seed Production

Pollination

While carrots are fully self-fertile, a majority of the seed produced is due to cross-pollination between plants in the population. Much of this cross-pollination is due to the fact that carrots are protandrous, *i.e.* the anthers on any particular flower shed pollen before the stigma of the same flower is receptive to fertilization. This decreases the chances of self-pollination as insects will fly from umbel to umbel in an attempt to find mature flowers. Protandry is an evolutionary mechanism to enforce a majority of cross-pollination in carrots as this species is more susceptible to potential inbreeding depression from excessive self-pollinations than most agricultural crops. Many species of pollinating insects are attracted to carrot flowers including wasps, syrphid flies (also known as hover flies), wild bees and domestic bees. In most climates, the best seed is harvested from the primary and secondary umbels as these have longest time to mature fully (seedalliance.org; www.biologydiscussion.com).

Emasculation

Emasculation is a laborious job and time consuming. As soon as the first bud in an umbel opens the whole umbel of the female parent is bagged in a muslin cloth bag. The flowers are removed daily until peak flowering has reached. Anthers are removed from the early opening outer flowers in the outer whorl of umbellets until sufficient flowers are emasculated. Unopened central florets in the emasculated umbellets and all late flowering umbellets are removed. Thus, only the emasculated flowers are left on the female inflorescence inside the bag. A pollen bearing umbel from previously protected male plant is inserted into the bag of the female parent along with some house-flies to ensure pollination. Daily for a few days in the morning, the male umbel is gently rubbed against the emasculated umbel to enhance artificial cross-pollination. Sometimes, 1-2 flowering umbels of both the parents are enclosed in the same cloth cage along with some house-flies and so for this the seeds from each parent is sown in adjacent rows. The hybrids and the parents could be identified (not always) and necessary roguing done to remove the selfed plants (www.biologydiscussion.com).

Another commonly used alternative for intercrossing male fertile plants is being practiced. A single isolated umbel will not develop seeds even though pollen is present in the flowers as this is protandrous. This umbel can serve as the seed parent in a cross if one week after anthesis the flowers of such an isolated umbel are sprinkled with water to flush out pollen. After it dries, pollen from the intended pollen parent can be introduced with a brush and the seed parent umbel again placed in isolation. This way the seeds produced are nearly always hybrids (www.biologydiscussion.com).

Sometimes two parents to be crossed are covered by a plastic or cloth screen pollination cage. Flies are released in the pollination cage to move pollen or pollen is moved by hand or brush. In this system, selfed and crossed seeds are harvested together. The selfed and crossed progenies need to be identified by phenotypic or molecular markers or by hybrid vigour when inbreds are crossed (www.biologydiscussion.com).

Seed Production Systems

Producing carrot seed on the farm is usually done across two seasons which is natural for the crop when growing in its native habitat. Carrot roots for plant breeding purposes are produced during one growing season which then at the end of the first season are dug up and evaluated for root characteristics, and then planted for seed production later in the year. That way the roots grow to marketable size as late as possible at the end of the first season, minimizing winter storage time. Overwintering can also be done *in situ* or in the field in milder climates but roots are required to be vernalized in either case which is necessary for flowering in biennial crops. In case of carrot, the root is exposed for 8-10 cumulative weeks of temperatures at 10°C or below. This is accomplished either by allowing the roots to overwinter in the ground in regions with mild winters or by storing the roots at ideal storage conditions of 2°C and 95% humidity through winter months (seedalliance.org).

The seed production of Asian carrots can be done successfully in the plains. For this, carrot seeds are sown in August to September and its roots are harvested along with the stems (leaves) in November to December. Immediately after digging the leaves are sheared off leaving only 5-7.5cm upper portion and then these are planted in the fully prepared field at 30cm distance in rows that are spaced 40-60cm apart. Light irrigation should be given immediately after planting the roots (krishijagran.com).

Carrot being a cross-pollinated crop, even then it is very necessary to keep isolation distance of 800 to 1000 metres between its two varieties for quality seed production. For good seed production, only breeder or basal or authenticated seed should be used and information of variety identification, certification, *etc.* should be ensured in advance. The field should always be free from weeds, pests and diseases. Carrot seeds are ready by the end of May. After seeds mature, the inflorescence should be harvested at right stage to avoid seeds being falling off and then the plants are stacked for 1-2 weeks for curing. On an average, about 10-20q/ha of seed is obtained from a good crop (krishijagran.com).

Classification and Cultivars

A wide variety of types and varieties of carrot exist across cultures that use this nutritionally important vegetable as an everyday staple of their diet. Worldwide, there are dozens of different market types of carrots, each with hundreds of unique varieties that have been important to some segment of the market.

The major root types used today include varietal groups as European (Nantes, Chantenay, Danvers, Paris Market, Flakkee, Berlicum, Amsterdam Forcing); Asian (Kuroda); North American (Imperator); South American (Brasilia), *etc.* (www.biologydiscussion.com).

Carrot cultivars for our convenience can be classified as 'Eastern carrots' and 'Western carrots' (Grubben, 2004); 'European', 'American' and 'Asian' carrots; 'Tropical' and 'Temperate' carrots; 'exotic' carrots, or they can be classified based on colours such as orange, red, purple, *etc.*

Eastern and Western Carrots

- **Eastern (with reference to European and American continent) carrots** were domesticated in Persia (now known as Iran) and Afghanistan within West Asia during 10th century or even earlier. Specimens of the Eastern carrots that survive to the present day are commonly purple or yellow and often have branched roots. The purple colour common in these carrots is due to the presence of anthocyanin pigments (Tiwari *et al.,* 2012).
- **Western carrot** is probably the result of Dutch cultivation in the 17th century when patriotic growers turned a vegetable which was then purple into orange, the colour of the national flag (http://news.bbc.co.uk). The orange colour of these cultivars is due to the presence of abundant carotenes (en.wikipedia.org/wiki/Carrot).
    - 'Chantenay', is an excellent cultivar for canning and storage, having vigorous foliage and greater girth, but the roots are shorter than other cultivars. Their roots are attractively orange, smooth and thick at the shoulders with gradual tapering towards a blunt and rounded tip. They have a pale-coloured core and are mostly used for processing. Their roots are more juicy and have less fibre content. Cultivars include 'Carson Hybrid' and 'Red Cored Chantenay' (nhb.gov.in/pdf/vegetable/carrot/car012.pdf; en.wikipedia.org/wiki/Carrot).
    - 'Danvers', developed in 1871 in Danvers (Massachusetts), having strong foliage and are somewhat shorter than 'Imperator' cultivars. Their conical-shaped roots with a well-defined shoulder, tapering to a point are longer than Chantenay types. They are more tolerant of heavy soil conditions. Danvers cultivars store well and are used both fresh and for processing. Cultivars include 'Danvers Half Long' and 'Danvers 126' (en.wikipedia.org/wiki/Carrot).
    - 'Imperator' carrots, a mid-season to late-maturing cultivar, have vigorous foliage and long, slender roots with tapering to slightly rounded end. The roots are white to deep orange, 15-17.5cm long and 2.4-4.5cm in diameter with high sugar content. It is the most widely cultivated by commercial growers. Cultivars include 'Imperator 58' and 'Sugarsnax Hybrid' (nhb.gov.in/pdf/vegetable/carrot/car012.pdf; en.wikipedia.org/wiki/Carrot).
    - 'Nantes' carrots have sparse foliage and cylindrical with short root having a blunter tip than Imperator types. These carrots give high yields in a range of conditions. Their skin is easily damaged and the core is deeply pigmented. They are brittle, high in sugar but

store less well than other types. Cultivars include 'Nelson Hybrid', 'Scarlet Nantes', 'Sweetness Hybrid', 'Nantes Half Long' and 'Early Nantes' (nhb.gov.in/pdf/vegetable/carrot/car012.pdf; en.wikipedia.org/wiki/Carrot).

'**Bolero**' is a nantes-type hybrid which can be planted early for delicious sweet carrots but when planted later (in June here), it will achieve a good length of 15 to 18cm with a 2.5-4cm diameter. After fall storage, this carrot will still be crisp and sweet. These can be harvested as late as possible even after frost but before the ground freezes (www.backwoodshome.com).

European, American and Asian Carrots

- **European and American varieties** require cool temperatures and do not tolerate heat. Their roots are cylindrical, medium-long, tail-tipped and dark orange in colour. Their average yield is 250q/ha. The main European varieties are 'Danvers', 'Nantes', 'Chantenay', 'Ball' (or Mini) (krishijagran.com), 'Paris Market', 'Flakkee', 'Berlicum', 'Amsterdam Forcing', 'Imperator' (North American), 'Brasilia' (South American) (www.biologydiscussion.com).
- **Asian varieties** can tolerate high temperatures. Some cultivars include 'Pusa Meghali', 'Carrot No-29', 'Pusa Kesar', 'Hisar Geric', 'Hisar Rasili', 'Hisar Madhur', 'Selection No-223', 'Pusa Rudhira', 'Pusa Ashita', 'Pusa Jamdagni', 'Pusa Vasuda' (F1 hybrid for tropical condition) and 'Pusa Vrishti' (heat tolerant). These are sown early, i.e. from August to September. However, they can be sown till October (krishijagran.com).

Carrot Types Grown in India

European Nantes types (cylindrical roots, orange in colour) and Japanese Kuroda type (conical roots, deep orange in colour) are quite common in India, and almost entire Asia. A dark orange selection from 'Gosun' by Mr. Kuroda in 1950s resulted in the 'Kuroda-Gosun' cultivar which is being grown widely in India with the name of 'Kuroda'. Most of private sector seed companies are marketing Kuroda carrot on a large scale (www.biologydiscussion.com).

The carrots in India have deep red to orange to purplish black colour. Although the red carrots (*desi gajar*) and the orange carrots are found all over India, purple carrots (*kali gajar*) is found only in some limited parts of North India. Purple carrots get their characteristic hue due to extra antioxidant anthocyanin pigments. The orange carrots are genetically engineered hybrid variety, while the red carrots are the non-hybrid original variety cultivated in higher numbers in the Asiatic region. Red and orange carrots derive their colour from the beta carotene (food.ndtv.com) which these possess.

New cultivars have been developed with a view to quantify large amounts of chemically-stable acylated pigments such as anthocyanins which enrich carrot colour based on the density and types of anthocyanin to produce different carrot colours (Iorizzo *et al.*, 2013, 2020).

Asian/Tropical types

- **Arka Suraj** was developed by IIHR. It flowers and sets seeds under tropical condition. Its deep orange conical roots are smooth having 15-18cm length and 3-4cm diameter with self core colour. It contains TSS 8-10 brix with carotene content of 11.27mg/g. It is tolerant to powdery mildew and nematodes (iihr.res.in).
- **Kuroda** is an old variety developed in Japan by Mr. Kuroda but is very popular in Indian seed companies for large scale sales. This orange-rooted tapered variety with blunt point, *vis-à-vis* wide shoulders has long, tender and sweet taste. Its roots are smooth, uniform and conical in shape with better storability.
- **No. 29** was developed by PAU, Ludhiana. Its light red roots are long tapering (nhb.gov.in/pdf/vegetable/carrot/car012.pdf).
- **Pusa Asita** is a first black-rooted obtriangular carrot having long, smooth, round shouldered black coloured self-core and is suitable for sowing from September to October under north Indian plains. It matures for harvesting during December-January, requiring 90-110 days from sowing where average root yield is 25t/ha.
- **Pusa Kesar**, evolved from a cross of Local Red and Nantes, was released in 1963 at IARI, New Delhi. It has red-coloured roots and self-coloured core. It can tolerate higher temperatures and is suitable for sowing from early August to October, crop maturing in 90-110 days with average yield of 25-30t/ha (nhb.gov.in/pdf/vegetable/carrot/car012.pdf). It is rich in carotene (38mg/100g edible portion) and the roots stay longer in field without bolting. Its seed production in north Indian plains is quite successful (www.biologydiscussion.com).
- **Pusa Kulfi** is a first self-core cream-coloured tropical variety, suitable for sowing in mid-September under north Indian plains. It matures in 90-100 days, yielding 25t/ha root yield where one root may weigh 100-150g.
- **Pusa Meghali** was developed in 1985 by IARI, New Delhi from a cross between Pusa Kesar and Nantes. It has short top, smooth roots with

orange flesh and self-coloured core. It is suitable for sowing from August to September. Its average yield is 25-30t/ha (nhb.gov.in/pdf/vegetable/carrot/car012.pdf). It is richer in carotene content (11571 IU/100g) than Pusa Kesar (7753 IU/100g) and produces seeds in plains. It is suitable for early sowing and takes 110-120 days to attain harvest maturity. Its root yield is 260-280q/ha (www.biologydiscussion.com).

- **Pusa Rudhira** is a long blood-red variety with self core colour and its shape is attractive obtriangular. It is suitable for sowing from mid-September to October and the roots are ready for harvesting during December-January, i.e. 90-110 days with average root yield of 30t/ha.
- **Pusa Vasuda** hybrid was developed using CMS system. Its roots attain marketable maturity in 80-90 days with average yield of 35-40t/ha. Its roots are attractive, vigorous, smooth, red with self core colour, juicy, sweet and rich in total carotenoids, lycopene, TSS & minerals. It is quite suitable for extraction of juice, preparing salad, cooking and for carotenoid extracrtion.
- **Pusa Vrishti** is a first heat-tolerant tropical variety in red colour with red colour core, suitable for sowing from July end to 1st half of August under north Indian plains, producing roots weighing 150-200 grammes each with an average root yield of 20-22t/ha. It attains marketable maturity in about 90 days as is ready for harvesting by mid-November.
- **Selection 223** was developed by PAU, Ludhiana. Its roots are orange with light orange flesh.

European/Temperate Types

- **Nantes** is an old introduction by IARI Regional Station, Katrain. Its seed production is possible only in the hilly areas. The roots are half long (12-15cm), slim, well-shaped, cylindrical with stumped end and forming a small thin tail though its keeping quality is poor. The cortex and core are deep orange. It ranks good in quality but top being brittle make root pulling a bit difficult. It is suitable for cultivation in cooler months and takes 90-110 days for root development (www.biologydiscussion.com). Its average root yield is 25-30t/ha.
- **Nantes Half Long** has cylindrical stumpy roots with orange-scarlet self-coloured core. Crop matures in 110-120 days (nhb.gov.in/pdf/vegetable/carrot/car012.pdf).
- **Early Nantes** is an excellent European cultivar with almost cylindrical roots terminating abruptly in small thin tail. It has a weak brittle top which makes pulling difficult. Due to thin skin and fine texture, it can

be stored for a limited period. Crop matures within 90-100 days (nhb.gov.in/pdf/vegetable/carrot/car012.pdf).

- **Chantenay** - see above.
- **Imperator** - see above.
- **Zeno** is a high yielding good quality variety popular in Nilgiri hills and emits a good flavour (nhb.gov.in/pdf/vegetable/carrot/car012.pdf).
- **Pusa Nayanjyoti** is a first temperate CMS-based orange carrot hybrid developed by public sector which yields some 39.6t/ha of roots. Its orange roots are smooth, uniform, cylindrical, stumpy with small indistinct self colour core. It is sown from April to August on the hills as well as in November-December under north Indian plains.
- **Pusa Yamdagini** (Selection 5) has been developed at IARI Regional Station, Katrain from an inter-varietal cross between EC 9981 and Nantes to combine earliness and root size of the former and self-coloured core character of the latter. It is early in maturity (90-100 days) and is suitable for growing only on the hills where it yields 20-25t/ha of roots. Its proper sowing time on the hills is April to August. It has orange flesh with self-coloured core. Crop matures early by 7-10 days than Early Nantes (nhb.gov.in/pdf/vegetable/carrot/car012.pdf).

Exotic Varieties

(nhb.gov.in/pdf/vegetable/carrot/car012.pdf)

- USA for processing purpose: 'Red Cored Chantenay', 'Danvers Half Long', 'Imperator'.
- New Zealand: 'Akaroa Long Red', 'Spring Market Improved', 'Wanganui Giant'
- Japan: 'Suko'
- Belgium: 'Belgian White'
- France: 'Chantenay', 'Nantes', 'Oxheart', 'Touchon'
- Australia: 'Red Elephant', 'Western Red', 'Yellow'
- Netherlands: 'Early Horn'

Nutritional Value

The roots contain high quantities of alpha- and beta-carotene, and are a good source of vitamin A, vitamin K and vitamin B6 (en.wikipedia.org/wiki/Carrot).

Beta carotene is the naturally occurring group of pigments that determines the colour of many fruits and vegetables. It is essential to make vitamin A, which are required for good eyesight. Although beta-carotene is most abundant in carrots, it is also found in pumpkins, apricots and nectarines. Another compound called lycopene, the one also found in tomatoes is also responsible for the reddish hue. Lycopene protects our body against free radicals for its high antioxidant properties, putting the risk of heart diseases at bay. The orange carrots also get their bright colour from beta-carotene which varies slightly from the red carrots. The red carrots definitely have extra antioxidant pigments naturally. But orange carrots too are loaded in vitamins A, E and K. It also contains the eye-healthy antioxidants lutein and lycopene. Purple carrots get their characteristic hue due to extra antioxidant anthocyanin pigments that may benefit the arthritis, many heart problems and many more.

Carrot dietary fibre comprises mostly cellulose, with smaller proportions of hemicellulose, lignin and starch (Rubatsky *et al.,* 1999). Free sugars in carrot include sucrose, glucose and fructose. The carrot gets its characteristic bright orange colouration from β-carotene as well as lesser amounts of α-carotene, γ-carotene, lutein, and zeaxanthin. α- and β-carotenes are partly metabolized into vitamin A, providing more than 100% of the Daily Value (DV) per 100g serving of carrots. Carrots are also a good source of vitamin K (13% DV) and vitamin B6 (11% DV), but otherwise have modest content of other essential nutrients.

Nutritional value per 100g				
	Carrots, boiled		**Carrots, raw**	
Energy	147 kJ (35 kcal)		173 kJ (41 kcal)	
Carbohydrates	8.22 g		9.6 g	
Sugars	3.45 g		4.7 g	
Dietary fiber	3 g		2.8 g	
Fat	0.18 g		0.24 g	
Protein	0.76 g		0.93 g	
Vitamins	**Quantity**	**%DV†**	**Quantity**	**%DV†**
Vitamin A equiv. beta-Carotene lutein zeaxanthin	852 μg 8330 μg 687 μg	107% 77%	835 μg 8285 μg 256 μg	104% 77%
Thiamine (B1)	0.066 mg	6%	0.066 mg	6%
Riboflavin (B2)	0.044 mg	4%	0.058 mg	5%
Niacin (B3)	0.645 mg	4%	0.983 mg	7%
Pantothenic acid (B5)	0.232 mg	5%	0.273 mg	5%
Vitamin B6	0.153 mg	12%	0.138 mg	11%
Folate (B9)	14 μg	4%	19 μg	5%

Nutritional value per 100g				
Vitamin C	3.6 mg	4%	5.9 mg	7%
Vitamin E	1.03 mg	7%	0.66 mg	4%
Vitamin K	13.7 μg	13%	13.2 μg	13%
Minerals	**Quantity**	**%DV†**	**Quantity**	**%DV†**
Calcium	30 mg	3%	33 mg	3%
Iron	0.34 mg	3%	0.3 mg	2%
Magnesium	10 mg	3%	12 mg	3%
Manganese	0.155 mg	7%	0.143 mg	7%
Phosphorus	30 mg	4%	35 mg	5%
Potassium	235 mg	5%	320 mg	7%
Sodium	58 mg	4%	69 mg	5%
Zinc	0.2 mg	2%	0.24 mg	3%
Other constituents	**Quantity**		**Quantity**	
Water	90.2 g		88 g	
Units: μg = micrograms; mg = milligrams; IU = International units †Percentages are roughly approximated using US recommendations for adults.				
***Table source:** https://en.wikipedia.org/wiki/Carrot*				

Phytochemicals

Phytochemicals contribute to the dietary value of carrots and comprise mainly of four types ~ phenolic compounds, carotenoids, polyacetylenes and ascorbic acid.

Phenolic compounds are secondary plant metabolites which constitute one of the most ubiquitous groups of plant metabolites and are an integral part of both human and animal diets. The role of polyphenols in the prevention of degenerative diseases such as cancer, cardiovascular and neurodegenerative diseases has been reported. Interest in food phenolics has increased greatly over the past two decades owing to their antioxidant capacity and their function as a defence against oxidative stress caused by excess reactive oxygen species (Tsao, 2010).

Phenolic compounds can be divided into different subgroups such as phenolic acids, flavonoids, tannins, lignans, stilbenoids and curcuminoids. Carrots are rich in phenolic acids such as p-hydroxybenzoic, caffeic and chlorogenic as well as in anthocyanins ~ a class of flavonoids (Gonçalves *et al.,* 2013). Isocoumarins and phenolic acids are the potentially bitter compounds found in carrot peels. The bitter taste in carrots is caused by terpenoids and water-soluble phenolics (Czepa and Hofmann, 2004).

Carotenoids are a group of isoprenoid molecules present in all photosynthetic plants including carrots. Some non-photosynthetic fungi and bacteria also possess carotenoids. Several conjugated double bonds of a polyene chain that function as a chromophore are responsible for the yellow, orange and red carotenoid colouration (Aguila Ruiz-Sola and Rodríguez-Concepción, 2012; Rodriguez-Concepcion and Stange, 2013). There are two types of carotenoids present in carrot such as carotenes and xanthophylls. The major carotenoids in carrot roots are β-carotene (75%); α-carotene (23%); lutein (1.9%); and β-cryptoxanthin such as lycopene and zeaxanthin (Soltoft *et al.*, 2011).

Polyacetylenes are a prominent group of non-volatile bioactive phytochemicals. Plants of the Apiaceae family contain aliphatic C17-polyacetylenes of the falcarinol type (Christensen, 2011). More than 1400 polyacetylenes have been identified in plants and 12 polyacetylenes were isolated from carrot. Out of the 12 ~ falcarinol, falcarindiol and falcarindiol-3-acetate are essential polyacetylenes predominately found in carrot roots. The other nine polyacetylenes that are isolated from carrot are (E)-isofalcarinolone, falcarindiol-8-acetate, 1,2-dihydrofalcarindiol-3-acetate, (E)-falcarindiolone-8-acetate, (E)-falcarindiolone-9-acetate, 1,2-dihydrofalcarindiol, (E)-1-methoxy-falcarindiolone-8-acetate, (E)-1-methoxy-falcarindiolone-9-acetate and panaxydiol (Dawid *et al.*, 2015; Schmiech *et al.*, 2009).

Polyacetylenes in carrots show cytotoxic activities (Zidorn *et al.*, 2005; Baranska *et al.*, 2005) and recent studies on the biological activity of polyacetylenes have indicated their potential to improve human health due to anticancerous, antifungal, antibacterial, anti-inflammatory and serotogenic effects (Dawid *et al.*, 2015). Two such compounds are falcarinol and falcarindiol (cis-heptadeca-1,9-diene-4,6-diyne-3,8-diol); falcarindiol compound being the main compound responsible for bitterness in carrots (Czepa and Hofmann, 2003) and shows antifungal activity towards *Mycocentrospora acerina* and *Cladosporium cladosporioides* (Garrod *et al.*, 1978). These findings suggest targeting vegetables with elevated levels of bisacetylenic oxylipins such as falcarinol; and due to the abundant availability, high diversity of cultivars, worldwide experience, and high agricultural yields, carrot genotypes are a very promising target vegetable (Dawid *et al.*, 2015).

l-ascorbic acid (vitamin C) is one of the most abundant water-soluble low molecular weight antioxidant known to play a central role in regulating the cellular redox potential in cells (Fotopoulos and Kanellis, 2013; Gest *et al.*, 2013). As humans and some other primates lack the ability to synthesize and store vitamin C, they depend on fresh fruits and vegetables to cover their daily requirements (75–90 mg RDA).

Other compounds such as pyrrolidine (present in the leaves), 6-hydroxymellein, 6-methoxymellein, eugenin, 2,4,5-trimethoxybenzaldehyde (gazarin) or (Z)-3-acetoxy-heptadeca-1,9-diene-4,6-diin-8-ol (falcarindiol 3-acetate) can also be found in carrot (en.wikipedia.org/wiki/Carrot).

Uses

Carrots are commonly consumed raw or cooked in various cuisines. Desi or red carrots are seasonal whereas the hybrid variety is found round the year. Desi carrots are more sweet, juicy and crunchy and hence preferred for preparations of desserts. Indians use the red carrots for a longer span of time in the year compared to the Europe or America where availability of carrots is low as they use more of the orange ones which are less sweet and are used widely for grilling, sauteing and in salads. Both red and orange carrots offer maximum nutritional benefit when eaten raw. A few carrot sticks a day can help strengthen and clean children's teeth and may also help encourage lower-jaw development. Juicing gives highest concentration of beta carotene. Orange carrots are made by altering the genes to receive the best characteristic nutrition of the vegetable. Hence nutritive output of orange carrots could be more upon juicing. Also there sugar content is lesser than the red *desi gajar* (food.ndtv.com).

Only 3 per cent of the β-carotene in raw carrots is released during digestion though this can be improved to 39% by pulping, cooking and adding cooking oil (Hedren *et al.*, 2002). Alternatively they may be chopped and boiled, fried or steamed, and cooked in soups and stews, as well as in baby and pet foods. The sweetness of carrots allows the vegetable to be used in some fruit-like roles. Carrots can also be used alone or blended with fruits in jams and preserves. Carrot juice is also widely marketed, especially as a health drink, either standing alone or blended with juices extracted from fruits and other vegetables (Shannon, 1998).

Though rarely eaten by humans (Rubatsky *et al.*, 1999), the carrot greens harvested young in high-density plantings before significant root development are edible as a leafy vegetable (en.wikipedia.org/wiki/Carrot; Richard, 2006) but some sources suggest that greens contain toxic alkaloids (Yeager, 2008; Brown, 2012). The leaves may typically be used stir-fried or in salads (Rubatsky *et al.*, 1999).

In India carrots are used in a variety of ways ~ as salads or as vegetables added to spicy rice or pulse dishes. A popular variation in north India is the carrot dessert (*Gajar Ka Halwa*) which has grated carrots added with sugar and cooked in milk until the whole mixture becomes solid, after which nuts and butter are added (Gupta, 2000). Carrot salads are usually made with sliced or grated carrots with or without a seasoning of mustard seeds and green chillies popped in hot oil. Carrots can also be cut in thin strips and added to rice, can form part of a dish of mixed roasted vegetables or can be blended with tamarind to make chutney (Chapman, 2007).

Processing

Baby carrots or mini-carrots (carrots that have been peeled and cut into uniform cylinders) have been a popular ready-to-eat snack food available in many supermarkets (Bidlack and Rodriguez, 2011). Carrots are pureed and used as baby food, dehydrated to make chips, flakes, and powder, and thinly sliced and deep-fried similar to potato chips (Rubatsky *et al.,* 1999).

Health Benefits

Carrots are rich in vitamins, minerals, antioxidant compounds and fibre. As part of a balanced diet, these can help support immune function, reducing the risk of some cancers and promoting wound healing and digestive health. Carrots contain vitamin A whose deficiency may result in xerophthalmia ~ a progressive eye disease that can cause night blindness or difficulty in seeing when levels of light are low (www.medicalnewstoday.com). Carrots are an excellent source of Vitamin A and Beta-carotene. Along with these nutrients, it is a good source of vitamin C, lutein, zeaxanthin, vitamin K, dietary fibre, *etc.* The antioxidants lutein and zeaxanthin in carrots protect the retina and lens, *vis-a-vis* improve eye health (pharmeasy.in).

Too many free radicals in the body may increase the risk of various types of cancer. The antioxidant effects of dietary carotenoids ~ yellow, orange and red organic pigments present in carrots and other vegetables may reduce this risk. Lutein and zeaxanthin are two examples of these carotenoids. Vitamin A and beta carotene are also quite beneficial (www.medicalnewstoday.com).

Carrots being rich in fibre so adequate fibre intake can assist in the optimal working of the entire digestive system. A high fibre diet may have a lower risk of colorectal cancer than those who consume little fibre. High fibre diet may also lower the risk of developing cardiovascular diseases and reducing the levels of low-density lipoprotein or 'bad cholesterol'. Moreover, as carrots contain less sodium but more potassium so potassium helps relaxing the blood vessels, reducing the risk of high blood pressure, *vis-a-vis* other cardiovascular issues. Carrots contain vitamins A & C that help in boosting the immune system.

Vitamin C in carrots also contributes to the production of collagen that is a key component of connective tissue and is essential for wound healing and keeping the body healthy. Carrots contain vitamin K and small amounts of calcium and phosphorus so all of these contributing to bone health and may help prevent osteoporosis (www.medicalnewstoday.com).

Carrots are though rich in nutrients but low in calories, and these nutrients assist to feel fuller and longer and so lowering quantity of energy eaten. Carrots in meal rotation may help in reducing weight. Low sugar *cum* high-fibre foods such as carrots may help in prevention of type II diabetes. For those who already have diabetes, the benefits of carrots can help in the management of blood sugar (pharmeasy.in).

Side Effect

Highly excessive consumption over a period of time can result in carotenemia, a yellow-orange discoloration of the skin caused by a build up of carotenoids (Amy and Anthony, 2011).

Allergy

Some people are allergic to carrots. In a 2010 study on the prevalence of food allergies in Europe, 3.6 per cent of young adults showed some degree of sensitivity to carrots (Burney *et al.*, 2010). Most carrot allergy sufferers were also allergic to pollen from these plants because the major carrot allergen, the protein Dauc c 1.0104, is cross-reactive with homologues in birch pollen (Bet v 1) and mugwort pollen (Art v 1) (Ballmer-Weber *et al.*, 2012).

Cultivation

Carrots are grown from seeds and will take 1 to 3 weeks to sprout. Germination in cold soil is quite slow than in warm soil. Depending on the cultivars being grown, the crop matures in 90-120 days after sowing though there are some early-maturing cultivars which take hardly 70 to 80 days under the right conditions (www.goodhousekeeping.com).

Climate

Carrots grow best in full sun but tolerate some shade (Elzer-Peters, 2014). Carrot is a cold season crop and the shapes of carrots have a great effect on temperature. Although carrots grow well and may form good shape and attractive root colour in the temperature range of 12 to 25°C (krishijagran.com), the optimum temperature is 16 to 21°C (Benjamin *et al.*, 1997).

Soil

The ideal soil for carrot cultivation is deep, loose and well-drained (krishijagran.com), sandy or light loamy with a pH of 6.3 to 6.8 (Cunningham, 2000). Growers should choose the right variety for root size and shape to suit their soil. Ball-type, Chantenay and Danvers carrots have blocky shapes that can handle heavy or shallow soil, while slender Nantes and Imperator carrots need deep and loose soil (www.goodhousekeeping.com). All these types are available in early and late cultivars, many of which are even disease-and crack-resistant (www.goodhousekeeping.com).

Sowing season and seed rate

Sowing of Asian varieties should be done from August to September and European varieties from October to November. Seed rate of 10 to 12kg per hectare is sufficient (krishijagran.com). Carrot is a cool-weather crop and the grower should start sowing it 3 weeks before the last expected frost, and then planting may again be done every 2 to 3 weeks after that, last being 2-3 months before the first expected fall frost. In Zone 8 and warmer zones, carrots are shown in fall or winter (www.goodhousekeeping.com).

Field preparation and Planting

Prior to sowing, field should be deeply ploughed 2-3 times and disked properly to eliminate debris and soil clods. If the soil is not rich in nutrients, 10-20 tonnes FYM, 1-2 tonnes vermicompost or 4-8 tonnes poultry manure should be mixed well while preparing the field after the last plooughing. For good yield and quality of roots, sowing should be done on light bunds. The distance between the bunds should be 30 to 45 cm where seeds can be sown 6 to 8 cm apart. Seeds should be sown by making a drain of 2 to 3 cm deep on top of the bunds (krishijagran.com).

Alternately, seeds can be broadcast or for easier weeding they are sown in rows by putting a pinch of about six seeds at a distance of 2.5cm. The plants should be thinned to 6.25cm apart when the tops are 5cm high since crowded carrots will produce crooked roots. Thinning should be performed again after about 2 weeks later to 8 to 10cm apart (www.goodhousekeeping.com).

Growers should cover the area with 0.75 to 1.25cm or a little more in warm and dry areas with strained compost and potting mix along with a little sand so that it becomes quite easy for the delicate seedlings to emerge. The soil must be kept moist for best germination where watering should be done gently with water can to avoid washing away of seeds (www.goodhousekeeping.com).

Manure and Fertilizers

Fertilizer should be applied according to soil type because the crop requires low levels of nitrogen, moderate phosphate and high potash. Rich or rocky soils should be avoided, as these will cause the roots to become hairy and/or misshapen (Abbott, 2012). If the soil is heavy, plenty of mature compost should be added (www.goodhousekeeping.com). In the average grade of land, about 20 tonnes of well-rotten cow dung manure or FYM per hectare is required at the time of soil preparation, and at the time of sowing 20kg each of nitrogen, phosphorus and potash should be applied in the field per hectare, and subsequently after about 3-4 weeks of sowing 20 kg of nitrogen is applied in the standing crop for better stand (krishijagran.com).

Irrigation

If there is less moisture while sowing the field, first light irrigation should be applied immediately after sowing. After this, irrigation should be carried out according to the weather condition and soil moisture level at an interval of 15 to 20 days (krishijagran.com). Grower should never let young carrot plants drying out. However, if the soil dries out completely between waterings, they should gradually remoisten the bed over a period of days as a sudden drenching may cause the roots to split (www.goodhousekeeping.com).

Interculture

Thinning

After sprouting, the crop is eventually thinned to a spacing of 8 to 10cm and weeded to prevent competition beneath and above the soil (en.wikipedia.org/wiki/Carrot).

Mulching

As the seedlings come up, mulching should gradually be applied to maintain even moisture level and reduce weed problems (www.goodhousekeeping.com).

Weeding

Two to three weedings may be required before root harvest. Carrot's feeder roots get easily damaged therefore any weed that pushes through the mulch should be hand-pulled or cut off just below the soil surface (www.goodhousekeeping.

com). If more weeds grow in the field, the grower can mix the pentamethylene 30 EC 3kg in 900 to 1000 litres of water and spraying it within 48 hours of sowing which will not allow growing of any weed for first 30 days of sowing (krishijagran.com).

Earthing up

Carrot crowns that push up through the soil as they mature must be covered with mulch or soil to prevent them from becoming green and bitter (www.goodhousekeeping.com).

Companion planting

Carrots make good use of space between rows of celery, onions, lettuce, spinach, Swiss chard, and mustard greens. These can be planted on raised 'peaks' with the 'valleys' planted with celery, lettuce, onions and other surface crops. Carrots do well alongside most plants, especially chives, tomatoes and sage. Dill, coriander and other members of the Apiaceae family should not be planted near carrots, especially if the growers are saving their own seed, because they may sometimes cross-pollinate (www.plantea.com/carrots.htm).

Carrots benefit from strongly scented companion plants such as onions, leeks and chives. The pungent odours of these plants help repel the carrot root fly (Cunningham, 2000). They can also be good companions for other plants. If left to flower, the carrots attract predatory wasps that kill many garden pests (Grubben, 2004).

Harvest

Harvesting can be started as soon as the carrots are big enough to eat or can be left to mature for a single harvest. However, these should be dug while these are in a soft state. The harvesting stage of the roots depends on the variety ~ usually fast-growing Asian varieties are harvested in 90-100 days of sowing the seed while slower maturing Asian varieties need a month longer, *i.e.* 120-130 days. However, European varieties require 70-80 days for root harvesting (krishijagran.com; www.goodhousekeeping.com).

Carrots are generally harvested before the first frost and on a day when the soil is moist but the air is dry. Since spading forks tend to bruise roots, they should be hand-pulled after loosening the soil with a trowel. Watering the bed before harvesting softens the soil and makes pulling easier (www.goodhousekeeping.com).

Post Harvest

Under normal conditions, carrots cannot be stored for more than 3-4 days but can be stored in perforated polyethene for at least 2 weeks while the same carrots packed in perforated polyethene can be stored for 1-2 months in cold store (krishijagran.com). Carrots can be stored for several months in the refrigerator or over winter in a moist and cool place. For long term storage, unwashed carrots can be placed in a bucket between layers of sand, a 50/50 mix of sand and wood shavings or in soil. Carrots can be easily preserved for a long time (2 to 3 months) at temperature range of 0-4°C and 98% humidity (www.plantea.com/carrots.htm; www.backwoodshome.com). During storage, carrots may be subject to the development of bitterness, white blush and browning which may lead to carrot losses. Bitterness can be prevented by storage in well-ventilated rooms with low ethylene content, *i.e.* without ethylene-producing fruits and vegetables. White blush and browning can be countered with application of edible films, heat treatment, application of hydrogen sulphide and ultraviolet irradiation (Papoutsis and Edelenbos, 2021).

Yield

The yield and quality of carrots depend on the variety, time of sowing, type of land and climatic conditions, lifting time, etc. Its early crop yields on an average about 20-25t/ha in August-sown crop, medium crop 30-40t/ha in September to October sown crop while the late-sown November crop 28-32t/ha (krishijagran.com).

Pests and Diseases

There are several diseases that can reduce the yield and market value of carrots. The most devastating carrot disease is *Alternaria* leaf blight, which has been known to eradicate entire crop. A bacterial leaf blight caused by *Xanthomonas campestris* can also be destructive in warm, humid areas. Root knot nematodes (*Meloidogyne* species) can cause stubby or forked roots, or galls. (Davis, 2004) Cavity spot, caused by the oomycetes *Pythium violae* and *Pythium sulcatum*, results in irregularly shaped, depressed lesions on the taproots (ipm.ucanr.edu).

List of Carrot Diseases

Source: *en.wikipedia.org/wiki/List_of_carrot_diseases*

Fungal Diseases

Disease	Pathogen
Alternaria leaf blight	*Alternaria dauci*
Black root rot	*Thielaviopsis basicola - Chalara elegans* [synanamorph]
Black rot (black carrot root dieback)	*Alternaria radicina = Stemphylium radicinum*
Blue mold rot (blue green mold)	*Penicillium expansum*
Brown rot (Phoma disease)	*Leptosphaeria libanotis- Phoma rostrupii* [anamorph]
Buckshot rot	*Typhula* spp.
Canker	*Thanatephorus cucumeris* • *Rhizoctonia solani* [anamorph]
Cavity spot	*Pythium violae* *Rhizoctonia solani*
Cercospora leaf spot	*Cercospora carotae*
Cottony rot	*Sclerotinia minor* *Sclerotinia sclerotiorum*
Crater rot	*Athelia arachnoidea* (anamorph = *Fibulorhizoctonia carotae*)
Crown rot	*Rhizoctonia solani*
Damping off	*Fusarium* spp. *Pythium* spp. *Rhizoctonia solani*
Dieback of carrots	*Pythium* spp. *Pythium debaryanum*
Downy mildew	*Plasmopara crustosa = Plasmopara nivea*
Forking, brown root	*Pythium* spp. *Pythium irregulare* *Pythium paroecandrum* *Pythium sylvaticum*
Fusarium dry rot	*Fusarium* sp.
Gray mold rot	*Botryotinia fuckeliana* *Botrytis cinerea* [anamorph]
Hard rot	*Fusarium* sp. *Gliocladium roseum*
Lateral root dieback	*Pythium* spp.
Leaf rot	*Typhula variabilis*
Leaf spot	*Ramularia* spp.
Licorice rot	*Mycocentrospora acerina = Centrospora acerina* *Sclerotinia sclerotiorum* *Thielaviopsis basicola* *Typhula* spp.
Phytophthora root rot	*Phytophthora megasperma*
Pink mold rot	*Trichothecium roseum*
Powdery mildew	*Erysiphe heraclei* *Erysiphe polygoni* *Erysiphe umbelliferarum f. dauci*
Pythium brown rot and forking	*Pythium irregulare* *Pythium paroecandrum* *Pythium sulcatum* *Pythium sylvaticum*
Pythium root dieback	*Pythium* spp.
Rhizoctonia canker	*Rhizoctonia solani*
Rhizoctonia seedling disease	*Rhizoctonia* spp.

Rhizopus wooly soft rot	*Rhizopus arrhizus = Rhizopus oryzae* *Rhizopus stolonifer = Rhizopus nigricans*
Root canker	*Rhizoctonia* spp.
Root dieback	*Pythium* spp.
Root rot	*Fusarium* spp. *Fusarium culmorum* *Rhizoctonia* spp. *Sclerotium rolfsii*
Phymatotrichum root rot (cotton root rot)	*Phymatotrichopsis omnivora = Phymatotrichum omnivorum*
Rubbery brown rot	*Phytophthora cactorum* *P. megasperma*
Rubbery slate rot	*Pythium debaryanum*
Rust	*Aecidium foeniculi* *Uromyces graminis* *Uromyces lineolatus subsp. nearcticus = Uromyces scirpi*
Rusty root	*Pythium* spp. *Olpidium brassicae*, as vector, tobacco necrosis virus
Sclerotinia rot	*Sclerotinia sclerotiorum* *Sclerotinia minor*
Seed mold	*Alternaria alternata* *Gibberella fujikuroi* *Fusarium moniliforme* [anamorph]
Sooty rot	*Aspergillus niger*
Sour rot	*Geotrichum candidum*
Southern blight	*Athelia rolfsii* *Sclerotium rolfsii* [anamorph]
Stem spot	*Diaporthe arctii*
Tip rot	*Numerous pathogens*
Umbel blight	*Colletotrichum gloeosporioides*
Violet root rot	*Helicobasidium brebissonii= Helicobasidium purpureum* *Rhizoctonia crocorum* [anamorph]
Watery soft rot	*Sclerotinia minor* *Sclerotinia sclerotiorum* *Sclerotium rolfsii* *Botrytis cinerea*

Bacterial Diseases

Bacterial leaf blight	*Xanthomonas campestris*
Bacterial soft rot	*Erwinia chrysanthemi* *E. carotovora* subsp. *carotovora = Pectobacterium carotovorum* subsp. *carotovorum* *E. carotovora* subsp. *atroseptica = Pectobacterium carotovorum* subsp. *atrosepticum*
Carrot bacteriosis	*Xanthomonas campestris* pv. *carotae*
Crown gall	*Agrobacterium tumefaciens*
Hairy root	*Agrobacterium rhizogenes*
Scab	*Streptomyces scabiei = Streptomyces scabi*

Nematodes, Parasitic

Cyst nematode	*Heterodera carotae*
Lance nematode	*Hoplolaimus uniformis*
Lesion nematode	*Pratylenchus penetrans* *Pratylenchus* spp.
Root knot	*Meloidogyne hapla*
Sting nematode	*Belonolaimus longicaudatus*

Virus and Viroid Diseases

Alfalfa mosaic	genus *Alfamovirus*, Alfalfa mosaic virus (AMV)
Carrot latent	genus Nucleorhabdovirus, Carrot latent virus (CtLtV)
Carrot mottle	genus Umbravirus, Carrot mottle virus (CMoV)
Carrot red leaf	genus Luteovirus, Carrot red leaf virus (CaRLV)
Carrot thin leaf	genus Potyvirus, Carrot thin leaf virus (CTLV)
Carrot yellow leaf	Carrot yellow leaf virus (CYLV)
Celery mosaic	genus Potyvirus, Celery mosaic virus (CeMV)
Cucumber mosaic	genus Cucumovirus, Cucumber mosaic virus (CMV)
Curly top	genus Hybrigeminivirus, Beet curly top virus (BCTV)
Motley dwarf	genus Luteovirus, Carrot red leaf virus (CaRLV) genus Umbravirus, Carrot mottle virus (CMoV)

Phytoplasmal and Spiroplasmal Diseases

Aster yellows	Aster yellows phytoplasma

Miscellaneous Diseases and Disorders

Crown rot disorder	No specific pathogen associated with this disorder
Heat canker	High soil surface temperature
Hollow black heart	Boron deficiency
Ozone injury	Ozone pollution
Root scab	Physiological
Speckled carrot	No pathogen; genetic disorder

It is important to recognize and distinguish the root diseases of carrots from other problems. Nematodes such as root-knot, root lesion and cyst cause some unique symptoms which can be easily confused with diseases. Forked and split carrots are a major problem in many areas.

These losses are often associated with inconsistent moisture patterns, soil compaction or disturbance patterns and sometimes due to herbicide residues. It is important to properly diagnose carrot root diseases before initiating a management strategy (www.ontario.ca).

Diseases and Their Control

Aster yellows symptoms on leaves	**Black root rot symptoms**	**Cavity Spot symptoms**	**Common Scab symptom**
Yellowed foliage Photo: Howard	**Black rot lesions** Photo: Pepin	**Many cavity spots** Photo: Ormrod	**Scab lesions on the surface** Photo: Evans
Lateral root dieback	**Leaves totally killed off following infection by the *Alternaria* leaf blight fungus**		**Cercospora leaf spot**
Fungus causes stubby carrots Photo: Howard	Photo: McDonald		**Lesions on carrot leaf stems** Photo: McDonald
Source: www.alberta.ca/diseases-of-vegetables-carrots.aspx			

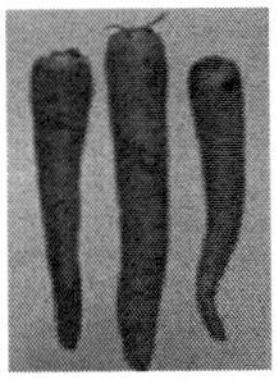

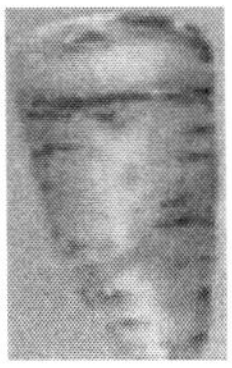

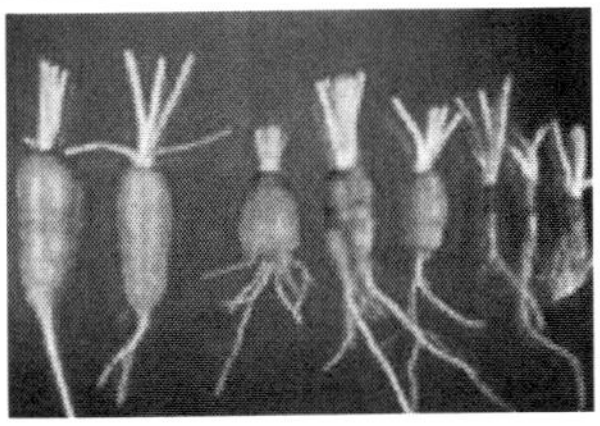

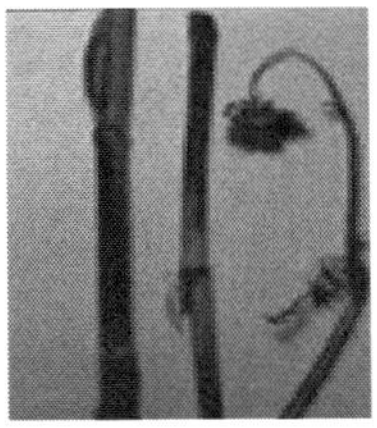

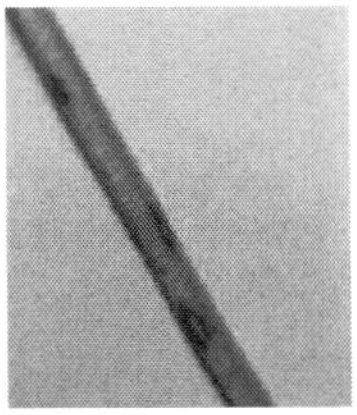

Aster yellows [mycoplasma (phytoplasma-like organism, PLO)] are common diseases in carrot production causing losses from trace levels to 10% or more in the unharvested crop. Affected carrots have leaf yellowing and the carrot taproot produces masses of fibrous side roots making it unmarketable. This disease is spread by leaf hoppers. Aster yellows affects canola, onions, parsnips and faba beans as well as flowers, forage and field crops (www.alberta.ca/diseases-of-vegetables-carrots.aspx). Dormant buds in the crown grow out into chlorotic shoots, giving rise to wither-broom appearance on the top. Size and quality of roots are reduced and malformed. Internal texture of roots show marked changes causing reduction in value of carrots for fresh market as well as for processing. Roots of infected plants have a bitter taste with astringent flavour (nhb.gov.in/pdf/vegetable/carrot/car002.pdf). The disease is transmitted by the 6-spotted leafhopper. It can be managed through effective weed control especially of those acting as alternate hosts. Insecticidal spraying for control of leafhopper vector will help reducing the disease incidence (nhb.gov.in/pdf/vegetable/carrot/car002.pdf). Chlorotic mottle disease is due to CMV, spreading through the vector *Aphis gossypii*. This causes mild chlorotic mottling, interveinal chlorosis, vein banding and stunted growth coupled with reduced leaf size. This disease can be minimized by spreading 'silver colour agri-mulch sheets', spray of mineral or neem oil, and through spraying at 10 days intervals of dimethoate or monocrotophos at 0.05% so that vectors are controlled. Yellow mosaic is symptomized by yellowing and marginal chlorosis of leaves which ultimately reduces the lamina size and causes wild curling.

This virus spreads through whitefly. As soon as such plants appear, these should be lifted and destroyed. Silver agri-mulch should be used and 2% mineral or neem oil should be sprayed on the crop. The vectors can also be killed through spraying the crop with acephate 0.15% or dimethoate 0.05% at every 10 days. Carrot red leaf virus causes plant stunting, sometimes with mottling of plants, *vis-à-vis* reddening and yellowing of carrot leaves. Its vector is carrot willow aphids, *Cavariella aegopodii*. Carrots should be grown by spreading the 'silver colour agri-mulch sheet in the field' to repel the aphids. Foliar spraying with 2% mineral oil or neem oil or 0.05% dimethoate or monocrotophos at 10-15 days interval will keep the vectors at bay.

Black root rot or black mould [*Thielaviopsis basicola* (*Chalara elegans*)] is though quite rare in Ontario (Canada) but elsewhere it can be a problem on carrots grown on muck or high organic soils. Root damage and high storage temperatures appear to favour disease development. Washed carrots may show scattered black lesions when stored in polythene bags (www.alberta.ca/diseases-of-vegetables-carrots.aspx).

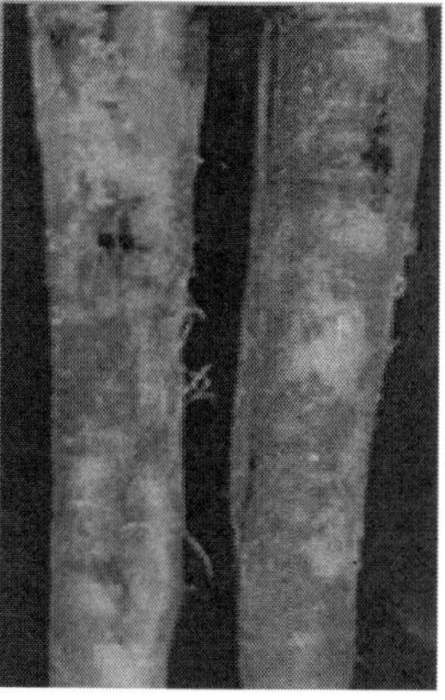

Black root rot
Source: www.ontario.ca

Field symptoms of this disease are not seen as it is a post-harvest disease associated with carrots stored at high temperatures (above 25°C) and at high humidity. It is symptomized as dark irregular lesions developing on the surface of the root. Roots damaged during packaging are more likely to become infected (www.ontario.ca). Careful harvest, handling and storage at proper temperatures from harvest to sale are the key management suggestions (www.ontario.ca). Cavity spot and lateral root dieback are caused due to several species of *Pythium*. Cavity spot and *Pythium* root dieback (formerly known as rusty root) are common diseases of carrots especially on muck soils. Cavity spot occurs on both parsnips and carrots grown on mineral or peat (muck) soils. In its infection though crop yield is not affected but the cavities make the roots unmarketable (www.alberta.ca/diseases-of-vegetables-carrots.aspx). Cavity spots are often seen at the time of harvesting though infections can occur early at the time of root development which is favoured usually by moist soil conditions (www.ontario.ca). Horizontal lesions from 1 to 10mm appear on root surface which enlarge into larger decayed areas if invaded by other fungi or bacteria (www.ontario.ca). Management of *Pythium* disease is difficult if the growing season remains wet. However, precision planting on raised beds, crop rotations with cole crops, onions or potato and improvements to field drainage may help combating this problem. There is evidence that high nitrogen levels in the soil may also contribute to the problem. Avoiding problem in the fields is helpful if that option is available (www.ontario.ca). However, as there are some tolerant varieties so only such varieties should be grown. Seed treatment before sowing as well as soil application of fungicides such as Captafol for managing cavity spots may be helpful, however, persisting soil moisture conditions and disease pressure levels may aggravate the situation (www.ontario.ca).

Lateral root dieback symptoms appear as rusty-brown lateral root formation, forking and stunting and such symptoms can easily be confused with damage from nematodes, soil compaction or soil drainage problems (www.ontario.ca).

Pythium root dieback symptoms on carrots
Source: *www.ontario.ca*

Many *Pythium* species are involved in this disease often referred to as rusty root in Ontario. The disease occurs in both mineral and muck soils. For managing this disease there should be sufficient drainage facilities in the field (www.alberta.ca/diseases-of-vegetables-carrots.aspx).

Leaf blight (*Cercospora dauci*) is the most common leaf disease of carrots and parsley. This fungus is seed- and residue-borne. Complete loss of foliage can take place during periods of prolonged wet and humid weather (www.alberta.ca/diseases-of-vegetables-carrots.aspx). This disease first appears as irregular dark-browning to black at leaflet margins. Lesions produced on the petioles and stems are dark brown and often coalesce and girdle the stems. As the disease progresses entire leaflets may shrivel and die appearing scorched. Leaf lesions are generally more prevalent on older foliage and plants. This occurs due to high leaf density *cum* poor air circulation among the older lower leaves in the canopy where there is more moisture even.

This disease can be managed through seed treatment, crop rotation and by the use of foliar fungicides such as 0.2% Foltaf or 0.3% copper hexachloride (nhb.gov.in/pdf/vegetable/carrot/car002.pdf). Seed treatment with Thiram at the rate of 3g per kilogramme of seed before sowing is also very effective. HWT of seeds at 50^{0}C for 15 minutes is also effective. However, suitable crop rotation is also the option provided there is ample provision of drainage facilities as in no case the water should stand in the field. Field sanitation and destruction of infected plant parts will minimize the disease incidence. Kavach 0.2%, difenaconazole/tricyclazole 0.05-0.06% or thiophanate methyl 0.1% spraying at 10-14 days is quite effective in controlling this disease.

Cercospora leaf spot (*Cercospora carotae*) is as destructive as *Alternaria* leaf blight. The disease is both seed- and residue-borne. Younger leaves are more susceptible than older leaves. Its control is similar to *Alternaria* leaf spot (www.alberta.ca/diseases-of-vegetables-carrots.aspx). Powdery mildew (*Oidium dauci; Erysiphe polygoni*) occurs as white powdery growth on the leaves and petioles causing the leaves to turn brown and wilt. Spraying of Bavistin or benomyl (0.1%) at an interval of 8-10 days controls this disease effectively.

Watery soft rot (*Sclerotinia sclerotiorum*) appears as white mould which is a quite common disease on most of the vegetables. Carrots are susceptible, especially late in the season and during storage. The disease is present in soil, storage boxes or storage areas and often shows up after the crop has been harvested. Symptoms appear in the field with characteristic white mould with black sclerotia on the crown of infected carrots (www.ontario.ca; nhb.gov.in/pdf/vegetable/carrot/car002.pdf).

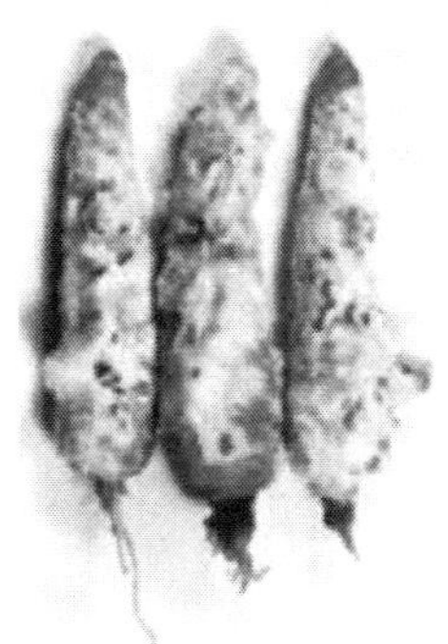

Sclerotinia mold covers carrots
Photo: Howard

It is the most destructive disease of carrots in Alberta (Canada). The infectious ascospores of this fungus are abundant as a consequence of the build-up of the disease on millions of acres of canola and bean crops. Carrots may show little or no damage incidence in the field but following washing and storage white mould outbreaks is observed often on the stored roots. Only a small percentage of the roots may be initially infected but mycelium can move rapidly from carrot to carrot. In a matter of weeks the whole storage container may become a mass of white mould and black sclerotia surrounding each and every carrot (www.alberta.ca/diseases-of-vegetables-carrots.aspx).

Crop rotation, raised bed sowing, weed control especially to improve air circulation, rapid cooling prior to storage and meticulous sanitation of all storage components are necessary to reduce losses from this disease. Field applied fungicides are available; however, heavy disease pressure in the field or placement of carrots into infected storages will still lead to severe losses (www.ontario.ca; nhb.gov.in/pdf/vegetable/carrot/car002.pdf). Frequent inspection in storage, low temperatures, aeration and washing in a final water of 2-5% diluted bleach solution [1 part bleach (sodium hypochlorite) to 20 parts water] may give adequate control (www.alberta.ca/diseases-of-vegetables-carrots.aspx).

Crown rot (*Rhizoctonia solani*) and dry rot (*Fusarium* spp.) may cause problems in stored carrots (www.alberta.ca/diseases-of-vegetables-carrots.aspx). Crown rot is caused by the same *Rhizoctonia* fungus which causes damping-off when seedlings are small. On carrot roots, early symptoms are horizontal dark brown lesions and as the crop matures the tops may die in patches in the field. Infections can occur early in the growing season during wet periods, however, symptoms may not be detected until later in the season. Near harvest the lesions join to form large, deep, rotten areas on the top part of the root. Control of this disease is nearly impossible if late summer and fall conditions are wet (www.

ontario.ca). Growers should store the roots as cool as possible at 0 ^{0}C and avoid planting on land where these diseases previously may have occurred (www.alberta.ca/diseases-of-vegetables-carrots.aspx). Earlier harvests, planting on ridges, crop rotation, careful handling during harvest, storage sanitation *cum* extremely good storage conditions might reduce losses due to this disease. Reducing the spread of infected soil on cultivation and harvesting equipment is also important (www.ontario.ca).

Rubbery brown rot (*Phytophthora porri*) as well as **violet root rot** only sometimes occur on carrot roots (www.alberta.ca/diseases-of-vegetables-carrots.aspx) whose control measures are the same as to that of *Rhizoctonia* rot. Violet root rot is caused by a closely related species of *Rhizoctonia*, however, this disease has characteristically unique symptoms. It is found in patches in the field, often as weak areas or areas where the carrot tops are dying. Characteristic root symptoms are soil clinging to partially rotten and leathery textured roots with purplish or violet colouration on roots (www.ontario.ca). The only management suggestions are to avoid infected areas for its cultivation, rotate grain crops, ensure good nutrition and drainage, and most importantly the reduction of the spread of infected soil through farm implements (www.ontario.ca).

Foliage symptoms & carrot decay | Lesions on harvested carrots

Crown rot

Lesions on harvested carrots | Muck soil field symptoms

Violet root rot

Source: www.ontario.ca

Black rot (*Alternaria radicina*) of carrots is different from black root rot mentioned above and is also of rare occurrence. It attacks the roots and can be a problem on celery, parsnip and dill. The disease can persist for up to 8 years in the soil if once infected (www.alberta.ca/diseases-of-vegetables-carrots.aspx). It is a seed- and soil-borne disease characterized by a shiny black decay at the crown area and a greenish-black mould on the taproot. The infected tissue is greenish-black to jet black due to presence of black spore masses. This disease affects the roots in the field as well as in storage (nhb.gov.in/pdf/vegetable/carrot/car002.pdf). To avoid infection of this pathogen, only clean and treated carrot seeds should be used by following proper management practices recommended similar to those of *Pythium* diseases. An eight year crop rotation away from carrots, parsnips, parsley and celery should be followed along with the deep ploughing of the field (www.ontario.ca). Proper field sanitation, use of resistant cultivars, low temperature storage (0^0C) at 95% RH of roots and use of foliar application of fungicides such as 0.1% Bavistin alternate with 0.2% Captafol alternatively at fortnightly intervals are quite effective (www.alberta.ca/diseases-of-vegetables-carrots.aspx; nhb.gov.in/pdf/vegetable/carrot/car002.pdf).

Common scab (*Streptomyces scabies*) is favoured by alkaline and dry organic soils. Both raised and sunken, dry, corky lesions may form on the carrot root, however, symptoms are rarely severe enough to cause major yield losses (www.ontario.ca). Carrot scab is caused due to soil bacterium *Streptomyces* which also attacks potatoes and other root crops. To manage this disease, crop rotation with non-host crops should be followed and soil pH should be brought down below 7 (www.alberta.ca/diseases-of-vegetables-carrots.aspx).

Bacterial leaf blight (*Xanthomonos campestris* pv. *carotae*) is a soil-borne disease causing leaf spotting similar to *Cercospora* or *Alternaria blights* (www.alberta.ca/diseases-of-vegetables-carrots.aspx).

Bacterial soft rot (*Erwinia carotovora* pv. *carotovora* and *E.c.* pv. *atroseptica*) is common on carrots both in the field and in storage (www.alberta.ca/diseases-of-vegetables-carrots.aspx). Bacterial tip rot can occur in the field. It occurs where drainage conditions in the field are utterly poor, soil conditions are wet, autumn season remains quite wet or storage conditions remain poor. Losses due to bacterial soft rot in storage usually occurs after other storage diseases have taken hold (www.ontario.ca). Soft water-soaked irregular lesions appear on the roots. Initially these lesions are superficial but soon spread to cover the inner tissues. The foliage may remain green until the disease on the root advances considerably. Entire plant wilts when complete rotting of the root takes place and then this emanates a foul odour from the decayed roots. Disease incidence is favoured by abundant moisture level on root surface (nhb.gov.in/pdf/vegetable/

carrot/car002.pdf). While harvesting, the roots should be handled carefully so that these are not injured. Handling, grading and sanitation are the only ways to reduce the problem. Root surface should be kept dry and these should be stored at 0 ^{0}C with 95% relative humidity. In poorly drained soils, these should be grown on raised beds (www.ontario.ca). Roots should finally be washed in 2-5% bleach solution [1 part bleach (sodium hypochlorite) to 20 parts water] for getting effective control (www.alberta.ca/diseases-of-vegetables-carrots.aspx).

Insect-Pests

Nematodes such as northern root-knot nematode (*Meloidogyne hapla*) has a very wide host range. This nematode can attack carrots and may render them unmarketable. (www.alberta.ca/diseases-of-vegetables-carrots.aspx) as the feeder roots are found galled and forked so these roots become unable to suck soil nutrients, *vis-à-vis* unmarketable. To control this, one should follow crop rotation with spinach, marigold and radish. Soil application of neem cake along with *T. harzenianum* (containing 1×10^6 spores@10g/2.5m^2) has also proved quite effective (Parvatha Reddy, 2019). Carrot var. 'Arka Suraj' is tolerant to root-knot nematode.

Cutworms are thick and dark caterpillars (the larvae of several species of night-flying moths) which can be a problem during seedling establishment. These generally attack young seedlings at night and may cut the stem off (sfyl. ifas.ufl.edu). These are the common pests of many vegetable crops including carrots, celery, lettuce, onion, tomato, pepper, eggplant, cole crops, rutabaga, beans, cucurbits, sweet corn and several others. Most species of cutworms are solitary feeders found in the soil, however, some species occasionally attack even the foliage and/or fruit of some vegetable crops. Commonly found cutworms include the black cutworm (*Agrotis segetum, A. ipsilon, A. exclamationis*), variegated cutworm (*Pedridroma saucia*), dark-sided cutworm (*Euxoa messoria, E. nigricans*), dingy cutworm (*Feltia jaculifera*), glassy cutworm (*Crymodes devastator*), red-backed cutworm (*Euxoa ochrogaster*), black army cutworm (*Actebia fennica*), large yellow underwing (*Noctua pronuba*), lesser yellow underwing (*N. comes*), and white cutworm (*Euxoa scandens*).

The black cutworm and the variegated cutworm are the most common ones attacking vegetables though other species may be present in some other areas. Recognizing damage from cutworms is relatively simple when plants are small. Seedlings are often 'cut off' at ground level and if the soil around the fresh severed plants is dug up to a depth of about 5cm the characteristic cutworm larvae can be found. Sometimes even outer bark of the plants are gnawed away. While feeding these make cavities on carrot roots. Generally these feed during night (www.ontario.ca/page/managing-cutworms-vegetable-cr). Preparing soil before planting may help control cutworms (sfyl.ifas.ufl.edu). Managing cutworms requires regular and consistent monitoring and diligent field management as part of an 'integrated crop management programme' (ICM). Field selection, field history and weed management are important methods to reduce damage from cutworms. Fields that had large weed population, especially grasses and chickweed may experience more cutworm problems the following season.

Crop rotation is also a helpful management tool. For black cutworm, treatment with insecticides is effective only if the larvae are very small (under 2.5cm) and soil conditions are favourable though older larvae may be tolerant. If damage is isolated to a small patch or field border area, spot treatments are recommended. Application of insecticide late in the day may kill these. Chlorpyriphos 20EC at 0.1% soil drenching is quite effective. Many natural predators, parasites and birds prey upon cutworms and reduce the population. Take steps to preserve such beneficial organisms in and around vegetable fields. Reducing unnecessary spraying for other pests and providing refuges may help to encourage beneficial organisms (www.ontario.ca/page/managing-cutworms-vegetable-cr). One of the pathogenic nematodes, *Steinernema carpocapsae* can be used in moist but well-drained soil but only during the summer months.

Wireworms are slender, shiny, hard and orange-brown click beetle larvae (*Agriotes lineatus, A. obscurus, A. sputator* and *Atbous haemorrhoidalis*) having three pairs of short legs near the head end. These feed on roots and bore into root vegetables including carrot roots. These can stay in the soil for 1-5 years. These are only up to 2.5cm long having a small protuberance on the underside of the rear end of the abdomen. These pests typically cause problems in organic soils rather than mineral soils. While living in the soil, these kill seedlings by biting into the stems just below soil level and during hot months these tunnel into carrot roots. Wireworms can attack developing carrots directly or provide entry points for secondary rot-causing pathogens. If wireworms are present in the soil, 4-5 bait stations can be set (approximately 15cm-deep holes with untreated wheat and corn) four to five weeks before planting (sfyl.ifas.ufl.edu). The crop at seedling stage is protected by dusting seed rows with pirimiphos methyl or lindane.

Tawny mole cricket (*Talpa europaea*) is one of the few mole cricket pests. Though mole crickets primarily feed on roots, they also feed on stems and leaves at night during warm and wet weather. They create meandering tunnels in, around and under-developing root systems, damaging young seedlings. The most useful control measure is to apply soil insecticide before planting. Parasitic nematode *Steinernema scapterisci* can also be an effective control agent (sfyl.ifas.ufl.edu). These can also be controlled by igniting smokes in the tunnel.

Female adult **leafminers** are small flies with yellow abdomens whose larvae cause leaf damage through mining the leaves. These insects insert their eggs in feeding punctures located on the plant's upper leaf surfaces. When the eggs become larvae, these start feeding through mining the leaves between the upper and lower leaf surfaces. Healthy plants can typically tolerate leafminer damage though heavy damage may cause leaf drop. Additionally, the exit holes in old mines could provide access for pathogens. Chemical control may be difficult because larvae are protected in the leaf during the feeding stage. The best option is to target small larvae (sfyl.ifas.ufl.edu). To get rid of this pest, affected leaves should be removed and burnt together with spraying of malathion or primiphos methyl.

Green peach aphids (*Myzus persicae*) are the most reported aphid species though even **melon aphids** (*Aphis gossypii*) also damage carrots. These small soft-bodied insects pierce plant tissues with their needle-like mouthparts and suck out the plant sap. Toxins in their saliva may cause foliage to curl and deform and together can invite black sooty mould to grow as these leave large amounts of honeydew on plant surfaces. Despite feeding on tender parts and on leaves, these also act as vectors in transmitting various viral diseases to the affected plants. Aphids are biologically controlled by natural predators such as lady bird beetles and lacewings (sfyl.ifas.ufl.edu). Chemically these are controlled by spraying with pirimicarb, heptenophos with permethrin, dimethoate, pirimiphos methyl or fenitrothion.

The wingless summer form of the **willow carrot aphid** (*Cavariella aegopodii*) is pale green with the cornicles, cauda (tail-like structure) and legs pale to slightly dusky. These are medium-size aphids with rather elongate oval-shaped bodies that are flattened front to back. The upper surface of the body may be roughened by numerous small depressions. The wingless forms have a second tail-like process called the supracaudal process directly above the cauda giving the aphid the appearance of having twin tails when viewed with a hand lens. This supracaudal process may be triangular or fingerlike in shape. The winged forms have a black head and thorax. The abdomen is pale green with dark areas on the sides and dark bands on the top. The antennae are black. The legs are pale and black at the tips. The cornicles are somewhat swollen near the tip (ipm.ucanr.edu/agriculture/carrot/willow-carrot-aphid/). The nymphs and adults of this insect feed on the plants by sucking the sap from their tender plant parts, and even transmit the viral diseases such as 'carrot motley dwarf', 'carrot red leaf', 'parsley yellow fleck', 'celery mosaic virus', 'sugarbeet mosaic virus' and 'cauliflower mosaic virus'. Predators such as green lacewing larvae, lady beetles and syrphid fly larvae prey on this aphid as well as on other aphid species (ipm.ucanr.edu/agriculture/carrot/willow-carrot-aphid/). These aphids are controlled similar to those of 'green peach aphid' as elaborated above.

Heat canker in carrots is caused by collapsing and dying of tissues of the root at or near the soil surface due to direct strong sunlight which causes even major disfigurement of the root. Black or dark soils are most conducive to this type of injury (www.alberta.ca/diseases-of-vegetables-carrots.aspx).

Physical damage can also reduce the value of carrot crops. Two main forms of damage are: i) splitting which consequently causes a longitudinal crack during root growth that can be a few centimetres to the entire length of the root, and ii) breaking which occurs postharvest. These disorders can affect over 30% of commercial crops. Factors associated with high levels of splitting include wide plant spacing, early sowing, lengthy growth durations and genotypic factors (Benjamin *et al.*, 1997). Increase or decrease of N in the soil increases or decreases the splitting. Ammonium compounds being more in the soil, splitting is also more, as compared to any other form of nitrogen.

Fanging or **root forking** in carrot roots affects its marketability. It is usually due to early damage of apical root meristem which forces the roots becoming short *cum* distorted and branched.

References

Abbott, Catherine, 2012. The Year-Round Harvest: A Seasonal Guide to Growing, Eating, and Preserving the Fruits and Vegetables of Your Labor. *Adams Media,* pp. 54–5. ISBN 978-1-4405-2816-3.

Aguila Ruiz-Sola, M. and M. Rodríguez-Concepción, 2012. Carotenoid biosynthesis in arabidopsis: A colorful pathway. *BioOne*, 1–28.

Amy, S. Paller and J. Mancini Anthony, 2011. SPEC - Hurwitz Clinical Pediatric Dermatology E -Book 12Month Subscription: A Textbook of Skin Disorders of Childhood and Adolescence. *Elsevier Health Sciences*, pp. 264.. ISBN 978-1-4377-3613-7.

Ballmer-Weber, B.K., Hansen K. Skamstrup, J. Sastre, K. Andersson, I. Bätscher, J. Ostling, L. Dahl, K.M. Hanschmann, T. Holzhauser, L.K. Poulsen, J. Lidholm and S. Vieths, 2012. Component-resolved in vitro diagnosis of carrot allergy in three different regions of Europe. *Allergy*, **67**(6): 758–66. doi:10.1111/j.1398-9995.2012.02827.x. PMID 22486768. S2CID 23864612.

Banga, O. 1957. Origin of the European cultivated carrot. *Euphytica*, **6**(1): 54–63. doi:10.1007/BF00179518. S2CID 39884390.

Baranska, Malgorzata, Hartwig Schulz, Rafal Baranski, Thomas Nothnagel and Lars P. Christensen, 2005. In situ simultaneous analysis of polyacetylenes, carotenoids and polysaccharides in carrot roots. *J. agric. and Food Chem.*, **53**(17): 6565–71. doi:10.1021/jf0510440. PMID 16104767.

Baranski, Rafal, Anna Maksylewicz-Kaul, Thomas Nothnagel, Pablo F. Cavagnaro, Philipp W. Simon and Dariusz Grzebelus, 2012. Genetic diversity of carrot (*Daucus carota* L.) cultivars revealed by analysis of SSR loci. *Gen. Res. and Crop Evol.*, **59** (2): 163–70. doi:10.1007/s10722-011-9777-3.

Benjamin, L. R., A. McGarry and D. Gray, 1997. The root vegetables: Beet, carrot, parsnip and turnip. The Physiology of Vegetable Crops. CABI, Wallingford, UK, pp. 553–80. ISBN 978-0-85199-146-7.

Bidlack, Wayne R. and Raymond L. Rodriguez, 2011. Nutritional Genomics: The Impact of Dietary Regulation of Gene Function on Human Disease, 321p. CRC Press, USA.ISBN 978-1-4398-4452-6.

Bradeen, James M. and Philipp W. Simon, 2007. Carrot. In: Vegetables. Genome Mapping and Molecular Breeding in Plants (ed. Cole, Chittaranjan), vol. 5, pp. 162-84. Springer, New York, USA. ISBN 978-3-540-34535-0.

Brown, Ellen, 2012. The Complete Idiot's Guide to Smoothies, 21p. DK Publishing. ISBN 978-1-4362-9393-8.

Burney, P., C. Summers, S. Chinn, R. Hooper, R. van Ree and J. Lidholm, 2010. Prevalence and distribution of sensitization to foods in the European Community Respiratory Health Survey: A EuroPrevall analysis. *Allergy*, **65**(9): 1182–8. doi:10.1111/j.1398-9995.2010.02346.x. PMID 20180791. S2CID 22718444.

Chapman, David, 2008. Exploring the Cornish Coast, 110p. Alison Hodg, Penzance. ISBN 9780906720561.

Chapman, Pat, 2007. India Food and Cooking: The Ultimate Book on Indian Cuisine, pp. 158-230. New Holland Publishers. ISBN 978-1-84537-619-2.

Christensen, P.L. 2011. Aliphatic C17-Polyacetylenes of the Falcarinol Type as Potential Health Promoting Compounds in Food Plants of the Apiaceae Family. *Rec. Patents Food Nutr. Agric.*, **3**: 64–77.

Crosby, D.G. and N. Aharonson, 1967. The structure of carotatoxin, a natural toxicant from carrot. *Tetrahedron*, **23**(1): 465–72. doi:10.1016/S0040-4020(01)83330-5. PMID 6037290.

Cunningham, Sally Jean, 2000. Great Garden Companions: A Companion-Planting System for a Beautiful, Chemical-Free Vegetable Garden. Rodale, pp. 195–6. ISBN 978-0-87596-847-6.

Czepa, A. and T. Hofmann, 2003. Structural and sensory characterization of compounds contributing to the bitter off-taste of carrots (Daucus carota L.) and carrot puree. *J. agric. Food Chem.*, **51**(13): 3865–73. doi:10.1021/jf034085+. PMID 12797757.

Czepa, A.and T. Hofmann, 2004. Quantitative studies and sensory analyses on the influence of cultivar, spatial tissue distribution, and industrial processing on the bitter off-taste of carrots (*Daucus carota* L.) and carrot products. *J. Agric. Food Chem.*, **52**: 4508–14.

Davis, R. Michael, 2004. Carrot diseases and their management. In: Diseases of Fruits and Vegetables: Diagnosis and Management (ed. Naqvi, S.A.M.H.), pp. 397-439. Springer. ISBN 978-1-4020-1822-0.

Dawid, C., F. Dunemann, W. Schwab, T. Nothnagel and T. Hofmann, 2015. Bioactive C 17-polyacetylenes in carrots (*Daucus carota* L.): Current knowledge and future perspectives. *J. Agric. Food Chem.*, **63**: 9211–22.

Elzer-Peters, K. 2014. Midwest Fruit & Vegetable Gardening: Plant, Grow, and Harvest the Best Edibles – Illinois, Indiana, Iowa, Kansas, Michigan, Minnesota, Missouri, Nebraska, North Dakota, Ohio, South Dakota & Wisconsin, 136p. Cool Springs Press, USA. ISBN 978-1-61058-960-4.

en.wikipedia.org/wiki/Carrot.

en.wikipedia.org/wiki/Daucus_carota.

en.wikipedia.org/wiki/List_of_carrot_diseases.

encyclopedia.pub/entry/11770.

Faulkner, Herbert Waldron, 1917. The Mysteries of the Flowers, 210p. Frederick A. Stokes, UK.

food.ndtv.com/food-drinks/red-carrots-desi-gajar-versus-orange-carrots-which-ones-would-you-choose-1766761.

Fotopoulos, V. and A.K. Kanellis, 2013. Altered apoplastic ascorbate redox state in tobacco plants via ascorbate oxidase overexpression results in delayed dark-induced senescence in detached leaves. *Plant Physiol. Biochem.*, **73**: 154–60.

Garrod, B., B.G. Lewis and D.T. Coxon, 1978. Cis-heptadeca-1,9-diene-4,6-diyne-3,8-diol, an antifungal polyacetylene from carrot root tissue. *Physiol. Pl. Path.*, **13** (2): 241–6. doi:10.1016/0048-4059(78)90039-5.

Gest, N., H. Gautier and R. Stevens, 2013. Ascorbate as seen through plant evolution: The rise of a successful molecule? *J. exp. Bot.*, **64**: 33–53.

Gonçalves, E.M., J. Pinheiro, M. Abreu, T.R.S. Brandão and C.L.M. Silva, 2013. Carrot (*Daucus carota* L.) peroxidase inactivation, phenolic content and physical changes kinetics due to blanching. *J. Food Eng.*, **97**: 574–81.

Grubben, G.J.H. 2004. Vegetables. Plant Resources of Tropical Africa, 282p. ISBN 978-90-5782-147-9.

Hedren, E., V. Diaz and U. Svanburg, 2002. Estimation of carotenoid accessibility from carrots determined by an in vitro digestion method. Europ. J. Clin. Nutrit., **56**(5): 425–30. doi:10.1038/sj.ejcn.1601329. PMID 12001013.

http://news.bbc.co.uk/2/hi/health/7188969.stm.

iihr.res.in/carrot-arka-suraj-0.

Iorizzo, Massimo, Julien Curaba, Marti Pottorff, Maroi G. Ferruzzi, Philipp Simon and Pablo F. Cavagnaro, 2020. Carrot anthocyanins genetics and genomics: Status and perspectives to improve its application for the food colorant industry. *Genes,* **11**(8): 906. doi:10.3390/genes11080906. ISSN 2073-4425. PMC 7465225. PMID 32784714.

Iorizzo, Massimo, Douglas A. Senalik, Shelby L. Ellison, Dariusz Grzebelus, Pablo F. Cavagnaro, Charlotte Allender, Johanne Brunet, David M. Spooner, Allen van Deynze and Philipp W. Simon, 2013. Genetic structure and domestication of carrot (*Daucus carota* subsp. *sativus*) (Apiaceae). *Amer. J. Bot.,* **100**(5): 930–38. doi:10.3732/ajb.1300055. PMID 23594914.

ipm.ucanr.edu/agriculture/carrot/cavity-spot/.

Jansen, Gabrielle Claire and Hans Wohlmuth, 2014. Carrot seed for contraception: A review. *Aust. J. Herbal Med.*, **26**(1): 10–17.

John M. Riddle, 2023. Contraception and Abortion from the Ancient World to the Renaissance, 58p. Retrieved onMarch 27, 2023 via en.wikipedia.org/wiki/Daucus_carota.

Kartika, Engla, Chandra Boy Zulharmita and Harrizul Rivai, 2001. *Int. J. Pharmaceut. Sci. & Med. (IJPSM),* **6**(1): 75-82.

krishijagran.com/agripedia/carrots-are-the-best-crops-to-grow-in-august-read-full-details-of-how-to-grow-this-nutritious-vegetable/.

Leja, M., I. Kaminska, M. Kramer, A. Maksylewicz-Kaul, D. Kammerer, R. Carle and R. Baranski, 2013. The content of phenolic compounds and radical scavenging activity varies with carrot origin and root color. *Pl. Foods Hum. Nutr.*, **68**: 163–70.

McClintock, David and R.S.R. Fitter, 1956. *The Pocket Guide to Wild Flowersi,* 108p. Collins. England.

nhb.gov.in/pdf/vegetable/carrot/car002.pdf.

nhb.gov.in/pdf/vegetable/carrot/car012.pdf.

Papoutsis, Konstantinos and Merete Edelenbos, 2021. Postharvest environmentally and human-friendly pre-treatments to minimize carrot waste in the supply chain caused by physiological disorders and fungi. *Trends in Food Sci. & Techn.*, **112**: 88–98. doi:10.1016/j.tifs.2021.03.038. S2CID 233620416.

Parvatha Reddy, 2019. Nematode Management in Horticultural, Plantation, Spice and Tuber Crops. In Handbook of Horticlture, vol. 2., pp. 1109-37. ICAR, New Delhi.

Peters, Emily, 2014. Wild Carrot Monograph (PDF). Ember Peters, RHP Wild Current Herbalism. Archived from the original (PDF) on October 20, 2020. Retrieved August 11, 2020.

Peterson, Roger Tory and Margaret McKenny, 1968. A Field Guide to Wildflowers: Northeastern and North-central North America, 48p. Houghton Mifflin Harcourt, England. Retrieved on 27 March 2017.

pharmeasy.in/blog/5-wonderful-benefits-and-uses-of-carrots/.

plants.ces.ncsu.edu/plants/daucus-carota/.

powo.science.kew.org/taxon/841063-1.

Richard, Michel 2006. Happy in the Kitchen, pp. 74, 76. Artisan Books, England. ISBN 978-1-57965-299-9.

Rodriguez-Concepcion, M. and C. Stange, 2013. Biosynthesis of carotenoids in carrot: An underground story comes to light. *Arch. Biochem. Biophys.*, **539**: 110–16.

Rubatsky, V. E., C.F. Quiros and P.W. Siman, 1999. Carrots and Related Vegetables, Umbelliferae, pp. 22-8 & 253-6. CABI Publishing, U.K. ISBN 978-0-85199-129-0.

Schmiech, L., A. Carole, B. Witulski and T. Hofmann, 2009. Structure determination of bisacetylenic oxylipins in carrots (*Daucus carota* L.) and enantioselective synthesis of falcarindiol. *J. Agric. Food Chem.* **57**: 11030–40.

seedalliance.org/publications/how-to-breed-carrots-for-organic-agriculture/.

sfyl.ifas.ufl.edu/archive/hot_topics/agriculture/carrot_pest_management.shtml.

Shannon, Nomi, 1998. The Raw Gourmet, 33p. U.K. Book Publishing Company. ISBN 978-0-920470-48-0.

Soltoft, M., A. Bysted, K.H. Madsen, A.B. Mark, S.G. Bügel, J. Nielsen and P. Knuthsen, 2011. Effects of organic and conventional growth systems on the content of carotenoids in carrot roots, and on intake and plasma status of carotenoids in humans. *J. Sci. Food Agric.*, **91**: 767–75.

Tiwari, B.K., Nigel P. Brunton and Charles Brennan, 2012. Handbook of Plant Food Phytochemicals: Sources, Stability and Extraction, 405p. John Wiley & Sons, USA. ISBN 978-1-118-46467-0.

Tsao, R. 2010. Chemistry and biochemistry of dietary polyphenols. *Nutrients*, **2**: 1231–46.

Umar, G., S. Kaur, S. Gurumayum, P. Rasane, 2015. Effect of hot water blanching time and drying temperature on the thin layer drying kinetics and anthocyanin degradation in black carrot (*Daucus carota* L.) shreds. *Food Technol. Biotechnol.*, **53**: 324–30.

Wurtele, E S., H. Wang, S. Durgerian, B.J. Nikolau and T.H. Ulrich, 1993. Characterization of a gene that is expressed early in somatic embryogenesis of *Daucus carota. Pl. Physiol.*, **102**(1): 303–12. doi:10.1104/pp.102.1.303. PMC 158776. PMID 8108498.

www.alberta.ca/diseases-of-vegetables-carrots.aspx.

www.backwoodshome.com/articles2/gist82.html.via web.archive.org/web/20110906104024/http://www.backwoodshome.com/articles2/gist82.html.

www.biologydiscussion.com/vegetable-breeding/carrot-origin-production-and-breeding-method-india/68528.

www.etymonline.com/word/carrot.

www.fao.org/faostat

www.first-nature.com/flowers/daucus-carota.php.

www.goodhousekeeping.com/home/gardening/a20705702/how-to-grow-carrots/.

www.medicalnewstoday.com/articles/270191.

www.ontario.ca/page/identification-and-management-carrot-root-diseases.

www.plantea.com/carrots.htm.

Yeager, Selene, 2008. The Doctors Book of Food Remedies: The Latest Findings on the Power of Food to Treat and Prevent Health Problems – From Aging and Diabetes to Ulcers and Yeast Infections, 366p. Rodale. ISBN 978-1-60529-506-0.

Zidorn, Christian, Karin Johrer, Markus Ganzera, Birthe Schubert, Elisabeth Maria Sigmund, Judith Mader, Richard Greil, Ernst P. Ellmerer and Hermann Stuppner, 2005. Polyacetylenes from the Apiaceae vegetables carrot, celery, fennel, parsley and parsnip and their cytotoxic activities. *J. agri. & Food Chem.*, **53**(7): 2518–23.

2

Beet (*Beta vulgaris* subsp. *vulgaris*)

Family: Amaranthaceae.

Subfamily: Betoideae.

Introduction and Origin

The generic name *Beta* is the ancient Latin name for beetroot (Gledhill, 2008), possibly of Celtic origin, becoming *bete* in Old English (en.wikipedia.org/wiki/Beetroot). *Beta vulgaris* has three recognized subspecies (Romeiras *et al.,* 2016; Kadereit *et al.,* 2006; Lange *et al.,* 1999):

- *B.v.* subsp. *vulgaris,* probably originated in Europe and includes all domesticated cultivars.
- *B.v.* subsp. *adanensis* (Pamukc. ex Aellen) Ford-Lloyd & J.T.Williams (Syn.: *Beta adanensis* Pamukc. ex Aellen), formed by a distinct group of semi-annual plants with specific morphological characteristics, and having a great decline in auto-fertilization. They occur in disturbed habitats and steppes of SE Europe (Greece) and W. Asia (Cyprus, Israel, W. Syria and Turkey).
- *B.v.* subsp. *maritima* (sea beet), the wild ancestor of all the cultivated beets which is formed by a large complex of morphological types that occur in a vast geographic area. It is distributed in SE, SW and N.Europe

along the Atlantic coasts and the Mediterranean Sea, in North Africa, Macaronesia, to W. Asia and naturalized in other continents. These grow at coastal cliffs, on stony and sandy beaches, in salt marshes or coastal grasslands, and in ruderal or disturbed places.

Domestication of beet as a leafy vegetable and root crop is known from prehistoric era but its production as a crop is new to many countries like India (Panella and Kaffka, 2011). Greeks described colourful chards, a special type of leaf beet with elongated, broad and fleshy leaf midribs and petioles. In Roman times, chard was called beta (Nottingham, 2004). Sweet *cum* swollen root types were probably selected from leafy beets which led first to table or red root beets for human consumption and later on the selection was for large white beets used as animal fodder. Today beetroot is grown all around the world for its edible leaves and taproot.

Beets are cultivated for fodder (e.g. mangelwurzel), for sugar (the sugar beet), as a leaf vegetable (chard or 'bull's blood') or as a root vegetable (beetroot/table beet/garden beet) (en.wikipedia.org/wiki/Beta_vulgaris).

Botany

Cultivated beets are herbaceous biennials but usually grown as annuals for the edible thick fleshy taproot and leaves that form during the first season (www.britannica.com). The plant grows up to a height of about 1.2m (rarely 2m) (en.wikipedia.org/wiki/Beta_vulgaris) with erect stems or in the wild often procumbent, may be simple or branched in the upper part (Shultz, 2003), and their surface is ribbed and striate (Zhu *et al.,* 2003). The basal leaves have a long and typically red petiole though may be white or yellow in some cultivars. The simple leaf blade is oblanceolate to cordate, dark green to dark red, slightly fleshy, usually with a prominent midrib, with entire or undulate margin, 5–20 cm long on wild plants and often much larger in cultivated plants. The upper leaves are smaller having rhombic to narrowly lanceolate blades (Shultz, 2003). Strongly swollen, globular to long, tapered and fleshy taproot usually has dark purplish-red skin and flesh though some cultivars are nearly white or orange (www.britannica.com). Wild subspecies have brown and fibrous taproots that are sometimes swollen and woody (en.wikipedia.org/wiki/Beta_vulgaris).

In the second season, a tall, branched and leafy stem arises to bear clusters of minute green (www.britannica.com) (tinged reddish in some cultivars) and urn-shaped hermaphrodite flowers consisting of five basally connate perianth segments (tepals), 3-5×2–3mm 5 stamens and a semi-inferior ovary with 2-3 stigmas (en.wikipedia.org/wiki/Beta_vulgaris). The perianths of neighbouring flowers are often fused (Flores Olvera *et al.*, 2008). The flowers are wind- to insect-pollinated (Free *et al.*, 1975) and develop into brown corky fruits (utricles/seedballs). The fruit immersed in the swollen and hardened perianth base is enclosed by the leathery and incurved perianth (Shultz, 2003). The lenticular seed is 2–3mm with a shiny red-brown seedcoat. Seed contains an annular embryo and copious perisperm (feeding tissue) (Zhu *et al.*, 2003).

Classification

All the cultivated beets belong to *Beta vulgaris* subsp. *vulgaris*. According to Lange *et al.* (1999), *Beta vulgaris* ssp. *vulgaris* cultivars can be subdivided into four cultivar groups:

- Garden beet group or Beetroot group [Conditiva group (www.plantnames.unimelb.edu.au) syn. *B. v.* ssp. *vulgaris* convar. *vulgaris* var. *vulgaris* (Hanelt *et al.*, 2001)]. This is the red root vegetable that is most typically associated with the word 'beet'.
- Sugar beet group [Altissima group (www.plantnames.unimelb.edu.au) syn. *B. v.* ssp. *v.* convar. *vulgaris* var. *altissima*) (Hanelt *et al.*, 2001)]: It is a white coloured strain grown in the US and Europe for sugar production.
- Leaf beet group is a cultivar with edible leaves, petioles and roots with no significantly increased diameter.
    - Cicla group (www.plantnames.unimelb.edu.au), spinach beet or chard (syn. *B. v.* ssp. *v.* convar. *cicla* var. *cicla*) (Hanelt *et al.*, 2001): The leaf beet group has a long history dating to the 2nd millennium BC. The first cultivated forms were believed to have been domesticated in the Mediterranean but were introduced to the Middle East, India and finally to China by 850AD. These were used as medicinal plants in Ancient Greece and Medieval Europe. Their popularity declined in Europe following the introduction of spinach. This variety is widely cultivated for its leaves which usually are cooked like spinach. It can be found in many grocery stores around the world.
    - Flavescens group (www.plantnames.unimelb.edu.au), Swiss chard (syn. *B. v.* ssp. *v.* convar. *cicla* var. *flavescens*) (Hanelt *et al.*, 2001):

Chard leaves have thick and fleshy midribs. Both the midribs and the leaf blades are used as vegetables often in separate dishes. Some cultivars are also grown ornamentally for their coloured midribs. The thickened midribs are thought to have arisen from the spinach beet by mutation.

- Fodder beet group or Crassa group (www.plantnames.unimelb.edu.au), Mangelwurzel (syn. *B. v.* ssp. *v.* convar. *vulgaris* var. *crassa*) (Hanelt *et al.*, 2001): This variety was developed in the 18th century for feeding herds.

Beetroot (*Beta vulgaris* subsp. *vulgaris* 'Conditiva Group)

Synonyms: *Beta vulgaris* subsp. *cicla* (L.) Arcang.; *B.v.* subsp. *rapacea* (Koch) Döll).

Other common names: Beet in North America but in British English: Beetroot, Common beet, Dinner beet, Garden beet, Golden beet, Red beet, Red beetroot and Table bee. In Hindi it is called 'Chukander'.

A bundle of beetroot	Concentric rings in cross section of taproot

Beetroot probably originated in Europe from *Beta vulgaris* L. ssp. *maritima* by hybridization with *B. patula*. Earlier types were with long roots like that of carrot (eagri.org). It is one of several cultivated varieties of *Beta vulgaris* grown for their edible taproots and leaves (called beet greens).

Botany

Shultz (2003) reports its chromosome number as $2n=2\times=18$. Beetroot and other group of cultivars produce a fleshy elongated hypocotyl and a rosette of leaves in first year and flowers in the second. Upper portion of fleshy root gets developed from hypocotyls and basal part from tap root. Concentric rings that can be seen in cross section of root are due to alternate formation of vascular tissues and storage parenchyma tissues. Root skin and flesh colour varies from orange red to dark purple red (eagri.org).

Rosette leaves are formed in a close spiral, with newer ones being inside.

Inflorescence is a spike bearing numerous sessile flowers in 3-4 clusters. Small bisexual flowers are inconspicuous without corolla but with green calyx. It produces abundant pollen and is wind- cum insect- pollinated.

Fruits are mostly aggregates formed by cohesion of two or more fruits and held together by swollen perianth (calyx) base forming an irregular dry cork like body called 'seedball' or the so called seed. If there is a single flower, a single germ seed (monogerm) alone gets developed in the seed. True seeds are small, kidney-shaped and brown. Varieties with monogerm seeds are more preferred since thinning is a costly and essential operation in beetroot cultivation (eagri.org).

A selection of different coloured beetroots

By Beet man - Own work, CC BY-SA 3.0,

Varieties

Varieties popular in India

A few of the improved varieties popular in India are (eagri.org):

- **Detroit Dark Red**: Roots perfectly round with smooth uniform deep red skin; flesh dark blood red with light red zoning; heavy yielder with a duration of 80-100 days.
- **Crimson Globe**: It produces round to flat round roots. Outer skin is medium red and flesh is crimson red without zonations; duration 55-60 days.
- **Early Wonder**: Roots flat globular with dark red skin and dark red flesh and light red zoning.
- **Ooty-1**: This TNAU variety has round roots with blood red flesh colour; yields 28 t/ha in 120 days; it sets seeds under Nilgiris conditions.

- **Crosby Egyptian**: Roots flat globe with dark purplish red flesh; duration 55-60 days; produces white zoning under warm weather.
- **Madhur**, **Ruby Queen** and **Ruby Red** are a few of the varieties marketed by private seed industry

Detroit Dark Red Beets

By BriannaWalther - Own work, CC BY-SA 3.0, https://commons.wikimedia.org/w/index.php?curid=7155494

Exotic varieties

- **UK**
  - **Table purpose:** 'Red Ace Beetroot', 'Exhibition Long Parsnip', 'Nandor F1 Hybrid Carrot', 'Long Carrot', 'Snow-White Beetroot'.
  - **Processing purpose:** 'Cylindra' or 'Formanova Beetoot', 'Detroit Globe Beetroot', 'Crimson Globe Beetroot'.
- **USA**
  - **Table purpose:** 'Early Wonder Beetroot', 'Red Beetroot'
- **Italy**
  - **Table purpose:** 'Barbabietola di Choggia', 'Burpee's Golden Beetroot', 'Cheltenham Green-Top Beetroot'.

Yellow beetroots or Golden beets

The yellow beetroot ~ a variant of regular beetroot was introduced from the Mediterranean region probably from North Africa. The Romans spread it throughout the Roman Empire around 2000 years ago. The intense colours of today's varieties, *i.e.* yellow, red and white beets, came from breeding, and are very different to the wild beetroot form (plantura.garden).

Yellow beetroot usually tastes much sweeter as compared to the red ones. Moreover, yellow beetroots do not 'bleed' during processing. These are round, circular or long, cylindrical taproots with very thin skin and yellow flesh that is juicy and crunchy. The elongated oval leaves of the yellow beetroot form a 15 to 30 cm-long leaf blade. They are arranged in rosettes and grow directly from the vegetable. The juicy green resembles the red beets but the leaves of the yellow variety are streaked with yellow veins. The edges usually curl with age. Like its red relatives, yellow beetroots too are the biennial herbaceous plant which develop the beets and leaf rosettes in the first year though in the second year an inflorescence about 1.5m tall. Its 5-petalled flowers are hermaphrodite, pollinated through wind (plantura.garden).

Some of the popular yellow beet varieties are 'Burpees Golden', 'Boldor F_1', 'Golden Eye', 'Touchstone Gold', 'Golden Detriot', *etc.* (plantura.garden; www.specialtyproduce.com).

- **Burpees Golden**: It has an intense orange skin and light yellow flesh. This fruity mild-flavoured and quite sweet variety is very productive and fast-growing. It is good to use raw or in smoothies, but also for cooking, baking or frying.
- **Boldor F_1**: This high-yielding variety has orange skin and golden-yellow flesh. It is a large beet containing a lot of sugar, therefore particularly suitable for baking. Young leaves are good for adding to salads.
- **Golden Eye**: It has a bright orange skin and golden-yellow flesh. This mildly sweet and earthy-flavoured variety is very uniformly round and good for fresh eating or canning.

Nutritional value

Beetroots are a good source of riboflavin as well as folate, manganese and the antioxidant betaine. Beet greens are a source of riboflavin, iron and vitamins A, C, and K (www.britannica.com).

Beetroot, Raw	
Nutritional value per 100 g	
Energy	180 kJ (43 kcal)
Carbohydrates	9.56 g
Sugars	6.76 g
Dietary fiber	2.8 g
Fat	0.17 g
Protein	1.61 g
Vitamins	**%DV† (Quantity)**

Beetroot, Raw	
Vitamin A equiv. beta-Carotene	0% (2 μg) 0% (20 μg)
Thiamine (B1)	3% (0.031 mg)
Riboflavin (B2)	3% (0.04 mg)
Niacin (B3)	2% (0.334 mg)
Pantothenic acid (B5)	3% (0.155 mg)
Vitamin B6	5% (0.067 mg)
Folate (B9)	27% (109 μg)
Vitamin C	6% (4.9 mg)
Minerals	**%DV† (Quantity)**
Calcium	2% (16 mg)
Iron	6% (0.8 mg)
Magnesium	6% (23 mg)
Manganese	16% (0.329 mg)
Phosphorus	6% (40 mg)
Potassium	7% (325 mg)
Sodium	5% (78 mg)
Zinc	4% (0.35 mg)
Other constituents	**Quantity**
Water	87.58g
Units: μg = micrograms, mg = milligrams, IU = International units †Percentages are roughly approximated using US recommendations for adults. *Source: USDA FoodData Central*	

en.wikipedia.org/wiki/Beetroot

Nutrient composition of beetroot and its byproducts (per 100ml or gramme)

Beetroot consists of multiple biologically active phytochemicals including betalains such as betacyanins and betaxanthins; flavonoids; polyphenols; saponins and inorganic nitrate (NO3). It is also a rich source of diverse minerals such as potassium, sodium, phosphorous, calcium, magnesium, copper, iron, zinc and manganese (Mirmiran *et al.,* 2020). It is commonly consumed in form of supplemental juice, powder, bread, gel, boiled, oven-dried, pickled, pureed or jam-processed across different food cultures (Wruss *et al.,* 2015; Morgado *et al.,* 2016; Guldiken *et al.,* 2016). As shown in the table below, 100ml of beetroot juice comprises of 95Kcal energy, 22.6g carbohydrates, 0.70g proteins, 0.16g total lipids, 0.91g total dietary fibre and 12g total sugars. As such, the micro-nutritional composition of 100ml beetroot juice is estimated as 8.8g sucrose, 0.86g fructose and 2.5g glucose.

	Raw	**Cooked, boiled**	**Canned**	**Fresh juice**
Water, g	87.58	87.06	90.96	–
Energy, kcal	43	44	31	30
Protein, g	1.61	1.68	0.91	1.02
Total fats, g	0.17	0.18	0.14	0
Carbohydrate, g	9.56	9.96	7.21	6.6
Fiber, g	2.8	2	1.8	0
Sugars, g	6.76	7.96	5.51	6.6
Calcium, mg	16	16	15	0
Iron, mg	0.8	0.79	1.82	0
Magnesium, mg	23	23	17	–
Phosphorus, mg	40	38	17	–
Potassium, mg	325	305	148	–
Sodium, mg	78	77	194	93
Zinc, mg	0.35	0.35	0.21	–
Vitamin C, mg	4.9	3.6	4.1	0
Thiamin, mg	0.031	0.027	0.01	–
Riboflavin, mg	0.04	0.04	0.04	–
Niacin, mg	0.334	0.331	0.157	–
Folate, µg	109	80	30	–
Total phenolic contenta	255	238	192	225
Total flavonoid contentb	260	261	173	126

[a]As mg gallic acid equivalent (GAE)/ 100 g; b as mg rutin equivalent (RE)/100 g sample

Source: Mirmiran et al. (2020)

Moreover, various commercial organic and conventional beetroot juices are reported to contain total sugar, vitamin C and total flavonoids within a range of 1.73–7.85g, 10.75–20.36mg and 2.02–2.36mg (per 100g), respectively (Kazimierczak *et al.*, 2016). Betalains make up to 70–100% of phenolic composition of beetroot, limited to 0.8–1.3g/l of fresh beetroot juice (about 60% betacyanins and 40% betaxanthins) (Wruss *et al.*, 2015).

Mineral component of beetroot juice

Oven drying method significantly increases the potassium value of the juice to 875mg compared to the fresh juice sample (230mg) with the least amount of potassium value and shows significant difference ($p>0.05$) in all the juice samples. The fresh juice has the highest iron content (21mg), followed by oven-dried sample with 16mg of iron and sun-dried 15mg of iron. The calcium content of the juice isn't affected by sun-drying method. With drying method, Mg value within the juices is increased. Oven-dried sample significantly increases to 102mg, sun dried sampled to 51mg compared to the fresh juice with Mg value of 34mg. It has been reported that by the method of drying, the nitrate content of juice sample was not influenced (Nguyen *et al.*, 2018).

Mineral Components (mg)	Fresh Juice Sample	Sun dried Juice Sample	Oven Derived Juice sample
Nitrate	0.04	0.04	0.04
Potassium	230	327	875
Iron	20	15	16
Calcium	3.1	3.6	3.9
Magnesium	34	51	100
Source: Nguyen *et al.* (2018).			

Phytochemicals and medicinal values

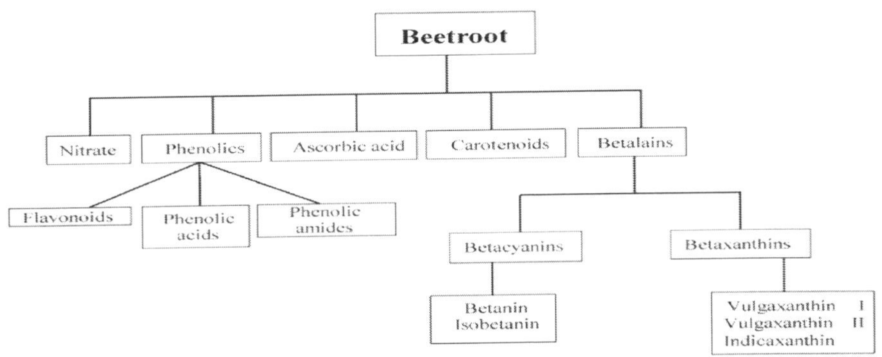

Potentially bioactive compounds in beetroot (Clifford *et al.*, 2015).

Beetroot is a rich source of phytochemical compounds that includes ascorbic acid, cartotenoids, phenolic acids and flavonoids (Georgiev *et al.*, 2010; Kujala *et al.*, 2002; Wootton-Beard and Ryan, 2011). Betalains (Lee *et al.*, 2005; Vulic *et al.*, 2014) is one of a highly bioactive pigment found in beetroot. Nitrosamine formed in beetroot juice can reliably be prevented by adding ascorbic acid (Kolb *et al.*, 1997). Most nitrosamines are carcinogenic in nonhuman animals (Yang *et al.*, 1990).

Betalain and Betaine

Unlike most other red plants which contain anthocyanin pigments such as red cabbage, the colour of red or purple beetroot is due to a variety of betalain pigments. The composition of different betalain pigments can vary resulting in strains of beetroot which are yellow or of other colours in addition to the familiar deep red (www.selfsufficientish.com).

Betaine (trimethylglycine or glycinebetaine) and betalain (red or yellow pigments), two health-beneficial compounds prevalent in *Beta vulgaris* are under basic research for their potential biological properties. Four predominant betalains, two betacyanins (betanin and isobetanin) and two betaxanthins (vulgaxanthin I and miraxanthin V) were isolated and quantified. Betanin and vulgaxanthin I are the major compounds in red and yellow beetroot extracts, respectively, and they comprised >90% of the betalain content in the tested cultivars. The contents of betalains and betaine in beetroots could be increased by modifying the growing conditions and that betanin and betaine extracted from beetroots had some anticancer effects against HepG2 cells (Lee *et al.*, 2014).

Beetroot, a major source of betalains, is classified as one of the 10 plants with the highest antioxidant activity as determined by the IC50 value, *i.e.* the extract's ability to reduce low-density lipoproteins (LDL) oxidation by 50% (Vinson *et al.*, 1998). Betalains are present in the taproot part of the plant, conferring its red-purple coloration (Ravichandran *et al.*, 2013). Members of the betalain family are categorised as either betacyanin pigments that are red-violet or betaxanthin pigments that are yellow-orange (Ninfali and Angelino, 2013). Betalains present in the taproot part of the plant have antioxidant (Ravichandran *et al.*, 2013), anti-inflammatory, hepatic protective and anti-cancer properties (Kapadia *et al.*, 2003; Winkler *et al.*, 2005). Betacyanin in beetroot may cause red urine in people who are unable to break it down and this is called beeturia (Eastwood and Nyhlin, 1995). Some of the betalains in beets are betanin, isobetanin, probetanin and neobetanin (the red to violet ones are known collectively as betacyanin). Other pigments contained in beet are indicaxanthin and vulgaxanthins (yellow to orange pigments known as betaxanthins). Indicaxanthin has been shown as a powerful protective antioxidant for thalassemia and prevents the breakdown of alpha-tocopherol (vit. E) (en.wikipedia.org/wiki/Beta_vulgaris).

The pigments are contained in cell vacuoles. Betalains are unstable and are degraded in the presence of light and oxygen and destroyed at high temperatures. This is why red beetroots leave a purple stain. Leaving the skin on when cooking, however, will maintain the integrity of the cells and therefore minimize the leakage (en.wikipedia.org/wiki/Beta_vulgaris).

Betaine is an important nutrient made from the B-complex vitamin, choline. Choline is key in regulating inflammation in the cardiovascular system since adequate choline is important for preventing unwanted build-up of homocysteine. Elevated levels of homocysteine are associated with inflammation and an increased risk of cardiovascular disease (www.theharvestkitchen.com).

Phenolic Compound

Beetroot contains several highly bioactive phenolics such as caffeic acid, pedicatechin and rutin which are also known to be excellent antioxidant (Frank *et al.*, 2005; Manach *et al.*, 2005). Phenolic compound composition and concentrations also vary according to the cultivar. Kujala *et al.* (2002) studied four distinct beetroot cultivars by high-performance liquid chromatography (HPLC), HPLC-electrospray ionisation-mass spectrometry (HPLC-ESI-MS) and NMR techniques and identified four flavonoids ~ betagarin, betavulgarin, cochliophilin A and dihydroisorhamnetin whereas Georgiev *et al.* (2010) identified the presence of catechin hydrate, epicatechin and rutin.

Ferulic acid, flavonoids and phenolic amides present mainly in beetroot skin confer strong antioxidant activity to this food (Kujala *et al.,* 2000, 2002).

Saponin

The occurrence of saponins in this species was characterized by Mroczek (2012) where 11 saponins derived from oleanoic acids were identified using reverse phase liquid chromatography coupled with electrospray ionization mass spectrometry (LC-ESI/MS/MS). Oleanoic acid is of importance among the saponins identified in the beet since it is capable of causing a marked hypoglycemic effect (Murakami *et al.,* 1999). Like flavonoids, the content of saponins may vary according to the cultivar with concentrations varying from 7.66 to 12.2 mg/g dry weight in three different *Beta vulgaris* cultivars. Saponin content may also vary according to beetroot processing. It was reported that saponin content was found to be almost three times higher in beetroot gel compared to juice (da Silva *et al.,* 2016).

Nitrate

Beetroot's effect on the vasculature is largely attributed to its high inorganic nitrate content, 250 mg·kg–1 as per fresh weight (Ormsbee *et al.,* 2013), which itself is not considered to mediate any specific physiological function and rather nitrate's beneficial effects are attributed to its *in vivo* reduction to nitric oxide (NO) ~ a multifarious messenger molecule with important vascular and metabolic functions (Hobbs *et al.,* 2013; Machha and Schechter, 2011).

Geosmin

The 'earthy' taste of some beetroot cultivars comes from the presence of geosmin. Researchers are yet not able to ascertain whether beets produce geosmin themselves or it is produced by symbiotic soil microbes living in the plant (Lu *et al.,* 2003). Breeding programmes can produce cultivars with low geosmin levels yielding flavours more accepble to consumers.

Uses

Usually, the deep purple roots of beetroot are eaten boiled, roasted or raw and either alone or combined with any salad vegetable. Beetroot can be roasted, boiled or steamed, peeled and then eaten warm with or without butter as a delicacy; cooked, pickled and then eaten cold as a condiment; or peeled, shredded raw and then eaten as a salad. Pickled beetroot is a traditional food in many countries (en.wikipedia.org/wiki/Beetroot). Beetroots are also commonly canned either whole or cut, and often are pickled, spiced or served in a sweet-and-sour sauce. The leaves of garden beets can be cooked like spinach or eaten fresh if picked young (www.britannica.com).

A large proportion of commercial production of beetroot is processed into boiled and sterilized beetroot or pickles. Beetroots are the main commercial source of betalains (concentrate or powder) in the food industry with the use of betanin restricted as a natural dye added to different foods like gelatins, desserts in general, confectionery, dry mixes, dairy products and beef and chicken-derived products (Delgado-Vargas *et al.,* 2000). When beetroot juice is used, it is most stable in foods with low water content, such as frozen novelties and fruit fillings. Beetroot can also be used to prepare wine (en.wikipedia.org/wiki/Beetroot).

Bioactive compounds present in beetroot have antioxidant, anti-inflammatory, hepatic-protective and anti-cancer properties. Beetroot juice can mediate cardiology effect and is anti-oxidant, anti-inflammatory and anti-tumourous, and results recommend that beetroot juice can provide treatment for angina, pain management as well as chemotherapeutic agents. The result of different studies indicates its need for implementation for therapeutic advantage for the near future.

As a source of nitrate, beetroot ingestion provides a natural means of increasing *in vivo* nitric oxide (NO) availability and has emerged as a potential strategy to prevent and manage pathologies associated with diminished NO bioavailability, notably hypertension and endothelial function (Clifford *et al.,* 2015). Consumption of nitrate-rich beetroot juice by athletes induces a number of beneficial physiological health effects that are linked to the formation of nitric oxide (NO) from nitrate. NO plays a fundamental role in the regulation of blood flow, muscle contractility, myocyte differentiation and glucose *cum* calcium homeostasis (Dejam *et al.,* 2004). However, following a secondary pathway, NO may also lead to the formation of N-nitroso compounds (NOCs) which are known to be carcinogenic in 39 animal species. The extent of the formation of NOCs is modulated by various other dietary factors such as vitamin C (Berends *et al.,* 2019).

Beeturia

The red colour compound betanin is not broken down in the body, and in higher concentrations may temporarily cause urine or stools to assume a reddish colouration. In case of urine it is a condition called beeturia (Frank *et al.,* 2005).

Health risks

When nitrite is digested in the acidic environment of the stomach, nitrous acid is formed (HNO_2). Two molecules of HNO_2 are converted to water and

N_2O_3, which reacts with amines to nitrosamines. HNO_2 can also be protonated and can react with amides to form nitrosamides. These NOCs can form reactive intermediates that may bind to DNA causing mutations that are potentially involved in the initiation of carcinogenesis (Vermeer and van Maanen, 2001).

Cultivation

They grow best in pH-neutral to slightly alkaline, deep and loose soils rich in plant nutrients including sodium and boron. Beets respond well to chemical fertilizers and manures. For growing these prefer relatively cool temperatures of 15 to 19°C. Leaf beets can thrive in warmer temperatures than beetroot. Beets are most extensively grown in temperate to cool regions or during the cooler seasons. They are grown worldwide in regions without severe frosts. The growing season is about 8-10 weeks (57-72 days) in favourable climates. Grown extensively under irrigation, beets tolerate salty soils and drought but are sensitive to high acidity and to a low content of boron. Boron deficiency retards growth and causes black lesions in the flesh of knobs (www.britannica.com; en.wikipedia.org/wiki/Beta_vulgaris).

Propagation

Beetroot is propagated from seeds, and for its sowing use of a precision seeder is recommended. Since beet seed is irregular in shape, it is pelleted to make it more for its use of usable ~ seeds are coated with inert clay materials moulded into a spherical shape before the precision seeder can use it. The inert clay dissolves in the soil (www.slideshare.net). Unfortunately, precision sowing machinery is not common in India, hence here sowing is done only manually (www.amrafarms.com).

Climate

Beetroots are cold hardy and can be grown year-round in areas with mild winters such as in the Nilgiri hills of Tamil Nadu. Although it prefers cool climate, beetroots can however be grown in warm weather but the root quality such as colour, texture, sugar content, *etc.* is much better only under cool weather conditions. High temperature causes zoning, *i.e.* appearance of alternate light and dark red concentric rings in the root. Extreme low temperature of 4.5-10°C for 15 days will result in bolting. It requires abundant sunshine for development of storage roots (eagri.org).

In India, beetroot is commonly cultivated as a seasonal crop in Haryana, Uttar Pradesh, Himachal Pradesh, Maharashtra and West Bengal during winters. It can however be cultivated anywhere when the weather conditions have a temperature ranging from 15-30°C (optimal being 18-25°C) and with minimal rainfall. Areas like Tamil Nadu and Kerala have started producing beetroot as they have found the weather conditions favourable for moderate yield. Temperature in these areas is usually slightly higher than 25°C during winter (www.amrafarms.com)

Season

Being a cool season crop, beetroot is raised during winter in the plains and as a spring-summer crop (March-April) in the hills. In the plains, the sowing is practiced during September-November. In India, beetroot is usually cultivated in the cooler months from mid-October to March. It can be grown in other months also but is not as profitable and the yield may be considerably lower. In Bengaluru and certain other similar areas, it may be sown in June-July also. Moreover the chances of pests and diseases are also higher when it is grown during summer (www.amrafarms.com).

Soil, Preparation of Land and Sowing

Deep, moisture-retentive but well-drained, rich in organic matter loamy or sandy-loam soil having pH of 6-7 is best for beet root cultivation. It is highly sensitive to acidic soils. Waterlogging should be avoided. Heavy clayey soils result in poor germination, *vis-à-vis* poor and misshapen development of roots. Beet root is one of a few vegetables which can be successfully grown in saline soils (eagri.org).

Land is ploughed to a fine tilth so that soil becomes utterly loose and friable. Clods, brick and stone pieces and rootstocks of perennial weeds, *vis-à-vis* the remnant of previous crops are to be removed completely. Well decomposed farmyard manure should be applied at the time of final ploughing @ some 10-15t/ha and thoroughly mixed in the soil before preparation of the beds. If there is no danger of waterlogging, the flat beds are prepared, however, for better growth and establishment the beds are prepared in ridges and furrows. Water-soaked 'seed balls' which contain 2-6 seeds are drilled 2.5cm deep on ridges at 30-45×15-25cm where some 5-6 kg of seeds is required for sowing one hectare of land. Staggered sowing at 1-2 weeks interval ensures steady supply of roots during the season (eagri.org).

If available, it would be better to use precision seeder otherwise individual seeds are sown manually but in a systemic manner by following plant to plant and line to line distances (www.amrafarms.com).

Manures and Fertilizers

Beetroot is a wonderful crop that requires minimal to no fertilizers if the land is prepared with enough organic matter and cow dung manure. However, if the soil is depleted of nutrients, growers may need to apply fertilizers. The fertilizers which are applied to beetroot include lesser content of nitrogen with more content of potash and phosphorus (www.amrafarms.com).

On sandy soils, organic manure @ 20-25t/ha is recommended. For an average soil, 60-70kg/ha of nitrogen, 100-120kg/ha of phosphorus and 60-70kg/ha of potash is recommended. Entire farmyard manure, half of N and full P and K should be applied as basal prior to sowing, especially at the time of land preparation and remaining at 30-45 days after sowing. Nitrate sources of N are preferred to ammonium sources. Beets have a relatively high boron requirement and its deficiency causes internal breakdown as black rot or dry rot (eagri.org).

Beets are one of the most boron-intensive of modern crops. On commercial farms, 60t/ha harvest requires 600g/ha of elemental boron for proper growth. A lack of boron causes the meristem and the shoot to languish eventually leading to heart rot (www.borax.com).

Irrigation

Moist soil is essential for seed germination and for further growth. There should be sufficient moisture in the soil at the time of seed sowing. Usually 5-6 irrigations are given during summer and three irrigations during winter in North Indian plains (eagri.org). Beetroots love moist soil but not waterlogged conditions. Growers should ensure that the soil is just moist and should leave it to dry an inch or 2 on the surface before watering again (www.amrafarms.com).

Intercropping

Beetroots can be intercropped with various other winter vegetables such as carrots, cabbages, cauliflowers and broccoli. Most of these plants have the same lifecycle and go from seed to harvest at almost the same time. The fertilizer and water requirements for these crops are almost same and they also do not attract many pests if grown in the right season. Some growers even grow beetroots along with brinjals, peas, onions, lettuce and French beans (www.amrafarms.com).

Cultural practices

Thinning is usually not required in monogerm seeds that are sown using precision seeder but multigerm seeds produce more than one plant/seed and require thinning to maintain proper distance between plants (Mall *et al.*, 2021). Overcrowding usually produces smaller roots and is not considered profitable for farmers. Therefore, thinning once the plants reach a certain stage is practiced to prevent overcrowding, especially if multigerm seeds are being used or if hand sowing is being practiced (www.amrafarms.com).

Field is usually kept weed-free by light hoeing at early stage of crop. Swollen roots are also to be covered with soil by earthing up (eagri.org).

Harvesting

The beetroot crop becomes ready for harvest within 60-75 days of sowing when the roots attain a diameter of 3-5cm. Roots should be harvested before development of spongy tissues within the rootknobs. The plants are usually harvested by pulling the top with hand. Later, the tops are removed, graded and marketed. In European countries where small-sized bunches are in demand, roots are tied in bundles of 4-6 with their tops. Over-matured and oversized roots become fluffy, woody and crack (eagri.org) so their harvesting should be done when still tender.

The oblong leaves typically have a red petiole and midrib and can be harvested continually throughout the first growing season, though production slows down in the heat of summer. Overharvesting of the leaves can negatively affect the size of the beetroot. In mild climates, the plant can overwinter, though the leaves turn bitter and unpalatable when the plant flowers in its second season (www.britannica.com).

Yield and Storage

Yield varies from 25-30t/ha and the knobs store well at 0°C temperature and 90% RH (eagri.org). This could differ depending on the variety being grown, climate and soil conditions.

Seed Production

Unlike other root crops, annual tropical types do not exist in beetroot. All beetroot cultivars are temperate biennial types and a low temperature of 4.4 to 7.7°C for 6-8 weeks is required for flowers to initiate. Seed production is possible in hills at 1400m a.s.l. An usual method of seed production is root to seed method. In this method, seeds are sown in July-August and well developed roots are dug out during November-December. After selection of roots, top is trimmed without injuring crown and then whole tuberous root of selected plants are transplanted at a spacing of 60×45-60cm in the well prepared field and irrigated. Seeds are harvesting during June-July. This being a cross-pollinated crop as it is self-incompatible it requires an isolation distance of 1000m for certified seed production and 1600m for breeder and nucleus seed production. Average seed yield is 2t/ha (eagri.org).

Pests and Diseases

Insect pests such as leaf miners, webworms, semi-loopers; fungal diseases such as *Cercospora* leaf spot and downy mildew; and viral diseases cause damage in beetroot cultivation.

Pests

- **Nematodes**
  - **Beet cyst nematode** (*Heterodera schachtii*): Nematodes are mainly transmitted by irrigation water, equipment, weed plants, etc. The symptoms may vary depending on the stage of the crop infested by the nematode. Symptoms are not much noticeable if the nematodes infest the older plants. Infested seedlings have stunted and reduced leaf growth, older leaves of seedlings become yellow and show wilting during hot days, roots appear stunted with lots of secondary roots and infested roots show yellow-brown cysts. If available, nematode-resistant varieties should be grown, field should be kept free from weeds, crop rotation with non-host crops should be followed and the field during summer should be solarized for two months to kill all sorts of germs including the nematodes.
  - **Root-knot nematode** (*Meloidogyne* spp.) (plantvillage.psu.edu): Galls can appear as quickly as a month of planting. In fact, nematodes prefer sandy soils and damage in areas with this type of soil is most likely. Galls on roots can be of any size but maximally up to 3.3cm. Root-knot infestation causes reduced plant vigour, yellowing of plants and their wilting during hot weather. The practices adopted in managing beet cyst nematode should be adopted in this case also.
  - **Spider mites** (*Tetranychus urticae* and others) (gardenerspath.com b): These pests being quite tiny are hard-to-spot. Numerous species of mites, such as strawberry spider mites and two-spotted spider mites attack the beets. Spider mites look for stressed plants, particularly those growing in hot and dry conditions. They use their needlelike mouthparts to suck the fluid out of plants. Spider mites weave webs and eat from the undersides of leaves which can cause white or yellow spots on beet foliage. This damage can eventually merge and cause leaves to drop. Beets should be kept well watered, Spraying

the webs off with a strong blast of cold water may help in reducing their population. Use of pesticides in the garden may also kill the beneficial insects like ladybugs, thrips, pirate bugs and lacewings that eat spider mites. However, spraying the infested plants with 0.2% Dicofol 18.5EC or 0.1% Spiromesifen 22.9EC will control these pests.

- **Insects**
  - **Aphids** (*Myzus* spp. and *Aphis* spp.) (gardenerspath.com b): Dozens of different species of tiny and pear-shaped aphids from the genera *Myzus* and *Aphis* love to feed on beet plants. Green peach, melon, potato, cowpea, and bean aphids are all common. Aphids range in colour from nearly black to nearly white, with some that are tan, orange, green, or red. Some have wings, and others don't. **Symptoms**: they suck on plant sap, leaving behind a sticky residue called honeydew, which also attracts ants and creates the perfect environment for fungi; large infestation may cause leaves to curl and become yellow and may cause stunted leaf and root growth. **Management**: blasting the plant with a strong spray of water from the hose; they won't return if knocked them loose; dusting plants with flour or wiping the stems and undersides of leaves with canola oil; allowing beneficial insects like parasitic wasps, ladybugs, lacewings, soldier beetles and syrphid flies to visit the garden.
  - **Beet webworms** are the moth larvae in the Crambidae family (gardenerspath.com b). The insect species attacking beet plants include the spotted beet moth webworm (*Hymenia perspectalis*), the Southern beet moth webworm (*Herpetogramma bipunctalis*), Hawaiian beet moth webworms (*Spoladea recurvalis*), the sugar moth webworm (*Loxostege sticticalis*), alfalfa moth webworms (*Loxostege cereralis*), and the garden moth webworm (*Achyra rantalis*). The webworms vary in appearance, some being olive-green while others yellow or bright green and sometimes with black stripes. These may range from just 2.2cm to 3.8cm long and are usually found in warmer regions. Fortunately, by the time the larvae hatch and mature, beets have attained enough growth though roots are still forming despite any damage to the foliage above ground. Although root damage is rare but severe infestation can almost defoliate the plants which may result into crop loss. The worms feed through the leaves completely skeletonizing the plants rapidly if their population is alarming. These use silky threads to make cozy nest-like webs to hide and cause curling or binding of leaves together. *Bacillus thuringiensis* can be used to parasitize these worms.

- **Cutworm** (*Agrots ipsilon, A. segetum*) larvae chew through the base of beet foliage at the soil level and kill the plants. Some cutworms feed on young leaves and stems in addition to cutting the stems down. The larvae are light gray to bright green or dark brown with greasy texture on their skin and measure about 2.5-5cm long with no hairs on their body. On disturbing, these curl up in a protective little 'C' shape. To manage this pest the field should be kept weed-free so that these may not get shelter to hide themselves anywhere nearby as for fear of being exposed they usually do not travel across a wide expanse of open ground. Moreover, the cardboard or plastic items can be used to create a barrier. When their infestation is first observed, nearby soil may be drenched with 0.1% Chlorpyriphos.
- **Leaf-eating caterpillar** (*Spodoptera litura*): Larvae after hatching from the clustered eggs, these start feeding gregariously on the plants skeletonizing the leaves and large larvae also bore in the knob. Whenever observed first, the clustered eggs and larvae should be killed manually. Neem seed-kernel extract at 4% should be used at the early stage of larvae. Methomyl or Thiodicarb @ 50g/kg of rice/wheat bran alongwith 100g of jaggery should be used as their poison bait to kill these.
- **Semi-loopers** (*Plusia* spp.): Green larvae of this lepidopterous insect feed on leaf surface and on disturbance these hang down through a self-made thread as a protection measure. In case of high population, the plant may be sprayed with Dichlorvos 76EC at 0.1%.
- **Darkling beetle or Rove beetle** (*Blapstinus* spp., *Staphylinid* spp.) (plantvillage.psu.edu): Generally beetles hide in organic debris during the day and become active at night. This insect is a dull blue-black or brown beetle about 0.6 cm long; tips of antennae are often enlarged, resembling a club. **Symptoms**: feeding damage on stems; death of seedlings; seeds dug up. **Management**: ditches filled with water can prevent spread of beetle to/from adjacent fields; all weeds from garden borders must be cleared off; use appropriate insecticides; insecticides are usually in the form of baits.
- **Leafminers** (*Lyriomyza* spp., *Pegomyia* spp.) (plantvillage.psu.edu): These insects may go through 7 to 10 generations per year and entire lifecycle can take as little as 2 weeks in warm weather. Mature maggots about 8mm long drop from leaves into soil to pupate. Its infestation can produce white, thin and winding trails on leaves though heavy mining can result in white blotches on leaves which results in leaf

dropping from the plant prematurely. Early infestation results in reduced fruit yield. Adult leafminer is a small black and yellow fly which lays its eggs in the leaves and its larvae after hatching start feeding on leaf interior. Transplants should be checked prior to planting for signs of leafminer damage; plants should be removed from soil immediately after harvest; insecticides should only be used when leafminer damage has been properly established as unnecessary spraying will also reduce natural enemy population. To manage this problem, fallen leaves should be destroyed and leaf undersides should be sprayed weekly with NSKE at 4% alongwith clipping and destroying of affected leaves.

- **Seed corn maggots** (*Delia platura*) (gardenerspath.com b): They attack numerous crops that grow underground and consume germinating seeds, totally destroying a crop before it even gets started. They are about a quarter-inch long and whitish-yellow. The maggots overwinter in the soil and emerge in the spring to eat and reproduce. There can be up to four generations per year. **Management**: if maggots are found, new seedlings should be started indoors, and transplanted outdoors when the plants are a few inches tall.
- **Vegetable maggots** (*Delia radicum*) (gardenerspath.com b): They overwinter in the soil and emerge in the spring to feed on roots. It looks almost identical to the seed corn maggot as well, about a quarter-inch long and cream-coloured. They not only do damage to the plants but can also introduce diseases. **Symptoms**: these maggots chew tunnels into the surface of beetroots and completely devour the finer root hairs; above ground, plants will look stunted, wilted and weak; leaves may turn yellow or mottled, or they may have purpling on the edges. **Management**: crop rotation with non-host crops; yellow sticky traps at ground level in the garden can be used to attract the maggots; growers can wait to transplant beets into the garden until late May to avoid the emerging maggots; use of floating row covers, especially in warmer areas, in the fall and spring prevents the adult flies from landing and laying eggs;

Diseases

- **Fungal Diseases**
  - **Alternaria leaf spot** (*Alternaria* spp.): The disease thrives in hot and humid conditions. In its infection small and round lesions appear on the leaves which after coalescing make the leaves brown and

necrotic along with shot holes and finally falling off. Crop rotation with non-host crops and at least three years gap should be maintained to grow the beets on the same land. Watering should be done only at the base of the plant so in such cases drip irrigation would be quite rewarding. The planted field should be kept weed-free throughout the season. Growers can apply the foliar spray of a product that contains hay bacillus (*Bacillus subtilis*). Mancozeb spraying is quite effective.

- **Beet rust** (*Uromyces betae*) (gardenerspath.com a) infects beets on their foliage and stems. The fungus thrives in cool and wet weather and can overwinter on plant debris or in seeds. It appears as small reddish-orange spots on leaves so foliage becomes weak which makes harvesting a bit difficult. Affected areas may discolour, wither and die off. To manage this problem infected leaves should be removed, air circulation in the cropped area should be improved, if available certified disease-free seeds should be used, in no case overhead irrigation should be applied and finally the fungicidal spray at 0.2% of penconazole, mancozeb, myclobutanil or bupirimate with triforine should be done at every 10 days interval.
- **Cercospora leaf spot** (*Cercospora beticola*) (plantvillage.psu.edu): This pathogen spreads through wind and rain. Fungus overwinters on crop residue or in seed. The infection is accelerated due to high temperatures and humidity. In its infection brown to gray spots surrounded by red-purple haloes on leaves are formed which cause leaves to yellow or becoming brown-necrotic. In this case the crop should be rotated every 2-3 years. Crop debris immediately after harvest should be destroyed to minimize the infection carry over. At first sign of infection, application of fungicide such as carbendazim and mancozeb at 0.2% spraying on the crop at first sign of disease should be applied.
- **Damping off** (*Rhizoctonia solani, Phoma betae, Pythium ultimum, Aphanomyces cochlioides*) (plantvillage.psu.edu): The disease emergence is favoured by warm and wet weather. In damping off infection the seedlings start collapsing, roots are blackened and constriction on the crown is formed. To manage this problem, while planting or sowing care should be taken that it is a well-drained soil and before planting the seeds should be treated with any of the contact fungicides. Moreover, soil should also be treated with a copper-based fungicide before sowing as well as treating the seedlings also with the same chemical. Sowing or planting is effected when

environment is still warm though not too much. Seeds should be sown thinly.

- **Downy mildew** (*Peronospora farinose*): Fungus overwinters in crop residues and infects new crop. It is usually a common disease of young plants and those growing in moist environments. In its infection the upper leaf surfaces become lighter green (yellow or discoloured) and the leaves small, puckered and thickened with fuzzy grey growth on both the leaf surfaces, vis-à-vis beneath the fungal growth is purplish. The plant growth is stunted. If possible only resistant varieties should be grown, infected crop debris must be removed and destroyed, air circulation around the plants should be increased through thinning, leaf removing and weeding and overhead watering should be avoided. After all, the crop should be sprayed at 0.2% with mancozeb.
- **Fusarium yellows and root rot** (*Fusarium oxysporum* f. sp. *spinaciae*, *F.o.* f. sp. *betae*) (plantvillage.psu.edu): The fungus survives on weeds such as pigweed, kochia and lamb's quarters during off season. Infected leaves exhibit yellowing between the larger veins and then entire leaves become dry, brittle and remain clustered around the crown. Usually only one side of the leaves gets affected and appear scorched. Vascular tissues become discoloured. Plant appears wilted during daytime though showing partial recovery at night. Tips of taproot become black due to rotting. Infected plants along with nearby soil should immediately be removed to minimize the spread of the disease. Five year rotation with non-host crops should be followed. Resistant varieties should be used if available. The field should be kept free from weeds. For this no chemical control is available.
- **Powdery mildew** (*Erysiphe betae*) (plantvillage.psu.edu): The spores are transmitted by air. If the weather conditions are good, dark brown to black, globular, sexual reproductive structures (chasmothecia = cleistothecia) can be seen, particularly on the upper surface of the older leaves. Symptoms initially appear on older leaves as small, scattered, circular, white mycelium mats on lower surface; later all the leaves of the plant gets infected and appear dusty white on both surfaces; in severe cases, the leaves become yellow and then turn purplish-brown. Management practices include use of resistant varieties; and spraying of suitable fungicide in severe infections.
- **Verticillium Wilt** (*Verticillium dahliae* and *V. albo-atrum*) (gardenerspath.com a): It gets into plants through the roots. It

causes older leaves to start yellowing and afterwards progressing to newer leaves so eventually the leaves can die back killing the entire plant. In its infection the leaves show interveinal discolouration, i.e. yellowing or browning between the veins and then die. No effective treatment is available for this disease. As the fungus survives in the soil by forming the resting bodies, and on plant debris so field should be kept clean, affected plants should be pulled and destroyed, and during summer such fields should be solarized for 60 days to kill all the germs present in the soil. After solarization, planting of beet in the same location should not be done again for at least four years. No chemical is so effective to control this disease if once infection has occurred.

- **Bacterial Diseases**
  - **Bacterial blight or bacterial leaf spot** (*Pseudomonas syringae pv. aptata*): The bacterium also infects other crops such as bean, eggplant, lettuce and pepper. The bacteria spread mainly by rainfall, mechanical and insect injuries and through wind & garden tools. In its infection, leaves show irregular-shaped spots with dark centres and black borders and sometimes the edges of the leaves may appear water-soaked which later turn yellow and necrotic. The spots can even grow and coalesce causing the leaf to dry up and then falling off the plant. To get rid of this problem only disease-free seeds should be used and the plants should be mulched around to prevent splashing water from landing on the leaves. Spraying the plants once a week with neem oil or 'Bonide Revitalize' ~ a biofungicide containing *Bacillus amyloliquefaciens* is also very effective.
  - **Scab** (*Streptomyces scabies*): This bacterium survives in the soil and the disease caused by this is more common during dry conditions. Its infection to beetroot causes small, round, raised gray, white or tan corky spots on root-knobs and ruptures the epidermis. To manage this disease, beet planting in the infected soil is avoided and effective crop rotation should be followed. Spraying of 0.3% copper oxychloride plus Bactosan or Bactrinashak will control this disease to a greater extent.
- **Viral Diseases**
  - **Beet curly top disease** [Beet curly top virus (BCTV), Beet severe curly top virus (BSCTV) and Beet mild curly top virus (BMCTV)]: The viruses infect many weed plants which act as a source of inoculum for the next cropping season. These are transmitted by

beet leafhopper, *Circulifer tenellus*. In virus infection the leaves become stunted, crinkled and rolled upward and inward, the veins become irregularly swollen on the lower surface, young roots do not develop properly, and rootlets become twisted, distorted and die which causes new rootlets to appear which ultimately become hairy-rooted. Root-knobs show discolouration of the vascular tissues. Only resistant varieties should be planted. Field should be kept weed-free throughout the growing season. Leafhopper-infested crop should be sprayed with mineral oil or neem oil at 2% or Acephate at 0.15% or Dimethoate at 0.05% at 10 days intervals.

- **Beet western yellows virus (BWYV)]:** The virus has an extensive host range and is transmitted by aphids ('green peach' and 'black bean aphids'). Older leaves start yellowing between the veins with possible small reddish brown spots which gives a distinct bronze cast on infected leaves. Later the leaves become thick, leathery and brittle. If available, one should grow resistant varieties. Field sanitation is a must even after the crop is over. To manage the disease, infected plants should immediately be removed from the field and destroyed and the remaining plants should be sprayed with 2% mineral or neem oil or the crop should be sprayed at every 10 days with 0.05% dimethoate or 0.15% Acephate.

Sugar beet (*B.v.* ssp. *vulgaris* var. *altissima or B.v.* ssp. *vulgaris* 'Altissima Group')

Synonym: *B.v.* subsp. *v.* convar. *vulgaris* var. *altissima* (Hanelt *et al.*, 2001).

Sugar beet was developed in Germany in the 18th century after the roots of beets were found to contain sugar in 1747. It is a major sugar producing commercial crop having high concentration of sucrose which is extracted to produce table sugar (en.wikipedia.org/wiki/Beta_vulgaris). Top 10 sugarbeet producing countries are Federation of Russia, France, Germany, United States of America, Turkey, Poland, China, Egypt, Ukraine and United Kingdom. Sugar beet was introduced in India during 1950s (Mall *et al.*, 2021).

Sugar beets have been bred for increased sugar content. Over these 200 years of breeding the sugarbeets have increased amount of sugar from then 8% to 18% by 2013. Together, there had been increase in resistance to viral and fungal diseases, taproot size, monogermy and less bolting. Discovery of a cytoplasmic male-sterility line has especially been useful in yield-breeding (Dohm *et al.*, 2013).

Sugar beets

By Stanzilla - Own work, CC BY-SA 3.0, https://commons.wikimedia.org/w/index.php?curid=23204161

Botany

Paesold *et al.* (2012) reported this beet containing 2n=2x=18 chromosome numbers. Sugarbeet has a conical, white and fleshy taproot with a flat crown and a rosette of numerous broad and rich brilliant green leaves some 35cm in length in a tuft, appearing from the crown. These leaves appear from the crown with or just above the ground surface (en.wikipedia.org/wiki/Sugar_beet). Through photosynthesis sugar is formed in the leaves and is then stored in the root. Average weight of one knob of sugarbeet root ranges from 0.5 to 1kg.

Varieties

The tropical sugarbeet varieties suitable for cultivation in Tamil Nadu (India) are 'Cauvery', 'Indus', 'Shubhra', 'Pasoda', 'H1 0064' and 'Doratea'. The duration of these tropical hybrids is 150-183 days depending on climatic conditions prevailing during crop growth period (agritech.tnau.ac.in; static.vikaspedia.in). Some of the common sugarbeet varieties being grown locally in Oregon (USA) are 'Emblem', 'Owyhee', 'Dillon', 'Canyon', 'Oasis', 'PM9', 'Sierra', 'Mustang', 'PM21', and 'Cassia'.

Constituents

The root of the beet contains 75% water, about 20% sugar (Dohm *et al.*, 2013) and 5% pulp (FAO/EBRD, 1999). Sugar content can vary between 12 and 21%, depending on the cultivar and growing conditions. Sugar is the primary value of sugarbeet as a cash crop. The pulp, insoluble in water and mainly composed of cellulose, hemicellulose, lignin and pectin is used as animal feed. Sugarbeet pulp is a pectin-rich (15–30% dry weight) byproduct of the sugar refining industry which is traditionally used as a raw material for livestock feeds (Knez *et al.*, 2018).

Phytochemicals

Phenolic compounds detected in methanolic extract of sugarbeet are given below. Data are presented as mg/100g dried powder (Arjeh *et al.*, 2022):

Compounds	Peak no.	RT (min)	Phenolic content	
			Peel	Flesh
Hydroxybenzoic acids				
Gallic acid	1	4.3	30.57 ± 2.69	22.95 ± 1.35
4-hydroxybenzoic acid	3	41.2	4.36 ± 1.02	6.02 ± 0.70
Vanillic acid	4	48.3	1.87 ± 0.49	1.52 ± 0.18
Hydroxycinnamic acids				
Chlorogenic acid	7	53.2	2.24 ± 0.21	Nd
Ferulic acid	8	59.9	3.20 ± 0.00	1.06 ± 0.03
Hydroxybenzoic acid				
Vanillin	5	51.1	0.65 ± 0.13	0.99 ± 0.20
Flavanol				
Epicatechin	2	6.3	31.16 ± 1.89	10.57 ± 1.27
Flavanone				
Eriodictyol	6	52.3	0.52 ± 0.00	0.26 ± 0.03
Flavonols				
Quercetin-3-O-rutinoside	9	64.6	30.14 ± 3.63	1.46 ± 0.06
Kaempferol	10	82.9	12.15 ± 1.62	Nd
Abbreviations: Nd= not detected; RT= retention time.				

The above table presents the concentration of the identified phenolics of the sugarbeet. Gallic acid, epicatechin and quercetin-3-O-rutinoside were the dominant phenolics detected in sugarbeet. The root peel extracts presented a higher amount of bioactive compounds and biological properties (antioxidant and chelating activity) compared to the flesh extract (Arjeh *et al.*, 2022). Phenolics existing in the methanolic extracts are quantified by well-defined peaks with maximum absorbance at 340nm for flavonoids and at 285nm for phenolic acids. The remarkable differences between phenolic compounds of peel and flesh extracts are due to the absence or presence of chlorogenic acid and kaempferol and to the relative concentration of them. Vanillin, vanillic acid, hydroxybenzoic acid, ferulic acid and kaempferol are the phenolic compounds previously identified in sugarbeet by-products (molasses and pulp) (Valli *et al.*, 2012), but other phenolic compounds (gallic acid, epicatechin, eriodictyol, chlorogenic acid and quercetin-3-*O*-rutinoside) were unambiguously identified for the first time among sugarbeet and its by-products (pulp and molasses). As can be seen, epicatechin, gallic acid and quercetin-3-*O*-rutinoside are the dominant phenolics that are detected in sugarbeet and presented average amounts of 31.16 ± 1.89, 30.57 ± 2.69, and 30.14 ± 3.63 mg/100g in peel extract, respectively. Gallic acid (3,4,5-trihydroxybenzoic acid) is a low molecular weight phenolic compound that is widely found in plants and can be either free or as part of hydrolyzable tannins. Nayeem *et al.* (2016) stated that gallic acid and its derivatives can exhibit their potential to prevent various diseases such as ischemic heart diseases, diabetes, cancer and other ailments. Rutin (quercetin-3-*O*-rutinoside) is one of the most prominent antioxidative flavonoids that scavenge free radical by its phenolic hydroxyl groups at the B-ring and the 3-position (Murakami *et al.*, 2008). Numerous studies have illustrated that quercetin derivatives have a high ability to scavenge the reactive species such as the hydroxyl and peroxynitrite (Amiri *et al.*, 2021; Boots *et al.*, 2008). Epicatechin is another dominant phenolic compound which is widely found in tea and belongs to the group of flavan-3-ols. Researchers have attributed a wide range of biological activities to epicatechin such as antioxidant, antibacterial, hypoglycemic, insulin resistance as well as preventing anti-inflammatory, enhancing immunity, lowering cardiovascular disease and so on (Tong *et al.*, 2018).

The fixed oil of the leaves of sugar beet contains undecane and undecane 2,6-dimethyl and that of the fruit contains hexadecanoic acid methyl ester and 9-octadecenoic acid (Afifi *et al.*, 2015) while the hexane extract of the seeds contains squalene, β-amyrine, cycloartenol, 24-methylenecycloartanol, cholesterol, 24-methylenecholesterol, campesterol, stigmasterol, ergost-7-en-3β-ol, stigmast-9(11)-en-3β-ol, spinasterol, sitosterol, stigmasterol, 22-dihydrospinasterol and avenasterol (Schmidt *et al.*, 1994). The sugarbeet root was found to contain inositol, galactinol and dehydrodiferulates (arabinan and galactan side chains esterified by ferulic acid residues (Brown and Serro, 1953; Ralet *et al.*, 2005). It was found that *B.v.* ssp. *vulgaris* var. *altissima* contains flavonoids as quercetin, quercetin 3-*O*-galactopyranoside, quercetin 3-*O*-β-D-glucopyranoside, vitexin, 3-hydroxytryptamine, betagarin, betavulgarin, 5, 2'-dihydroxy 6,7-methylene dioxy-isoflavone, dihydrooroxylin A,3,5-dihydroxy-6,7-methylene dioxy-flavanone, 2', 5-dihydroxy-6,7-methylene dioxy-isoflavone and 5-hydroxy-6,7-methylene dioxy-flavone (El-Gengaihi *et al.*, 2016; Geigert *et al.*, 1973; Gardner *et al.*, 1967; Chui *et al.*, 1985; Takahashi *et al.*, 1987).

In addition to phenolic acids this also recorded gallic, caffeic, vaneltic, coumaric, ferulic and cinnamic acids (El-Gengaihi *et al.*, 2016). Many saponins were identified as 3-*O*-ß-D-glucopyranosyl-3ß-hydroxy-olean-12-en-28-oic acid, the methyl ester of oleanolic acid, 3-*O*-ß-D-glucuronopyranosyl-3ß-hydroxy-olean-12-en-28-oic acid, 3-*O*-[ß-D-glucopyranosyl-(1-2)-(ß-D-xylopyranosyl-

(1-3)]-ß-D-glucuronopyranosyl-3ß-hydroxyolean-12-en-28-oic acid, 3-*O*-[ß-D-glucopyranosyl-(1-2)-(ß-D-xylopyranosyl-(1-3))-ß-D-glucurono-pyranosyl]-28-O-ß-Dglucopyranosyl-3ß-hydroxy-olean-12-en-28-oic, betavulgarosides I, II, III, IV, V, VI, VII, VIII, IX and X (Yoshikawa *et al.*, 1996; Marsh and Levvy, 1956; Bauserman and Hanzas, 1957; Ridout *et al.*, 1994; Massiot *et al.*, 1994; Yoshikawa *et al.*, 1995; Yoshikawa *et al.*, 1998). The saponin fractions of the leaves and roots of BVVA decreased the glucose in the blood of rats (Yoshikawa *et al.*, 1996, Yoshikawa *et al.*, 1995; Yoshikawa *et al.*, 1998; Kushiro *et al.*, 1998). *Beta vulgaris* is also known for having antioxidant, antihepatotoxic and anti-inflammatory activity (El-Gengaihi *et al.*, 2016). The precise nature of the bioactive components of the non-flowering aerial parts of BVVA responsible for its antioxidant activity still remains unclear.

Eight phenolic compounds (apigenin, luteolin, apigenin 7-*O*-di-glucoside, luteolin 4'-*O*-rhamnoside, caffeoyl6-(3,4-dihydroxy benzoyl) β-D-glucoside, apigenin 5-*O*-rhamnoside, isoscutellarein 7-*O*-glucosyl, 8-Oxyloside and 5, 7, 5'-trihydroxy-3'-methoxy-flavone 4'-*O*-glucoside) were isolated for the first time from the non-flowering aerial parts of *Beta vulgaris* var. *altissima* (BVVA) (sugarbeet) cultivated in Egypt. The GLC analysis of the unsaponifiable matter of the non-flowering aerial parts of BVVA revealed the presence of 18 compounds in which the main compound found was campesterol (23.85%). The gas liquid chromatography (GLC) analysis of the fatty acid methyl esters of the non-flowering aerial parts of BVVA revealed the presence of 13 compounds in which the major acid found was eicosenoic acid C20: 1 (41.38%) while the GLC of the unsaponifiable matter substantiated the presence of a mixture of 28 components in which campesterol (23.85%), n-triacontane C30 (12.71%) and cholesterol (8.77%) were the major components. The highest antioxidant activity was exhibited by the ethyl acetate fraction (62.65%) with effective concentration of 57.2 μg/ml using radical scavenging activity. The highest antiproliferative activity against HepG2 was exhibited by the methanol extract of BVVA. The per cent proliferation of human hepatocellular carcinoma was reduced to 69.5% and 72.4% on treatment with the ethyl acetate and the methanol extracts, respectively, at concentration of 100μg/ml (El-Hawary *et al.*, 2017).

Uses

Sugarbeet is used as a vegetable and food garnish in Japanese-style dishes (Yoshikawa *et al.*, 1996). The roots of sugarbeets have been used industrially as a raw material for sugar and the aerial parts of sugarbeets are used in feeding animals (Al-Dosari *et al.*, 2011). Along with sugarcane, sugarbeets are among the most common plants used in the production of white sugar. They are also used to produce other types of refined sugar such as molasses and brown sugar.

Unrefined thick sugary syrup can be produced directly from shredded sugarbeet by cooking for several hours, pressing the resulting mash and concentrating the juice produced until it has a consistency similar to that of honey.

In traditional Chinese medicine, the roots of sugarbeet have been listed as a medicinal herb which has been known to exhibit sedative and emmenagogue-like effects (Murakami *et al.*, 1999).

Sugarbeets can also be used to produce alcohol and biobutanol fuel. It requires 6.22kg of sugarbeet to produce 1kg of ethanol (approximately 1.27 litres at room temperature). Molasses can serve to produce monosodium glutamate (MSG). The waste left after extracting the sugar can be used as fodder for cattle. In New Zealand, sugarbeet is widely grown and harvested as feed for dairy cattle. It is regarded as superior to fodder beet because it has a lower water content (resulting in better storage properties). Both the beet knob and the leaves (with 25% protein) are fed to cattle (en.wikipedia.org/wiki/Sugar_beet).

Beet Sugar vs. Cane Sugar

Both beet sugar and cane sugar are composed of sucrose, a compound composed of glucose and fructose molecules. Although beet sugar and cane sugar are nearly identical in nutritional value, their distinct taste can alter the flavour of dishes. Beet sugar has an earthy, oxidized aroma and faint burnt-sugar aftertaste whereas cane sugar is characterized by a sweeter aftertaste and fruitier aroma. Cane sugar is said to caramelize more easily and result in a more uniform product than beet sugar. On the other hand, beet sugar can create a crunchier texture and has a unique taste that works well in certain baked goods.

Cultivation

Climate and Season

Sugarbeets grow exclusively in the temperate zone though sugarcanes in the tropical and subtropical zones. In India, the favourable time of sugarbeet cultivation starts from the end of September to mid-October (Deol and Kanwar, 1975) and is usually harvested in 150-180 days (Mall *et al.*, 2021). Both early and late sowing of sugarbeet causes a negative impact on root yield (Mathuria and Bhoj, 1977). Sugarbeet is grown in climates that are too cold for sugarcane growing (en.wikipedia.org/wiki/Sugar_beet) and owns a wider consistency with resistance to water stress (Mall *et al.*, 2018a), salt stress (Misra *et al.*, 2020) and frost (Webster *et al.*, 2016). Sugarbeet crop flourishes well in a fairly cool climate with an adequate amount of rainfall and bright sunshine during different phases of growth (Pathak *et al.*, 2014). Optimum temperature for seed germination is around 25°C and for root growth is 20°C (Dubetz and Russell, 1964). Low soil temperatures during early growth phase have positive impact on knob weight (Smit, 1983).

Soil

Sugarbeet grows well in clayey loam, loam or sandy-loam soils that are neutral to slightly alkaline in nature with a pH 7–8.5 though it can be grown at high pH of 8–9.6 due to its salt tolerance property. It can be grown on black cotton soils as is found in the Deccan tracts of Maharashtra, Karnataka, Andhra Pradesh and Tamil Nadu (Mall *et al.*, 2021; Pathak *et al.*, 2014).

Lang Preparation and Planting

Land is ploughed to a fine tilth by thorough ploughing and the clods and weeds are removed. Seeds can be sown directly in ridges at a spacing of 10cm. After thinning in multi-germ seeds, a distance of about 20cm between plants should be maintained (Mall *et al.*, 2021). However, row to row spacing is maintained at 30-45cm apart.

Fertilizers and Manures

Before sowing of beetroot, in July-end green manuring crop can be raised by sowing sunnhemp or *Sesbania* seeds which by 3rd week of September when the crop is some 50-60cm high, are knocked down in the field and then field is filled with water for its decomposition. After about one week farmyard manure some 10t/ha is thoroughly mixed in the soil and the soil is brought to a fine tilth. Now raised beds of convenient sizes are prepared. Before sowing the beetroot seeds some 50-60kg of nitrogen together with 60kg of P_2O_5 are mixed in the soil. Nitrogen @ 50-60kg/ha is again applied during earthing up of the beetroot crop. Late usage of nitrogen affects the canopy size and amount of nitrogen content in the plant (Malnoua *et al.*, 2008).

Irrigation

Irrigation supply usually depends on factors such as soil type, moisture content, rainfall, *etc.* About 10-12 irrigations supplied at 75 and 50mm cumulative pan evaporation (CPE) results into highest yield of this crop. Care must be given that water should not flow over the ridges while irrigating the crop (Mall *et al.*, 2021).

Weeding and earthing up

Sugarbeet crop faces competition with weeds from germination to harvest and requires weeding almost throughout the growing season. At the early growth stage, one manual weeding followed by two/three shallow hoeings/weedings favours good growth of the crop, however, with the first manual weeding, in early

December earthing up is also followed for harvesting a good yield of root-knobs (Mall *et al.*, 2021). Herbicide spray twice of Rifit @ 2 kg ha−1 or Alachlor @ 1 litre ha−1, once about 25 days of sowing and other at 50 days of sowing may suffice (Mall *et al.*, 2021). However, if need be afterwards, again one more spraying may be repeated but if weeds are sporadic these are lifted out manually.

Thinning

Thinning is usually not required in sugarbeet monogerm seeds, but multigerm seeds produce 2-3 plants/seed and require thinning to maintain a distance of about 20cm between plants. The most suitable time for singling and thinning is after 3–4 weeks of sowing, *i.e.* 4–6-leaf stage of crops (Mall *et al.*, 2021).

Harvesting

Harvesting can be done after around 5-6 months from day of sowing (Mall *et al.*, 2021). The yellowing of lower leaf whirls of matured plant and tuber brix reading of 15 to 18% indicate the maturity of taproots for harvest. Harvesting should be timed so that the roots reach the factory for processing within 48 hours. Till such time the sugarbeet roots should be retained in the field. The harvested taproot should be handled as gently as possible to remove soil and trash to minimize the beet breakage and bruising to get quality root (static.vikaspedia.in).

Propagation and Yield

Sugarbeets are propagated through seeds that start germinating within a week after sowing (Mall *et al.*, 2021). The average yield of beetroot is 30 to 35 tonnes/hectare (static.vikaspedia.in).

Seed Production

Sugarbeet requires thermal induction for flowering and seed production (Kapur *et al.*, 1986). This requirement lacks in sugarbeet grown in plains which prevents it to flower under such a condition. Due to this reason, seed production has been standardized only at higher altitudes (> 1500m) where favourable weather conditions prevail (Srivastava *et al.*, 1983). These places are Srinagar (Jammu & Kashmir), Mukteswar & Ranichauri (Kumaon Hills), Auli (Garhwal Hills), Darjeeling (West Bengal) as well as Shimla, Kalpa and Katrain (Himachal Pradesh). At all these places seed production technology is well established (Pathak *et al.*, 2011) and successful seed production of 'Ramonskaya-06' (R-06) ~ a Russian variety was performed followed by the development and production of indigenous varieties, *viz.* 'IISR Comp-1' and 'LS-6' (developed by IISR Lucknow, U.P.) and 'Pant S-10' (developed by GPUAT Pantnagar, Uttarakhand) (Pathak *et al.*, 2014). Kapur *et al.* (2000) stated that at these places high-temperature tolerant varieties were also developed and maintained along with the production of diploid varieties recommended for general cultivation areas. Kalpa centre (H.P.) was routinely producing foundation seeds of multigerm unpelleted varieties with mass multiplication by National Seeds Corporation (Pathak *et al.*, 2014). Presently ICAR-IISR Sugarbeet Outpost, Mukteshwar is producing and maintaining sugarbeet seeds.

Sugarbeet seeds are produced either through *in situ* method or through transplanting method. *In situ* method requires overwintering conditions in the field for the beet to blossom and produce seed. Due to its overwintering requirement, it is often termed as an overwintering method. In the transplanting method, selected stecklings (or young roots) are uprooted in the first season and stored in trenches for overwintering conditions followed by thermal induction. The stecklings are replanted during overwintering conditions in the next season for production of flowers and seeds. This practice is generally being performed in sugarbeets grown in the plains of India (Kapur *et al.*, 1986).

Flowering sugar beet

By Rasbak - Own work, CC BY-SA 3.0, https://commons.wikimedia.org/w/index.php?curid=252034

An important practice is to cut one-third of the top and root portion prior to planting these retaining only 10cm of leaves. Root weight of 7–16g with a pair of 5–10 leaves along with the height of the plant as 22–26cm is the requirement for transplanting method. Development of seed takes place inside the fruit of sugarbeet. Seed comprises of pericarp and operculum with an ovary cap. The sugarbeet seed comprises of two or five embryos supported by diploid perisperm rather than triploid one, due to which it is termed as multigerm seed. Conversely, the production of monogerm seeds (single embryo) occurs through mechanical means.

Inspected crop	Other crops	Minimum isolation distance required
Foundation	Crops planted with Foundation seed of the same pollen source	3 meters (10 feet), provided the pedigree of the Foundation seed used can be established and the prescribed isolation distance is free from harmful contamination (i.e. other species which can cross pollinate with the inspected crop)
	Non-pedigreed Sugarbeet pollen source	1525 meters or more as specified by the Breeder
	Other or unknown pollinator of genus Beta (including fodder beet, mangel, red beet, swiss chard)	3110 meters
Foundation -varieties with monogerm pollinator	- Monogerm pollinator sources	1525 meters
Certified	Crops planted with Foundation seed of the same pollen source	3 meters, provided the pedigree of the Foundation seed used can be established and the prescribed isolation distance is free from harmful contamination (i.e. other species which can cross pollinate with the inspected crop)
	Non-pedigreed sugarbeet pollen source	975 meters or more, as specified by the breeder
	Other or unknown pollinator of genus Beta (including fodder beet, mangel, red beet, swiss chard)	2440 meters
Certified -varieties with monogerm pollinator	- Monogerm pollinator sources	1525 meters

Seedgrowers.ca/wp-content/uploads/Circ6-SECTION-14.12_Sugar-Beet_English_Rev01.13-2018_20180108-1.pdf.

Pests and Diseases

The major insect-pests that affect the sugarbeet crop are aphids, tobacco caterpiller and diamond-back moth. Integrated pest management programme has to be adopted to control these insect- pests. To control aphids, spray 3% neem oil or dimethoate 2ml/l with teepol 0.5 ml/l; for tobacco caterpiller spray endosulfon 2ml/l or carbaryl 2g/l of water (static.vikaspedia.in).

The major diseases that affect the sugarbeet crop are *Rhizoctonia* wilt, powdery mildew, *Cercospora* leaf spot and *Fusarium* yellows. To control *Rhizoctonia* wilt, spot drenching with 1% Bordeaux mixture and for *Fusarium* wilt drenching the soil with carbendazim @0.1% (static.vikaspedia.in).

Sugarbeet plants are susceptible to *Rhizomania* ('root madness') which turns the bulbous taproot into many small roots, making the crop economically unprocessable. 'Beet necrotic yellow vein virus' (BNYVV) ~ a member of the genus *Benyvirus* (Tamada, 1999), is responsible for *Rhizomania* disease. It is transmitted by the plasmodiophorid *Polymyxa betae*. All the subspecies of *Beta vulgaris* suffer infections from BNYVV. Strict controls are enforced in European countries to prevent the spread though it is already present in certain areas there. It is also susceptible to both the beet leaf curl viruses which cause crinkling and stunting of the leaves and beet yellows (en.wikipedia.org/wiki/Beet_necrotic_yellow_vein_virus; en.wikipedia.org/wiki/Sugar_beet).

Mangelwurzel (*B.v.* ssp. *vulgaris* var. *crassa* or *B.v.* ssp. *vulgaris* 'Crassa Group')

Other names: Mangold wurzel, Mangold & Mangel beet (Wright, 2001), Field beet (Raynbird, 1851), Fodder beet and Root of scarcity (Chisholm, 1911; www.complete-herbal.com).

Mangelwurzel, a variety of *Beta vulgaris* (Chisholm, 1911), was developed in the 18th century as a fodder crop for feeding livestock. The name Mangelwurzel or Mangold wurzel may be from German Mangel/Mangold meaning 'chard' and Wurzel meaning 'root' (en.wikipedia.org/wiki/Mangelwurzel). It originated in the Middle East and was being used as cattle feed in ancient Greece in 500 BC (Henry, 2010). It then spread to Europe, and introduced into the USA in 1800 and is now cultivated worldwide in the cooler climates (Northern America, New Zealand and at higher altitudes (above 600-1000m) in the tropics (Heuze *et al.*, 2020). Though mangelwurzel isn't a well-known root vegetable, its leaves and roots provide a nutritious food so deserves some recognition.

Botany

Mangelwurzel is a biennial plant grown for its fleshy and swollen root. It develops seeds in its second spring. The external appearance of the mangelwurzel plant and its seed resembles very much to the beet-root. Its large white, yellow or orange-yellow swollen roots (en.wikipedia.org/wiki/Mangelwurzel) can grow much larger than beetroot, particularly in cooler climates. Its sweet and juicy roots are similar to a whitish-yellow beetroot when harvested young though their eating quality drops off in older roots as they become progressively woodier with size (www.theseedcollection.com.au).

The dark green, heart-shaped leaves that are borne in a rosette, lie horizontally to catch as much light as possible. After the first year, if the root is not harvested and after exposure to cold, the rosette turns into a 50-80cm tall floral stalk that bears small, green and bisexual flowers without petals (Henry, 2010). Fodder beet crops that are intended for fodder (rather than for seed production) are cultivated as annual crops and the roots must be harvested before winter since they do not withstand frost (Heuze *et al.,* 2020).

Nutritional value

Nutritional value per 100g	
Carbs	4.3 g
Dietary Fiber	3.7 g
Sugars	0.5 g
Fat	0.1 g
Saturated	0 g
Polyunsaturated	0 g
Monounsaturated	0 g
Trans	0 g
Protein	2.2 g
Sodium	226 mg
Potassium	762 mg
Cholesterol	0 mg
Vitamin A	126.5 %
Vitamin C	50 %
Calcium	11.7 %
Iron	14.3 %
Percentages are based on a diet of 2000 calories a day.	

Source: www.myfitnesspal.com/food/calories/mangel-wurzel-raw-118893642.

Uses

Though young tender roots and leaves can be eaten by humans, it is mainly grown for cattle, pig and other stock fodder (Chisholm, 1911). Both its leaves and roots provide a nutritious food for animals. For human consumption, the young and tender roots can be prepared boiled like potato for serving mashed, diced or in sweet curries. Young leaves can be prepared like other greens.

Cultivation

Climate and Season

The mangelwurzel cultivation is essentially the same as beetroot cultivation. They don't like tropical temperatures and tend to rot in high humidity or wet soil (morningchores.com). In temperate zones, seeds are generally sown in spring when the last frost has passed. Soil temperature should be no less than 10°C and an average of 15 to 18°C during the day. In subtropical zones, these are planted in late summer to early fall when nights become cool (morningchores.com).

Mangelwurzel is very susceptible to damage from frost and is suited to southern parts of England where climate is too warm and dry for the successful cultivation of turnip (Chisholm, 1911). Mangelwurzel isn't heat-tolerant too and will need protection in climates with hot summers. An area with morning sun but shade from midday onward is ideal for its cultivation but if such area is not available shade cloth or heavy netting will be needed for the hotter parts of the day (www.theseedcollection.com.au).

Soil

Loose, loamy or sandy soil with a pH range of 6-7 is best to allow the root to form and grow. It is recommended to use a lot of well-rotted compost made of rich organic matter and manure when planting (morningchores.com).

Planting Distance

In the cooler southern areas of Australia, plants should be spaced at least 50cm each way as the roots can reach up to 20kg in ideal conditions. Whereas, in more northerly parts, roots are more likely to reach only a few kilos if left to fully mature, so spacing of 30cm is more common (www.theseedcollection.com.au).

Fertilizer

Once the roots are approximately the size of a thumb, growers can use either a well-balanced fertilizer or well-rotted poultry manure (morningchores.com). Mangelwurzel may require supplementary potassium for optimum yields, flavour and texture. Foliage readily displays potassium deficiency as interveinal chlorosis. This can be corrected with either organic or inorganic sources of potash (en. wikipedia.org/wiki/Mangelwurzel).

Similar to most root vegetables, mangelwurzel leaves are also sensitive to fertilizer burn. Therefore growers should avoid lodging of any fertilizer or manure on the leaves. Also, growers should be careful in digging of its roots when still these are tender. The foliage can be sprayed each week with fish or seaweed-based spray after ensuring that these are suitable for foliage (morningchores.com).

Irrigation, Weeding and Mulching

Like other root vegetables, mangelwurzel too requires regular watering and weed control. The soil should not become too dry otherwise the root will split when it takes up the water. Drip irrigation may provide better results. In case of manual watering, a small amount should be watered at regular intervals ~ an inch a week is ideal but based on amount of rainfall. Mangelwurzel faces competition with weeds and requires weeding almost throughout the growing season. Mulching with straw or similar to a depth of about 2.5cm helps not only to suppress the weeds and retain the moisture but also protects the plant from extreme variations in weather (morningchores.com).

Companion Planting

Mangelwurzel can be planted along with chard, brussel sprouts, kohlrabi, onions, cauliflower, broccoli, lettuce, potato, tomato, corn, eggplant and pepper. It should not be planted with pole beans, crocus or mustard (morningchores.com).

Harvest

Harvesting of leaves can begin when these are large enough to eat. In this case a few of the outer leaves are removed leaving the inner ones to fuel the plant's growth. They are cut just above the ground or the top of the swollen root, using a clean and sharp knife (en.wikipedia.org/wiki/Mangelwurzel).

It takes five months for mangelwurzel to be ready for harvest of roots (morningchores.com). The roots are best harvested on the younger side if they're intended for the kitchen, with the larger and woodier specimens better reserved as fodder. In either case, it's important to harvest before the weather turns cold, as even a light frost can turn the roots into a mushy mess (en.wikipedia.org/wiki/Mangelwurzel).

Yield

Usually root production in a one hectare of land is around 50-75 tonnes.

Pests and diseases

These are the same as described above under the 'Beetroot'.

References

Afifi, M.S., M.A. Amer, S.R. Gedara and S.M. Elghandour, 2015. *J. environm. Sci.*, **44**: 161.

agritech.tnau.ac.in/agriculture/sugarcrops_tropicalsugarbeet.html.

Al-Dosari, M., S. Alqasoumi, M. Ahmed, M. Al-Yahya, M.N. Ansari and S. Rafatullah, 2011. *Farmacia,* **59**: 669.

Amiri, S., Saray F. Roshani, L. Rezazad-Bari and S. Pirsa, 2021. Optimization of extraction and characterization of physicochemical, structural, thermal, and antioxidant properties of mucilage from hollyhock's root: A functional heteropolysaccharide. *J. Food Measurem. & Characteriz.*, **15**(3): 2889– 2903.

Arjeh, E., S.M. Khodaei, M. Barzegar, S. Pirsa, I. Karimi Sani, S. Rahati and F. Mohammadi, 2022. Phenolic compounds of sugar beet (*Beta vulgaris* L.): Separation method, chemical characterization and biological properties. *Food Sci. & Nutr.,* **10**: 4238–46. https://doi.org/10.1002/fsn3.3017.

Bauserman, H.M. and P.C. Hanzas, 1957. *J. Amer. Soc. Sugar Beet Techn.,* **9**: 295.

Berends, J.E., L.M.M. van den Berg, M.A. Guggeis, N.F.T. Henckens, I.J. Hossein, M.E.J.R. de Joode, H. Zamani, K.A.A.J. van Pelt, N.A. Beelen, G.G. Kuhnle, T.M.C.M. de Kok and S.G.J. van Breda, 2019. Consumption of nitrate-rich beetroot juice with or without vitamin C supplementation increases the excretion of urinary nitrate, nitrite, and N-nitroso compounds in humans. *Int. J. Mol. Sci.*, **20**(9): 2277. doi: 10.3390/ijms20092277. PMID: 31072023; PMCID: PMC6539140.

Boots, A.W., G.R. Haenen and A. Bast, 2008. Health effects of quercetin: From antioxidant to nutraceutical. *Eur. J. Pharmacol.*, **585**(2–3): 325–37.

Brown R.J. and R.F. Serro, 1953.. *J. Amer. Chem. Soc.*, **75**: 1040.

Chisholm, Hugh, ed. 1911. Encyclopædia Britannica,.vol. 17 (11th ed.), 571p. Cambridge University Press, U.K..

Chui, H., Y. Arakawa, S. Ueda, M. Kuroda. And M. Izawa, 1985. *Phytochem.*, **25**: 281.

Clifford, T., G. Howatson, D.J. West and E.J. Stevenson, 2015. The potential benefits of red beetroot supplementation in health and disease. *Nutrients*, **7**(4):2801-22. doi: 10.3390/nu7042801. PMID: 25875121; PMCID: PMC4425174.

da Silva. D.V.T., P.O. Silva, D. Perrone, A.P.T.R. Pierucci, C.A. Conte-Junior, T.S. Alvares, E.M.D. Aguila and V.M.F. Paschoalin, 2016. Physicochemical, nutritional, and sensory analyses of a nitrate-enriched beetroot gel and its effects on plasmatic nitric oxide and blood pressure. *J. Food & Nutr. Res.,* **60**:1-9. DOI: 10.3402/fnr.v60.29909.

Dejam, A., C.J. Hunter, A.N. Schechter and M.T. Gladwin, 2004. Emerging role of nitrite in human biology. *Blood Cells Mol. Dis.*, **32**:423–9. doi: 10.1016/j.bcmd.2004.02.002.

Delgado-Vargas, F., A.R. Jimenez and O. Paredes-Lopez, 2000. Natural pigments: Carotenoids, anthocyanins, andbetalains - characteristics, biosynthesis, processing, and stability. *Crit. Rev. in Food Sci. & Nutr.*, **40**:173-289. DOI: 10.1080/10408690091189257.

Deol, D.S. and R.S. Kanwar, 1975. Effect of the dates of sowing and harvesting on the yield and quality of sugarbeet. *Crop Improv.*, **2**: 105–11.

Dohm, J.C., A.E. Minoche, D. Holtgrawe, S. Capella-Gutierrez, F. Zakrzewski, H. Tafer, O. Rupp, T.R. Sorensen, R. Stracke, R. Reinhardt, A. Goesmann, T. Kraft, B. Schulz, P.F. Stadler, T. Schmidt, T. Gabaldon, H. Lehrach, B. Weisshaar and H. Himmelbauer, 2013. The genome of the recently domesticated crop plant sugar beet (*Beta vulgaris*). *Nature Portfolio*, **505** (7484): 546–9. Bibcode:2014Natur.505..546D. doi:10.1038/nature12817. ISSN 0028-0836. PMID 24352233.

Dubetz, S. and G.C. Russell, 1964. Soil temperature and nitrogen effect on yield and phosphorus uptake by sugarbeet. *J. Amer. Soc. Sugar Beet Tech.*, **13**: 238–43.

eagri.org/eagri50/HORT281/pdf/lec30.pdf.

Eastwood M.A. and H. Nyhlin, 1995. Beeturia and colonic oxalic acid. *QJM: An Intern. J. Med*, **88**(10): 711–7. PMID 7493168.

El-Gengaihi, S.E., M.A. Hamed, D.H. Aboubaker and H.M. Abdel-Tawab, 2016. *Intern. J. Pharm. Pharmaceut. Sci.*, **8**: 281.

El-Hawary, S.S., F.M. Hammouda, W.A. Tawfik, H.A. Kassem, K.A. Abdelshafeek and S.S. El-Shamy, 2017. Investigation of some chemical constituents, cytotoxicity and antioxidant activities of *Beta vulgaris* var. *altissima* cultivated in Egypt. *Rasayan J. Chem.*, **10**(4): 1391-1401. ISSN: 0974-1496|e-ISSN: 0976-0083. https://rasayanjournal.co.in/admin/php/upload/291_pdf.pdf.

en.wikipedia.org/wiki/Beetroot.

en.wikipedia.org/wiki/Beet_necrotic_yellow_vein_virus.

en.wikipedia.org/wiki/Beta_vulgaris.

en.wikipedia.org/wiki/Mangelwurzel.

en.wikipedia.org/wiki/Sugar_beet

FAO/EBRD, 1999. Agribusiness Handbooks (vol. 43), pp. 35-43. Sugar Beets/White Sugar (PDF). Archived from the original (PDF) on 6 October 2014. www.responsibleagroinvestment.org/sites/responsibleagroinvestment.org/files/FAO_Agbiz%20handbook_White%20Sugar_0.pdf.

Flores Olvera, H., E. Smets and A. Vrijdaghs, 2008. Floral and inflorescence morphology and ontogeny in Beta vulgaris with special emphasis on the ovary position. *Ann. Bot.*, **102**(4): 643–51. doi:10.1093/aob/mcn140. PMC 2701786. PMID 18694878.

Frank, T., F.C. Stintzing, R. Carle, I. Bitsch, D. Quaas, G. Straß, R. Bitsch and M. Netzel, 2005. Urinary pharmacokinetics of betalains following consumption of red beet juice in healthy humans. *Pharmacol. Res.*, **52**(4): 290–97. doi:10.1016/j.phrs.2005.04.005. PMID 15964200.

Free, J.B., Ingrid H. Williams, P.C. Longden and M.G. Johnson, 1975. Insect pollination of sugar-beet (*Beta vulgaris*) seed crops. *Ann. Appl. Biol.*, **81**(2): 127–34. doi:10.1111/j.1744-7348.1975.tb00529.x. ISSN 1744-7348.

gardenerspath.com b: gardenerspath.com/how-to/disease-and-pests/common-beet-pests/.

gardenerspath.com a: gardenerspath.com /how-to/disease-and-pests/common-beet-diseases/.

Gardner, R., A. Kerst, D. Wilson and M. Payne, 1967. *Phytochem.*, **6**: 417.

Geigert, J., F. Stermitz, G. Johnson, D. Maag and D. Johnson, 1973. *Tetrahedron*, **29**: 2703.

Georgiev, V.G., J. Weber, E.M. Kneschke, P.N. Denev, T. Bley and A.I. Pavlov, 2010. Antioxidant activity and phenolic content of betalain extracts from intact plants and hairy root cultures of the red beetroot *Beta vulgaris* cv. Detroit dark red. *Pl. Foods Hum. Nutr.*, **65**:105-11. DOI: 10.1007/s11130-010-0156-6.

Gledhill, David, 2008.. The Names of Plants. Cambridge University Press, 70p. ISBN 9780521866453 (hardback), ISBN 9780521685535 (paperback).

Guldiken, B., G. Toydemir, K. Nur Memis, S. Okur, D. Boyacioglu and E. Capanoglu, 2016. Home-processed red beetroot (*Beta vulgaris* L.) products: changes in antioxidant properties and bioaccessibility. *Int. J. Mol. Sci.*, **17**(6):858.

Hanelt, Peter, R. Buttner, Rudolf Mansfeld and Ruth Kilian, 2001. Mansfeld's Encyclopedia of Agricultural and Horticultural Crops, pp. 235-41. Springer, Holland. ISBN 978-3-540-41017-1.

Henry, K. 2010. Fodder Beet. In: Handbook of Plant Breeding (ed. Bradshaw, J. E.), vol 7, pp. 221-44.

Heuze, V., G. Tran and D. Sauvant, 2020. Fodder beet roots. Feedipedia, a programme by INRAE, CIRAD, AFZ and FAO. https://feedipedia.org/node/534 Last updated on February 27, 2020, 16:42.

Hobbs D.A., T.W. George and J. A. Lovegrove, 2013. The effects of dietary nitrate on blood pressure and endothelial function: A review of human intervention studies. *Nutr. Res. Rev.*, **26**: 210–22. doi: 10.1017/S0954422413000188.

Kadereit G., S. Hohmann and J.W. Kadereit, 2006. A synopsis of Chenopodiaceae subfam. Betoideae and notes on the taxonomy of Beta (PDF). *Willdenowia*, **36**: 9–19. doi:10.3372/wi.36.36101. S2CID 84677159.

Kapadia, G.J., M.A. Azuine, R. Sridhar, Y. Okuda, A. Tsuruta, E. Ichiishi, T. Mukainake, M. Takasaki, T. Konoshima, H. Nishino and H. Tokuda, 2003. Chemoprevention of DMBA-induced UV-B promoted, NOR-1-induced TPA promoted skin carcinogenesis and DEN-induced phenobarbital promoted liver tumors in mice by extract of beetroot. *Pharmacol. Res.*, **47**:141-8. DOI: 10.1016/S1043-6618(02)00285-2.

Kapur R., B.L. Srivastava and H.M. Srivsatava, 1986. A new technique for vegetative propagation of sugarbeet seed crop. *Indian Sugar Crops J.*, **6**:16–7.

Kapur, R., H.M. Srivastava, V.K. Saxena and S. Bhatnagar, 2000. Adaptability studies for high temperature tolerance of sugarbeet varieties in subtropical India. *Crop Improv.*, **27**: 178–86.

Kazimierczak, R., A. Siłakiewicz, E. Hallmann, D. Srednicka-Tober and E. Rembiałkowska, 2016. Chemical composition of selected beetroot juices in relation to beetroot production system and processing technology. *Notulae Botanicae Horti Agrobotanici Cluj-Napoca*, **44**(2):491–8.

Knez, Z., M.K. Hrncic, M. Colnik and M. Skerget, 2018. Chemicals and value added compounds from biomass using sub- and super-critical water. *J. Supercrit. Fluids,* **133**: 591–602.

Kolb, E., M. Haug, C. Janzowski., A. Vetter and G. Eisenbrand, 1997. Potential nitrosamine formation and its prevention during biological denitrification of red beet juice. *Food & Chem Toxicol.*, **35**(2): 219–24. doi:10.1016/s0278-6915(96)00099-3. PMID 9146735.

Kujala, T.S., J.M. Loponen, K.D. Klika and K. Pihlaja, 2000. Phenolics and betacyanins in red beetroot (Beta vulgaris) root: Distribuition and effect of cold storage on the content of total phenolics and three individual compounds. *J. agric. Food Chem.*, **48**:5338-42. DOI: 10.1021/jf000523q.

Kujala, T.S., M.S. Vienola, K.D. Klika, J.M. Loponen and K. Pihlaja, 2002. Betalain and phenolic compositions of four beetroot (*Beta vulgaris*) cultivars. *Eur. Food Res. Techn.*, **214**: 505-10.

Kushiro, T., M. Shibuya and Y. Ebizuka, 1998. *Excerpta Medica International Congress Series,* pp. 421-8.

Lange, W., W.A. Brandenburg and T.S.M. Bock, 1999. Taxonomy and cultonomy of beet (*Beta vulgaris* L.). *Bot. J. Linn. Soc.*, **130**: 81-96. DOI: 10.1006/bojl.1998.0250.

Lee, C.H., M. Wettasinghe, B.W. Bolling, L.L. Ji and K.L. Parkin, 2005. Betalains, phase II enzyme-inducing components from red beetroot (*Beta vulgaris* L.) extracts. *Nutr. Cancer,* **53**: 91-103.

Lee, E.J.; D. An, C.T. Nguyen, B.S. Patil, J. Kim and K.S. Yoo, 2014. Betalain and betaine composition of greenhouse- or field-produced beetroot (*Beta vulgaris* L.) and inhibition of HepG2 cell proliferation. *J. agric. Food Chem.*, **62** (6): 1324–31. doi:10.1021/jf404648u. PMID 24467616.

Lu, G., C.G. Edwards, J.K. Fellman, D.S. Mattinson and J. Navazio, 2003. Biosynthetic origin of geosmin in red beets (*Beta vulgaris* L.). *J. agric. Food Chem.*, **51**(4): 1026–9. doi:10.1021/jf020905r. PMID 12568567.

Machha, A. and A.N. Schechter, 2011. Dietary nitrite and nitrate: A review of potential mechanisms of cardiovascular benefits. *Eur. J. Nutr.*, **50**:293–303. doi: 10.1007/s00394-011-0192-5.

Mall, A.K., V. Misra, Santeshwari Pathak, A.D. Pathak and S. Srivastava, 2021. Sugar beet cultivation in India: Prospects for bio-ethanol production and value-added co-products. *Sugar Tech.*, **23**: 1218–34. doi.org/10.1007/s12355-021-01007-0.

Mall, A.K., V. Misra, M. Kumar and A.D. Pathak, 2018a. Assessment of ethanol recovery and juice quality parameters in sugarbeet in drought conditions of sub-tropical India. In: 2nd International Conference on Advances in Agricultural, Biological and Applied Sciences for Sustainable Future (ABAS-2018)- Theme I (eds Singh, J., R. Nigam, W. Hasan, A. Kumar, R. Singh, H. Naz, N. Kapoor, S.L. Bairwa, A. Rani and M. Kumar), held at Swami Vivekanand Subharti University, Meerut (U.P.), India during 20–22 Oct. 2018, 99p.

Malnoua, C.S., K.W. Jaggarda, and D.L. Sparkes, 2008. Nitrogen fertilizer and the efficiency of the sugar beet crop in late summer. *Eur. J. Agron.*, **28**(1): 47–56. doi.org/10.1016/j.eja.2007.05.001.

Manach, C., G. Williamson, C. Morand, A. Scalbert and C. Remesy, 2005. Bio-availability and bio-efficacy of polyphenols in humans. I. Review of 97 bioavailability studies. *Amer. J. Clin. Nutrit.*, **81**: 230S-42S.

Marsh C.A. and G.A. Levvy, 1956. *Biochem. J.*, 63: 9.

Massiot, G., M.-G. Dijoux, C. Lavaud, L. Le Men-Olivier, J.D. Connolly and D.M. Sheeley, 1994. *Phytochem.*, **37**: 1667.

Mathuria, R.S. and R.L. Bhoj, 1977. A note on sugarbeet cultivation in western Uttar Pradesh. Cane Grower's Bull., **3**: 12–4.

Mirmiran, P., Z. Houshialsadat, Z. Gaeini, Z. Bahadoran and F. Azizi, 2020. Functional properties of beetroot (*Beta vulgaris*) in management of cardio-metabolic diseases. *Nutr. Metab. (Lond),* 17. Article number: 3. . doi.org/10.1186/s12986-019-0421-0.

Misra, V., A.K. Mall and A.D. Pathak, 2020. Sugarbeet: A sustainable crop for salt stress conditions. In: Agronomic Crops (ed. Mirza Hasaanzuman), pp. 40–62. Springer Nature Singapore Pte Ltd.

Morgado, M., G.V. de Oliveira, J. Vasconcellos, M.L. Monteiro, C. Conte-Junior and A.P.T.R. Pierucci, 2016. Development of a beetroot-based nutritional gel containing high content of bioaccessible dietary nitrate and antioxidants. *Int. J. Food Sci. Nutr.*, **67**(2): 153–60.

morningchores.com/growing-mangold/.

Mroczek, A. 2012. Triterpenesaponin content in the roots of red beet (*Beta vulgaris* L.) cultivars. *J. Sci. Food Agric.*, **60**:12397-402. DOI: 10.1021/jf303952x.

Murakami, A., H. Ashida snd J. Terao, 2008. Multitargeted cancer prevention by quercetin. *Cancer. Letters*, **269**(2): 315–325.

Murakami, T., H. Matsuda, M. Inadzuki, K. Hirano and M. Yoshikawa, 1999. Medicinal foodstuffs. XVI. Sugar Beet. (3): Absolute stereostructures of beta vulgarosides II and IV, hypoglycemic saponins having a unique substituent, from the roots of *Beta vulgaris L. Chem. & Pharmaceut. Bull.*, **47**: 1717-24. DOI: 10.1248/cpb.47.1717.

Nayeem, N., S. Asdaq, H. Salem and S. Ahel-Alfqy, 2016. Gallic acid: A promising lead molecule for drug development. *J.Appl. Pharm.*, **8**(2): 1– 4.

Nguyen, H., M. Nguyen, V. Chandran, Q.H. Nguyen, K.T. Nguyen and C. Nguyen, 2018. Phytochemical, cardiovascular effect, antioxidant, antiinflammation and anti-tumor properties by *Beta vulgaris* (beet) root juice. *J. Transl. Sci.*, 5. DOI: 10.15761/JTS.1000293. www.oatext.com/phytochemical-cardiovascular-effect-antioxidant-anti-inflammation-and-anti-tumor-properties-by-beta-vugaris-beet-root-juice.php.

Ninfali, P. and D. Angelino, 2013. Nutritional and functional potential of *Beta vulgaris* 'Cicla' and 'Rubra'. *Fitoterapia*, **89**:188–99. doi: 10.1016/j.fitote.2013.06.004.

Nottingham, S. 2004. Beetroot. Archived from the original (e-book) on 2012-02-09. web.archive.org/web/20120210091426/www.stephennottingham.co.uk/beetroot1.htm.

Ormsbee, M.J., J. Lox and P.J. Arciero, 2013. Beetroot juice and exercise performance. *J. int. Soc. Sports Nutr.*, **5**:27–35.

Paesold, Susanne, Dietrich Borchardt, Thomas Schmidt and Daryna Dechyeva, 2012. A sugar beet (*Beta vulgaris* L.) reference FISH karyotype for chromosome and chromosome-arm identification, integration of genetic linkage groups and analysis of major repeat family distribution. *Pl. J.*, **72**(4): 600–11. doi:10.1111/j.1365-313x.2012.05102.x. ISSN 0960-7412. PMID 22775355.

Panella, L. and S.R. Kaffka, 2011. Sugar beet (*Beta vulgaris* L.) as bio-fuel feedstock in the United States. In: Sustainability of the Sugar and Sugar Ethanol Industries (ed. Eggleston, G.). ACS Symposium Series, pp. 163–75. American Chemicsal Society, Washington DC.

Pathak, A.D., R. Kapur, R. Kumar and M.K. Vishwakarma, 2011. Impact of different vernalization treatments on flowering and seed production in sugarbeet (*Beta vulgaris* L.). *Indian J. Sugarc. Techn.*, **26**(1): 24–7.

Pathak, A.D., R. Kapur, S. Solomon, R. Kumar, S. Srivastava and P.R. Singh, 2014. Sugar beet: A historical perspective in Indian context. *Sugar Tech.*, **16** (2): 125–32. doi.org/10.100 7/s12355-014-0304-7.

plantura.garden/uk/vegetables/beetroot/yellow-beetroot.

plantvillage.psu.edu/topics/beet/infos.

Ralet, M-C., G. Andre-Leroux, B. Quemener and J.-F. Thibault, 2005. *Phytochem.*, **66**: 2800.

Ravichandran, K., N.M.M.T. Saw, A.A.A. Mohdaly, A.M.M. Gabr, A. Kastell, H. Riedel, Z. Cai, D. Knorr and I. Smetanska, 2013. Impact of processing of red beet on betalain content and antioxidant activity. *Food Res. Intern.*, **50**:670-75. DOI: 10.1016/j.foodres.2011.07.002.

Raynbird, Hugh, 1851. On the cultivation of mangold-wurzel or field-beet. *Trans. Highland & Agric. Soc. Scotland* (New Series), pp. 534–38.

Ridout, C.L., K.R. Price, G. Parkin, M.G. Dijoux and C. Lavaud, 1994. *J. Agric. & Food Chem.*, **42**: 279.

Romeiras, M.M., A. Vieira, D.N. Silva, M. Moura, A. SantosGuerra, D. Batista, M.C. Duarte and O.S. Paulo, 2016. Evolutionary and Biogeographic Insights on the Macaronesian *Beta patellifolia* species (Amaranthaceae) from a time-scaled molecular phylogeny. *Plos One*, **11**(3): e0152456. Bibcode:2016PLoSO..1152456R. doi:10.1371/journal.pone.0152456. PMC 4816301. PMID 27031338. Pages 1-16 of www.ncbi.nlm.nih.gov/pmc/articles/PMC4816301/pdf/pone.0152456.pdf.

Schmidt, J., C. Kuhnt and G. Adam, 1994. *Phytochem.*, **36**: 175.

seedgrowers.ca/wp-content/uploads/Circ6-SECTION-14.12_Sugar-Beet_English_Rev01.13-2018_20180108-1.pdf.

Shultz, L.M. 2003. Beta vulgaris. In: Flora of North America, North of Mexico., vol. 4: Magnoliophyta ~ Caryophyllidae, Part I. Edited by Flora of North America Editorial Committee. Oxford University Press, New York, pp. 266-7. ISBN 0-19-517389-9.

Smit, A.L. 1983. Influence of External Factors on Growth and Development of Sugar Beet (*Beta vulgaris* L.). Centre for Agricultural Publishing and Documentation, Wageningen, the Netherlands, pp. 1–117.

Srivastava, H.M., B.L. Srivastava and V.K. Saxena, 1983. Sugarbeet seed production in the Kumaon hills. II. Effect of steckling size and time of transplanting on yield of sugarbeet seed. *Indian Sugar Crops* J., **9**: 21–4.

static.vikaspedia.in/media/files_en/energy/energy-production/bio-energy/sugarbeet.pdf.

Takahashi, H., T. Sasaki. And M. Ito, 1987. *Bull. Chem. Soc. Japan*, **60**: 2261.

Tamada, T. 1999. Benyviruses. In:Encyclopedia of Virology (eds Webster, R. and A. Granoff), 2nd ed., pp. 154-60. Academic Press, New York.

Tong, G., X. Fu.,Y. Yang, L. Feng, Z. Xiang, Y. Pu and J. Fan, 2018. Advances in research on the distribution and pharmacological activities of epicatechin. *J. Yunnan agric. Univers.*, **33**(2), 343– 9.

Valli, V., A.M. Gomez-Caravaca, M. Di Nunzio, F. Danesi, M.F. Caboni and A.J. Bordoni, 2012. Sugarcane and sugarbeet molasses, antioxidant-rich alternatives to refined sugar. *J. agric. & Food Chem.*, **60**(51): 12508– 15.

Vermeer, I.T. and J.M. van Maanen, 2001. Nitrate exposure and the endogenous formation of carcinogenic nitrosamines in humans. *Rev. Environ. Health.*, **16**:105–16. doi: 10.1515/REVEH.2001.16.2.105.

Vinson, J.A., Y. Hao, X. Su and L. Zubik, 1998. Phenol antioxidant quantity and quality in foods: Vegetables. *J. agric. Food Chem.*, **46**:3630-34. DOI: 10.1021/jf980295o.

Vulic, J.J., T.N. Cebovic, J.M. Canadanovic-Brunet, G.S. Cetkovic, V.M. Canadanovic and S.M. Djilas, 2014. In vivo and in vitro antioxidant effects of beetroot pomace extracts. *J. Funct. Foods,* **6**: 168-75.

Webster, T.M., T.L. Grey, B.T. Scully and W.C. Johnson. 2016. Yield potential of spring harvested sugar beet (Beta vulgaris) depends on autumn planting time. *Indust. Crops & Products,* **83**: 55–60.

Winkler, C., B. Wirleitner, K. Schroecksnadel, H. Schennach and D. Fuchs, 2005. In vitro effects of beet root juice on stimulated and unstimulated peripheral blood mononuclear cells. *Appl. Biochem. & Biotech.*, **1**:180-85. DOI: 10.3844/ajbbsp.2005.180.185.

Wootton-Beard, P.C. and L. Ryan, 2011. A beetroot juice shot is a significant and convenient source of bioaccessible antioxidants. *J. Funct. Foods,* **3**: 329-34.

Wright, Clifford A. 2001. Mediterranean Vegetables: a cook's ABC of vegetables and their preparation in Spain, France, Italy, Greece, Turkey, the Middle East, and North Africa with more than 200 authentic recipes for the home cook, 52p. Boston, Massachusetts: Harvard Common Press. ISBN 1-55832-196-9.

Wruss, J., G. Waldenberger, S. Huemer, P. Uygun, P. Lanzerstorfer and U. Muller, 2015. Compositional characteristics of commercial beetroot products and beetroot juice prepared from seven beetroot varieties grown in Upper Austria. *J. Food Compos Anal.*, **42**:46–55.

www.amrafarms.com/beetroot-cultivation-and-farming-in-india.html.

www.borax.com/agriculture/files/beets.pioneer.pdf.

www.britannica.com/plant/beet.

www.complete-herbal.com/culpepper/rootofscarcity.htm : "Root of Scarcity". Culpeper's Complete Herbal. Retrieved 26 April 2023.

www.myfitnesspal.com/food/calories/mangel-wurzel-raw-118893642.

www.plantnames.unimelb.edu.au/Sorting/Beta.html.

www.selfsufficientish.com/beetroot.htm.

www.slideshare.net/PRAVINABARDE/production-technology-of-beetroot.

www.specialtyproduce.com/produce/Gold_Beets_260.php.

www.theharvestkitchen.com/top-5-health-benefits-beets/.

www.theseedcollection.com.au/blog/mangel-wurzel-a-versatile-and-productive-heirloom.

Yang, Chung S., Jeong-Sook H. Yoo, Hiroyuki Ishizaki and Junyan Hong, 1990. Cytochrome P450IIe1: Roles in nitrosamine metabolism and mechanisms of regulation. *Drug Metabol. Rev.*, **22**(2–3): 147–59. doi:10.3109/03602539009041082. PMID 2272285.

Yoshikawa, M.., T. Murakami, M. Kadoya, H. Matsuda, O. Muraoka, J. Yamahara and N. Murakami, 1996. *Chem. & Pharmaceut. Bull.*, **44**: 1212.

Yoshikawa, M., T. Murakami, M. Kadoya, H. Matsuda, J. Yamahara, O. Muraoka and N. Murakami, 1995. *Heterocycles*, **8**: 1621.

Yoshikawa, M., T. Murakami, M. Kadoya, J. Yamahara and H. Matsuda, 1998. *Chem. & Pharmac. Bull.*, **46**: 1758.

Zhu, G., S.L. Mosyakin and S.E. Clemants, 2003. Beta vulgaris. In: Zhengyi, W., Raven, P.H., & Hong, D. (eds.): Flora of China (eds Zhengyi, W., P.H. Raven and D. Hong), vol. 5, 345p. Ulmaceae through Basellaceae. Science Press/Missouri Botanical Garden Press, Beijing/St. Louis. ISBN 1-930723-27-X.

Radish [*Raphanus sativus* L.)

Family: Brassicaceae.

Introduction and Origin

The generic name *Raphanus* derives from *raphanis* ~ the ancient Greek name for a radish. *Raphanus sativus* (cultivated radish) is a subspecies of *Raphanus raphanistrum* L. [wild radish also called white charlock, jointed charlock (npgsweb.ars-grin.gov), jointed radish, jointed wild radish] which is native to western Asia, Europe and parts of Northern Africa, from where this has been introduced into most parts of the world. It spreads rapidly and is regarded as a habitat threatening invasive species in many areas such as Australia (en.wikipedia.org/wiki/Raphanus_raphanistrum). Resistant to several herbicides, wild radish is a naturalized species in Canada where sometimes it hybridizes with cultivated radish, *R. sativus* (Warwick and Francis, 2005).

All tender parts of wild radish including the leaves, flowers, seedpods, root and outer skin of the root are edible and have spicy taste though it has a distinct slender taproot which does not swell like that of the cultivated radish.

R. raphanistrum has several known subspecies including 'sea radish' [*R. raphanistrum* ssp. *landra* (Moretti ex DC.) Bonnier & Layens], *R. raphanistrum* ssp. *rostratus* (DC.) Thell. and the cultivated radish, both are simply called radish [*R. raphanistrum* subsp. *sativus* (L.) Domin, syn. *R. sativus*]. Subspecies *R. sativus* includes a diverse variety of cultivated radishes (en.wikipedia.org/wiki/Raphanus_raphanistrum).

Wild Radish in Behbahan

By Mysubarum- Own work, CC By -SA 4.0, com m ons. wikim edia. org/w/index. php?curid= 76234546

Wild Radish Flowers	
Yellow form	**White-flowered**
By Didier Desouens-Own work CC By-SA 4.0, commons.wikimedia. org/w/index.php?curid-104109640	**By Alex Lockton-Own work, CC By -SA.4.0, commons.wikimedia. org/w/index.php?curid-127862932**

Radish/cultivated radish (*Raphanus sativus*) is an edible root vegetable grown and consumed throughout the world mostly as a crispy salad vegetable with a pungent and slightly spicy flavour varying in intensity depending on varieties and the growing environment. The characteristic pungent flavour is due to the presence of volatile isothiocynates (4-methyl thio-3-butenyl isothiocyanate – MTB-ITC) (ecoursesonline.iasri.res.in). Domesticated in Asia prior to Roman times, radish today has numerous varieties that are broadly distributed around the world, but almost no archaeological records are available to help determine its early history and domestication (Daniel and Maria, 2000).

SW Asia is believed to be the origin of *Raphanus sativus* as this is the only region where true wild forms have been discovered. India and Middle Asia seem to have been secondary centres where differing forms were developed (www.missouribotanicalgarden.org2). Greeks wrote of their radishes in the 3rd century BC (Lewis-Jones, 1982) and Romans of the 1st century AD gave details of small, large, round, long, mild and sharply pungent varieties. Radish seems to have been one of the first European crops introduced to the Americas. A German botanist reported radishes of 45 kilogrammes and roughly 90cm in length in 1544 though only variety of that size today is the 'Japanese Sakurajima' radish (www.missouribotanicalgarden.org2). The large, mild, and white East Asian form was developed in China though it is mostly associated in the West with the Japanese daikon owing to Japanese agricultural development and larger exports (en.wikipedia.org/wiki/Radish). Black radish probably originated from *Raphanus maritimus* while spring radish varieties from *Raphanus landra* (Wein, 1964). Being easy to grow and quick to harvest it is being grown even as forage crop (Fitzgerald and Black, 1984). These are often planted by novice gardeners either as a cover or catch crop in winter (Price Andrew and Norsworthy, 2013).

Botany

The radish is a diploid species having 2n=18 chromosomes (Dixon, 2007). It is estimated that the radish genome contains between 526 and 574 Mb (Nishio, 2017). Radishes are annual or biennial crops grown for their swollen succulent taproots which as per varieties and growing condition vary in shape, size and colour. Depending on varieties, the taproots range in shape from spherical to long and cylindrical or tapered, and the outside skin can be white, yellow, pink, red, purple, green or black but the flesh is usually white. The roots acquire their colouration from anthocyanins as the red varieties usually have pelargonidin anthocyanin as a pigment whereas purple ones usually cyanidin (Nishio, 2017). Radishes vary in size from a few grammes in the most-popular early American and European varieties up to 1kg in the Japanese daikon radish.

The flesh of timely harvested radishes is crisp and sweet but becomes bitter and tough if left in the ground too long (Bradley *et al.*, 2016). The leaves form a basal rosette that emerge from the top of the root and have a lyrate shape, *i.e.*

divided pinnately with an enlarged terminal lobe and smaller lateral lobes. The flowers are borne on a racemose inflorescence (Gopalakrishnan, 2007). Flower stalks usually appear in the first season, bearing white or lilac-veined white flowers with four petals. The fruits are small pods called a silicle which can be eaten when young (Brickell, 1992) which is also sold in the vegetable market at a premium.

Smaller varieties have round taproots up to 2.5cm in diameter or more slender taproots up to 7cm long, *vis-a-vis* usually fewer leaves that are up to about 13cm long, Both of these are normally eaten raw in salads. Varieties with a longer root form such as oriental radishes, daikon or mooli, and winter radishes grow up to 60cm long with foliage about 60cm high and a spread of 45cm (Brickell, 1992). Small but quick-growing spring varieties have a mild, crisp, moderately firm flesh whereas the large, slow-growing summer and winter types have pungent and firm flesh.

Classification of Cultivated Radish

There are many named varieties of radish that are able to supply edible roots all year round. Over the centuries a number of distinct groups have evolved through cultivation and have been classified by botanists in various ways. Although the classification is subject to change, a separate record has been made for each of the following group (tropical.theferns.info):

- **Red radish** (*R. sativus*) are fast maturing plants with small roots that can be round or cylindrical and usually have red skins. They are grown primarily for their roots which in some varieties can be ready within three weeks from sowing the seed and are used mainly in salads. These are mainly grown for spring, summer and autumn use and can produce a crop within a few weeks of sowing.
- **White radish** (*R. sativus* subsp. *longipinnatus*) are also called *mooli*, deckon, winter radish, Oriental radish and long white radish. They are generally mild-flavoured winter radishes characterized by fast-growing leaves and a long white napiform root.
- **Seed pod varieties** [or rat-tailed radishes (*R. sativus* var. *caudatus*)] group does not produce roots of good quality so is cultivated mainly for the edible young seedpods which are usually harvested in summer.
- **Oriental or Spanish radishes** (*R. sativus* L. var. *niger* J. Kern) are grown for their larger edible root which can be round or cylindrical and can be available throughout the winter.

- **Fodder radishes** (*R. sativus* var. *oleiformis* Stokes Pers.) are grown mainly for their leaves and oil-rich seeds. They are used as a green manure or stock feed though they can also be eaten by people. The roots of these plants soon become fibrous though these make acceptable eating when young.

Radishes can also be classified as summer and winter varieties or as Asiatic/ Tropical and European/Temperate types. Common smaller varieties become ready for consumption within a month while larger daikon varieties take several months. Some radishes are grown for their seeds while others such as daikon for oil production, and certain others even for sprouting.

Seed Pod Varieties

Seedpods of almost all the radish varieties are edible but some are grown specifically for their seeds or seedpods rather than their roots. One such variety is 'rat-tailed radish' that is commonly found in India and SE Asia though is thought to have originated in China (commons.wikimedia.org/w/index. php?curid=59306701). It has long, thin, curly pods which can exceed 20cm in length. Pods are juicy *cum* crunchy having mild radish flavour which may be used fresh in salads, steamed, boiled or stir-fried. The pods should not be cooked too much as it reduces its fiery strength. The 'Munchen Bier' variety supplies seed pods that are sometimes served raw as an accompaniment to beer in Germany.

Rat-tail radish, Serpent radish or Mogri (*R. sativus* var. *caudatus*)

Other names: Green radish, Pod radish, Mogro, Mongri, Mogri, Mungori and Tail-pod radish.

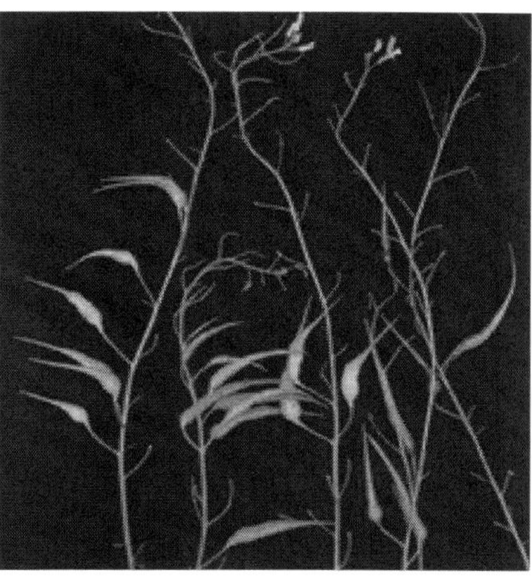

By MerielGJones - Own work, CC BY-SA 4.0, commons.wikimedia.org/w/index. php?curid=92115080

Synonyms: *Raphanus caudatus* L.; *R. sativus* L. var. *mougri* H. W. J.; *R. sativus* Rat-Tailed Radish Group.

Mogri has a relatively insignificant root but is cultivated as a vegetable in India and SE Asia for its edible foliage and relatively long seedpods which are eaten raw, cooked or pickled. Plants grow very tall, reaching 1m quickly with an eventual height of up to 1.5m (www.missouribotanicalgarden.org1). These plants do not grow a thickened root. Many flowers, usually yellow or white with purple veins and about 2cm in diameter appear on the plant quickly and are soon followed by a great number of edible seedpods (www.missouribotanicalgarden.org1). The pods themselves can be from a few centimetres to over 20cm long and either green or tinged with red or dark purple.

It was described as the species *Raphanus caudatus* by Linnaeus though now it is treated as a variety of the common radish (*R. sativus*, *R. caudatus* or *mougri*; npgsweb.ars-grin.gov).

Cultivation

Mogri is similar to mooli and grows best in warm climates with occasional rain. It needs 2 to 3 months of high temperatures and good light and does not grow well if temperatures fall below 10^0C. Seeds are sown 1-2cm deep and about 15cm apart with 45 to 60cm distance between rows under full sun and in well-drained fertile soil (tropicalfruitandveg.com).

Intercultures include mulching, regular fertilization and regular watering as per need to reduce bitterness. Keeping them in tall tomato cages or otherwise restraining them will greatly facilitate harvesting the edible seedpods (www.missouribotanicalgarden.org1).

It will be ready to harvest in 2-3 months (tropicalfruitandveg.com). Pods must be harvested young and small, as early as 5 weeks after planting and continuously after that (www.missouribotanicalgarden.org1). The texture of young pods is crispy, tender and very mild. As the pods mature they will become hard, spicy, hot and bitter so becoming unsuitable for eating. Flea beetles can be a problem as well as other pests typically found on Brassicaceae crops. Floating row covers can be very useful (www.missouribotanicalgarden.org1).

Summer Radishes

Summer radishes are sometimes called as European radishes or spring radishes if they are planted in cooler weather. They are generally small and have a relatively short 3-4-week cultivation time (Brickell, 1992). Generally these have relatively small and mildly pungent roots with quick quality deterioration (ecoursesonline.iasri.res.in). Its varieties are:

Easter egg radishes

Photo courtesy: By Self, en:User:Jengod - Own work, CC BY-SA 3.0, https://commons.wikimedia.org/w/index.php?curid=6292330

- '**April Cross**' is a giant white radish hybrid that bolts very slowly.
- '**Bunny Tail**', an heirloom variety from Italy is slightly oblong, mostly red and with a white tip.
- '**Cherry Belle**', familiar in North American supermarkets, is a bright red-skinned round variety with a white flesh (Faust, 1996).
- '**Champion**' is round and red-skinned like the 'Cherry Belle' but with slightly larger roots, *i.e.* up to 5cm with milder flavour (Faust, 1996). It never likes poorly drained soil.
- '**White Icicle** or **Icicle**' is a medium short European table variety, dating back to the 16th century, and is most suitable for kitchen garden. It matures in 25-30 days. The carrot-shaped roots measure 12-15cm with 2-3cm in diameter. Its skin is pure white while icy-white flesh is juicy and sweet-flavoured. Roots are free from green top. It slices easily and has better than average resistance to pithiness (Faust, 1996; Peterson, 1999). Sowing time is from mid-October to February.
- '**Red King**' has a mild flavour and with good resistance to club root (Faust, 1996).
- '**Sicily Giant**' is a large heirloom variety from Sicily. It can attain up to 5cm diameter (en.wikipedia.org/wiki/Radish).
- '**Snow Belle**' is a 'Cherry Belle' shape all-white variety (Faust, 1996).

- **'French Breakfast'** is a red-skinned elongated variety with a white splash at the root end. It is typically slightly milder than other summer varieties but is among the quickest to turn pithy (Peterson, 1999).
- **'Plum Purple'** is a purple-fuchsia radish which tends to stay crisp longer than average (Peterson, 1999).
- **'Gala'** and **'Roodbol'** are two varieties popular in the Netherlands in a breakfast dish, thinly sliced on buttered bread (Faust, 1996).
- **'Easter Egg'** is a mix of varieties with different skin colours, typically including white, pink, red and purple radishes. Sold in markets or seed packets under the name, the seed mixes can extend harvesting duration from a single planting, as different varieties may mature at different times (Peterson, 1999).

Winter Varieties

Winter varieties are slow-growing and late-maturing (50-90 days). These are strongly flavoured and generally have large roots and better shelf life (ecoursesonline.iasri.res.in).

Black Radish (*Raphanus sativus* L. var. *niger* J. Kern)

Black radishes which probably originated from *Raphanus maritimus* (Wein, 1964) are hot-flavoured, round, elongated or irregularly pear-shaped varieties with rough black or dull brown skin and white flesh (McIntosh, 1828; www.nature-and-garden.com). It is also called Black Spanish radish or Erfurter radish (en.wikipedia.org/wiki/Black_radish). The French name for these varieties is *Gros Noir d'Hiver* (Aiton, 1812). The black radish varieties are usually bigger than spring radish varieties and grow around 7-10cm in diameter or length. The plants can grow up to a height of 20 to 30cm depending on the variety (www.nature-and-garden.com; pfaf.org). These larger rooted types are cultivated mainly in Asia but still locally important in Germany.

Black radish can be consumed raw or cooked, or even in the form of juice. Ready to swallow capsules of black radish are also available. It is a very low-calorie food and has high fibre content. It has high levels of vitamins B and C, phosphorus and magnesium. It also has diuretic properties. It enhances gall bladder juices which help cleaning and detoxifying the liver and imparting tonic benefits for digestion. It is frequently used against kidney stone and gall stone (www.nature-and-garden.com; pfaf.org).

Black radish juice contains glucosinolates ~ more abundant being glucoraphasatin and glucoraphanin. It has been studied for its potential use in the prevention of cholesterol gallstones (Castro-Torres, 2013).

Black radish, round variety	Black radish, long variety
	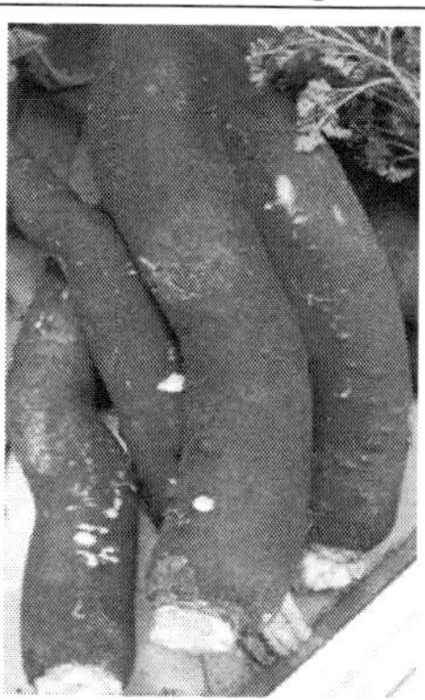
By goldlocki - of my own work, CC BY-SA 3.0, commons.wikimedia.org/w/index.php?curid=3106434	By goldlocki - of my own work, CC BY-SA 3.0, https://commons.wikimedia.org/w/index.php?curid=3106421

Cultivation

These varieties can be raised throughout full summer (April to October) through direct sowing in the ground depending on the climate. Sowing distance is usually 3-4cm in furrows 2-3cm deep. These require to be watered regularly with a fine drizzle. The weak sprouts should be thinned out and only most vigorous ones should be kept maintaining a gap of 15-20cm. Seeds can be sown indoors and the plants can later be transferred to the field. Frequent watering is required in summer to ensure that the soil remains cool. Mulching is recommended to sustain soil coolness (www.nature-and-garden.com; pfaf.org).

Harvesting though is dependent on the varieties but generally it takes after 2-4 months of sowing and is preferably targeted before the first frost spells though these can be harvested during entire winter. However, harvesting before the ground freezes over makes the pulling of the roots easier. A good hint for harvest time is when the tip of the black radish starts poking out of the ground. Once harvested, they should be dried outdoors but must be kept out of the rain. After drying, they are kept in a dry, cool and well ventilated spot. The 'Noir Long Maraicher' can be stored for a very long time in winter (www.nature-and-garden.com; pfaf.org).

Varieties

Some European black radish varieties are (www.nature-and-garden.com; pfaf.org):

- **'Noir Gros Rond d'Hiver'** ('Black Spanish Round'): Its growth is very rapid and has tolerance to short frost spells. It has good keeping quality.

- **'Noir Gros Long d'Hiver'** ('Black Spanish Long'): Its roots are usually longer than width and can grow up to 30cm. It has good storability.
- **'Noir Long Poids d'Horloge'** ('Noir Long Maraicher'): Its skin is quite black, root is some 20-25cm long and cylindrical.
- **'Serpent' or 'Rattail':** Its roots are often violet-coloured and as thin as a pencil, and it is harvested before maturity when initial swelling starts. Nibble when fresh or pickle with vinegar.
- **'Violet de Gournay'** ('Gournay'): One of the radishes with the mildest flesh. It has pinkish-purple skin and its keeping quality is good.

Some Asian varieties are (www.nature-and-garden.com; pfaf.org):

- **'Okhura'**: Its roots are round and white with mild pungency.
- **'China Rose'**: It can be sown in fall and in spring. Its roots are long and pink.
- **'Ilka':** It produces 10-15cm across dark pink and round roots that are quite juicy with mild pungency,
- **Mooli or 'Daikon'**: It includes *R. sativus* ssp. *longipinnatus* and certain other similar radishes.

'Daikon' in Japanese stands for *dai* meansing 'big' and *kon* meaning 'root' whereas in India 'mooli' meaning 'the root' and the name 'mooli' has been adopted in English. Sometimes, it is also called the 'Japanese radish', 'radish', 'white radish', 'winter radish', 'Oriental radish', 'long white radish' (en.wikipedia.org/wiki/Daikon), 'fodder radish' or 'oil radish' (commons.wikimedia.org/wiki/Category:Raphanus_sativus_var._oleiformis). Mooli or Daikon is native to continental East Asia and refers to a wide variety of winter radishes from Asia. They are generally mild-flavoured winter radishes characterized by fast-growing leaves and a long, white napiform root (Larkcom and Douglass, 1994). Varieties in this group are harvested and consumed throughout East Asia, the South Asia and worldwide. In some locations, daikon is planted for its ability to break up compacted soils and recover nutrients but there it is mostly not harvested (en.wikipedia.org/wiki/Daikon).

In North India, daikon is a popular ingredient used to make bhujiya or sabzi, stuffed paranthas, pakodas, salads, pickles and as garnish. The plant's leaves are used to make sabji, dal and kadhi, among other dishes. In South India, daikon is the principal ingredient in a variety of sambar.

Varieties

Mooli commonly has elongated white root, although many varieties of daikon exist. Several non-white mooli varieties also occur. Larkcom and Douglass

(1994) divide the term 'oriental radish' into two categories, which they label 'white mooli types' and 'coloured types' (Larkcom and Douglass, 1994). One well-known variety is 'April Cross' with smooth white roots (Faust, 1996; Peterson, 1999). The New York Times describes 'Masato Red' and 'Masato Green' varieties as extremely long, well-suited for fall planting and winter storage. The 'Sakurajima' radish is a hot-flavoured variety that typically grows to around 10kg though it can grow up to 30kg when left in the ground as such (Faust, 1996).

From left to right: Watermelon radish, Daikon, Lobak, Mu, Yeolmu Ponytail radish, varietals of *Raphanus Sativus* Photo courtesy: By KimYunmi - Own work, CC BY-SA 4.0, commons.wikimedia.org/w/index.php?curid= 115411117

Certain varieties of daikon, often named 'tillage radish' can be grown as a winter cover crop and green manure. These varieties have a quite large penetrating root that effectively performs deep cultivation. The roots bring nutrients from lower soil profile to higher reaches, hence are good nutrient scavengers, therefore these become good partners with legumes instead of grasses. If harsh winters occur the root decomposes releasing nitrogen in the spring (en.wikipedia.org/wiki/Daikon).

The 'Cantonese Lobak', 'Lo Pak', *etc.* names are sometimes referred to the usual Chinese form, *vis-à-vis* even to daikon which have a light green colouration on the top area of the root around the leaves. The '**Heirloom Watermelon**' radish is another Chinese variety of daikon with a dull green exterior but a bright rose or fuchsia-coloured flesh. The **Korean radish** (*mu*), a variety of white radish group, has a similar pale green shade halfway down from the top and is generally shorter, stouter and sturdier with denser flesh and softer leaves. Both are often spicier than the long white radishes (en.wikipedia.org/wiki/Daikon; www.honestfoodtalks.com).

'Gegeolmu or Gegeol radish', a speciality crop of Icheon and Yeoju in Korea, is a round, pungent variety of Korean white radish. Its thick rind and firm flesh do not get soft even after a few years of storage as *kimchi* ~ a traditional Korean side dish of salted and fermented vegetables, such as napa cabbage and Korean radish. It is smaller and firmer than regular Korean radishes and weighs about 500g, including the taproot (6-7cm long and up to 20cm in circumference) and the greens. Its greens tend to grow outward rather than upward (en.wikipedia.org/wiki/Gegeol_radish). Gegeol radish has a lower water content but more protein, fibre and minerals (magnesium, potassium, and calcium) as compared to regular Korean radishes. It also has a higher content of a sulphur compound that is responsible for its sharper taste. With regard to enzymes, the protease and myrosinase activities of the Gegeol radish are higher but the α-amylose activity is lower than regular Korean radishes (Kim *et al.,* 2007).

Gegeolmu (Korean radish)

http://cafe.daum.net/handheldgarden/4e0D/27, CC BY 2.5, https://commons.wikimedia.org/w/index.php?curid=54348847

Cultivation

The Indian and Chinese varieties of daikon can tolerate higher temperatures than the Japanese varieties. These varieties also grow well at lower elevations in East Africa. Under low moisture its flesh becomes overly tough and pungent (Grubben, 2004). The variety 'Long White Icicle' is available as seed in Britain and will grow very successfully in Southern England, producing roots resembling a parsnip by midsummer in good garden soil in an average year (en.wikipedia.org/wiki/Daikon).

The roots can be stored for weeks without leaves if lifted and kept in a cool and dry place. If left in the ground, the texture tends to become woody though storage life of whole untreated roots is not long (en.wikipedia.org/wiki/Daikon).

Daikon/Mooli vs. red radish, Korean radish and Chinese radish

The red radishes are much smaller and sharper in flavour than the radishes used in Japanese cuisine. Red radishes are peppery whereas the white radish is mild and slightly sweet. 'Mu' (Korean radish) is similar to long white Japanese radish, but it is shaped more like a potato. Watermelon radish, a variety of Chinese radish, has the same texture and crunch as Japanese and red radishes but is green on the outside, pink on the inside and has a mellower flavour (www.eatingwell.com).

Varieties Grown in India

Radish varieties grown in India are usually of daikon type.

Asiatic or Tropical types

Asiatic types don't require chilling temperature for bolting and set seeds freely under tropical conditions. Some Asiatic or tropical types are 'Arka Nishant', 'Chinese Pink', 'CO-I', 'C-2', 'Early Miino White', 'Hisar Mooli No. 1', 'Japanese White', 'Kalyanpur No. 1', 'IPVR-I', 'IPVR-II', 'Jaunpuri Mooli', 'Kalyani White', 'Punjab Pasand', 'Punjab Safed', 'Pusa Chetki', 'Pusa Desi', 'Pusa Reshmi'.

- '**Arka Nishant**' has long marble-white roots. It is resistant to pithiness and pre-mature bolting.
- '**Chinese Pink**' has 12-15cm long pink coloured roots but with white lower end of roots. It matures in 50-55 days. Roots are crisp and mildly pungent.
- '**CO 1**' is a selection from 'RS 44', developed at TNAU, Coimbatore. Its white roots are cylindrical with tapered end. It takes 40-45 days for maturity with an average root yield of 20-25t/ha and in Tamil Nadu it can be grown throughout the year. It is highly susceptible to aphids and white rust.
- '**Hisar Mooli No 1**' was developed by CCS, HAU, Hisar. Roots are long, soft, white and straight. It requires 50-55 days to mature.
- '**Japanese White**' is an introduction from Japan and recommended by IARI Regional Station, Katrain. The roots are cylindrical and blunt at the tip, 25-30cm long and 5cm in diameter, skin pure white, flesh crisp, solid and mildly flavoured. Top is medium large with deeply cut leaves. Suitable for growing between October- December in the plains and July- September in the hills. Crop matures in 60-65 days.

- '**Jaunpuri Mooli**' is a local variety of Jaunpur district (U.P.). Its white roots are 40-45cm long and 7-10cm thick. On an average, each root weighs 2.5-3 kg.
- '**Kalyani White**' requires 45-50 days for maturation. Roots are pure white, 25-30cm long, uniform in growth and blunt tipped.
- '**Kalyanpur No. 1**' is a selection from local material released in 1982 in Uttar Pradesh. Stem and foliage are green. Top is heavy, leaves are long, broad and less lobed. Roots are 22-23cm long, smooth, crispy, white with green shoulder, thick and tapering.
- '**Kashi Hans**' (**IIVR-2**) was developed through selection at IIVR, Varanasi. It is suitable for sowing from September to February, maturing in 40-50 days with straight and tapering roots some 35-40cm long *cum* 3.5-4.2cm diameter. Even if its harvesting is delayed by 10-15 days of its edible maturity, no pithiness in the roots is developed. Its foliage is soft and smooth.
- '**Kashi Mooli-40**' (**VRRAD-203**), suitable for summer cultivation, is tolerant to high temperature (35-42°C) and recommended for cultivation in Uttar Pradesh. Other features include delayed bolting and less pithiness. It is recommended for sowing from mid-September to mid-December, and from Mid-January to mid-April where for one hectare of land some 5-6kg seeds are required. After thinning the spacings maintained are 30-45×4-10cm so that about 6-8 lakhs/ha plants are accommodated to get optimum yield. In this case 60:40:40 kg/ha of NPK is usually applied along with 8-10kg/ha borax as basal dose. Its root yield during winter season is around 30-35t/ha though during summer 20-23t/ha.
- '**Kashi Sweta**' (**IIVR-1**) was developed through selection at IIVR, Varanasi which is suitable for sowing throughout the year where its roots become ready for harvesting in 30-35 days. Its roots are 25-30cm long with 3.5-4cm in diameter, straight, tapering and with blunt tip.
- '**Nadauni**' is a popular variety in Himachal Pradesh, developed through selection from local material. Roots are long, slim, light pink and tapering to a point.
- '**Palam Hriday**' is characterized by pink heart and green shoulder. It does not become pithy and its roots are crispy. It is rich in vitamin C. Its average root yield is 25-30t/ha.
- '**Punjab Ageti**' was developed by P.A.U., Ludhiana. Roots are red-skinned at the top and white at the lower half, tapering long (25.5 cm), medium thick (2.9 cm), less pungent with a few smooth hairs.

- '**Punjab Pasand**' is an early maturing variety developed by P.A.U., Ludhiana. Its pure white roots are long, semi-stumped and free from hairs. It is suitable for sowing in main season.
- '**Punjab Safed**' is a quick growing type with roots remaining edible for 10 days after attaining full size. Roots are white, tapering, smooth, mild in taste, free of forking, 30-40cm in length and 3-5cm thick. Crop matures in 50-60 days. It is suitable for sowing in September-October.
- '**Pusa Chetki**', an early maturing tropical type which was developed through selfing and massing at IARI, New Delhi through seeds collected from Denmark in 1966. Its roots have good tolerance to high temperature and humid weather conditions. Roots are pure white, smooth, medium long (12.5cm in summer and 20.5cm in rainy season), mildly pungent and almost stumpy. It matures in 40-45 days. It is grown as a summer and monsoon crop from mid-March to early September in northern and central parts of India, the best being from July to early September. In southern and western parts of India, it can be grown throughout the year. It yields 200-350 q/ha depending on the season.
- '**Pusa Desi**', an Asiatic type, is the first variety of radish released by IARI, New Delhi in 1965. It is a subtropical cultivar, suitable for sowing from middle of August to October in the northern plains. Roots are pure white, 30-35cm long, tapering with green stem end. It matures in 50-55 days after sowing. Suitable for sowing in northern India from mid-August to mid- October.
- '**Pusa Gulabi**' is the first entirely pink-fleshed nutritionally rich radish producing cylindrical roots with an average yield of 60t/ha. It contains high levels of total carotenoids, anthocyanins and optimal ascorbic acid contents.
- '**Pusa Jamuni**' is a nutritionally rich purple-fleshed radish developed in IARI, New Delhi with quite higher yield, *i.e.* 50t/ha.
- '**Pusa Reshmi**' is also an Asiatic variety with 30-45cm long roots having white with green tinge on top. Suitable for early sowing from mid-September to mid-November but it can tolerate slightly higher temperature. Roots are ready for harvesting in about 55- 60 days.
- '**Pusa Shweta**' is suitable for sowing from September to November under North Indian tropical conditions which reaches marketable maturity in 50-55 days. Its roots are medium long, cylindrical and ultra white. One root weighs some 200g with about 40t/ha root yield. Its 100g fresh roots contain about 4.20^{0}B TSS, 424.8ppm phosphorus, 8.84ppm calcium, 497.86 ppm sodium, 1.94ppm zinc, and 34.84mg ascorbic acid.

European or Temperate types

Temperate type radishes are quick growing with short duration (25-30 days) producing smaller and less pungent roots of good quality. Their seed production is limited to high hills. Some of these varieties are 'French Breakfast', 'Kvarta' (short duration suitable for protected cultivation), 'Palam Hriday', 'Pusa Himani', 'Rapid Red White Tipped', 'Scarlet Globe', 'Scarlet Long', 'Silver Queen', 'White Icicle', *etc.*

- '**Pusa Himani**' was developed by hybridizing Black and 'Japanese White' at IARI R.S., Katrain. It is the only variety which can be grown throughout the year in the hills barring three winter months (November to January). In the plains, it is suitable for sowing from December to February when no other variety can form such good roots. Its pure white roots with whitish green shoulder are semi-stumpy, 30-35cm long, mildly pungent, crispy and sweet-flavoured. Tops are short with green cut semi-erect leaves. Crop matures in 55 days. It is the only variety which can be grown throughout the year in the hills except for winter months from November-January.
- '**Rapid Red White Tipped** (RRWT) is an extra early European table type variety which matures in 25-30 days. Its bright red with white tipped roots having pure white flesh are small, smooth, round, crisp and snappy. It is sown from mid-October to February and becomes ready within 28 days.
- '**Scarlet Globe**' becomes ready for harvesting in 25-30 days after sowing. Its roots are round, 2 cm in diameter, red in colour with pure white and crispy flesh. Roots become pithy if they are not harvested at maturity.
- '**Scarlet Long**' becomes ready for harvesting in 30-40 days after sowing. Its light green leaves are 15-20cm long. Its long and point-tapered roots are red with soft and white flesh.
- '**White Icicle**' ('Icicle') is a medium-short duration variety maturing within 30 days of sowing. This produces icicle-shaped straight and tapering roots with small tops. Externally as well as internally the roots are complete white, crisp, juicy and sweet-flavoured.

Nutritional Value

Source: en.wikipedia.org/wiki/Radish	*Source*: en.wikipedia.org/wiki/Daikon
Overall, radish is low in calories but a good source of calcium, magnesium, copper, manganese, potassium, vitamin B6, vitamin C and folate.	Daikon is mostly watery and very low in food energy. Daikon also contains the active enzyme myrosinase (Shikita *et al.*, 1999).

Nutritional value per 100g			
Raw radishes		**Raw radishes**	
Energy	66 kJ (16 kcal)	Energy	76 kJ (18 kcal)
Carbohydrates	3.4 g	Carbohydrates	4.1 g
Sugars	1.86 g	Sugars	2.5
Dietary fiber	1.6 g	Dietary fiber	1.6 g
Fat	0.1 g	Fat	0.1 g
Protein	0.68 g	Protein	0.6 g
Vitamins	**%DV† (Quantity)**	**Vitamins**	**%DV† (Quantity)**
Thiamine (B1)	1% (0.012 mg)	Thiamine (B1)	2% (0.02 mg)
Riboflavin (B2)	3% (0.039 mg)	Riboflavin (B2)	2% (0.02 mg)
Niacin (B3)	2% (0.254 mg)	Niacin (B3)	1% (0.2 mg)
Pantothenic acid (B5)	3% (0.165 mg)	Pantothenic acid (B5)	3% (0.138 mg)
Vitamin B6	5% (0.071 mg)	Vitamin B6	4% (0.046 mg)
Folate (B9)	6% (25 μg)	Folate (B9)	7% (28 μg)
Vitamin C	18% (14.8 mg)	Vitamin C	27% (22 mg)
Minerals	**%DV† (Quantity)**	**Minerals**	**%DV† (Quantity)**
Calcium	3% (25 mg)	Calcium	3% (27 mg)
Iron	3% (0.34 mg)	Iron	3% (0.4 mg)
Magnesium	3% (10 mg)	Magnesium	5% (16 mg)
Manganese	3% (0.069 mg)	Manganese	2% (0.038 mg)
Phosphorus	3% (20 mg)	Phosphorus	3% (23 mg)
Potassium	5% (233 mg)	Potassium	5% (227 mg)
Zinc	3% (0.28 mg)	Sodium	1% (21 mg)
Other constituents	**Quantity**	Zinc	2% (0.15 mg)
Water	95.3 g	**Other constituents**	**Quantity**
		Water	94.6 g

Units: μg = micrograms; mg = milligrams; IU = International units
†Percentages are roughly approximated using US recommendations for adults.
Source: USDA FoodData Central

Phytochemicals

Phytochemical investigation of *Raphanus sativus* L. shows the presence of alkaloids, reducing sugar, flavonoids, glycosides, cardiac glycosides, tannins, saponin, protein, amino acid, terpenoids, steroids, *etc*. Leaves show most of the phytoconstituents present in quath, aqueous and methanolic extracts. Alkaloids are present in all the extracts, *i.e.* aqueous, methanolic and quath. Carbohydrate is absent by Barfoed's test in all the extracts while Molisch test shows positive results only in quath but absent in both aqueous and methanolic extracts. Reducing sugar is present in all the extracts by Benedict's test while it is absent by Fehling's test. Flavonoids are present in radish leaves while roots show negative results. Test for glycosides shows variable results by different tests. Cardiac glycosides are present in aqueous and methanolic extract while absent in quath. Tannin and phenolic compounds are present in all the extracts by ferric chloride test in radish leaves though absent in roots. Saponin is present in all the extracts. Amino acid and protein are present by ninhydrin test in all the extracts except quath while it is absent in biuret test. Terpenoids is present in all the extracts while steroids are present only in methanolic extract. Ninhydrin test for amino acids and protein shows mostly positive results. However, Biuret test shows negative results in all the extracts of roots and leaves. Tests for terpenoids and steroids present variable results (Kumar and Roshni Patwa, 2018).

Nutrients and phytochemicals were reported mainly in roots and leaves. Reported data showed largest categories of flavonoids (38.8%), non-flavonoid polyphenols (8.4%), terpenes and derivatives (8.2%), fat and fat-related compounds (6.4%) as well as glucosinolates and breakdown products (5.6%). Leaves have high concentrations of macronutrients ~ calcium, potassium and sodium as well as fiber, fatty acids, and non-flavonoid polyphenols while sprouts are a major source of flavonoids, specifically anthocyanins, β-carotene and vitamin C. Roots are rich in non-flavonoid polyphenols together with terpenes, its derivatives and glucosinolates, the latter also highly concentrated in seeds. Leaves and sprouts could be considered part of a healthy diet, and together with roots these could be explored as raw material for development of nutraceuticals (Magda Gamba *et al.*, 2021).

Anthocyanin pigments provide red colouration of roots while the sharp flavour of radishes is due to various chemical compounds produced by its plants including glucosinolate, myrosinase, and isothiocyanate (en.wikipedia.org/wiki/Radish). Anthocyanin concentrations of the mature radish taproot are significantly greater than in the sprouts of red, pink, and purple varieties. The primary anthocyanidins present in red and pink radish varieties are pelargonidin and delphinidin while primary anthocyanidin in the purple radish found was cyanidin. Radish sprouts were found 9- and 59-fold more potent than the corresponding mature taproot at activating the antioxidant response element (ARE) in a stably transfected hepatoma cell line. The ARE activity of the radish sprouts and mature taproots was significantly correlated with the total isothiocyanate concentration of the radishes (Hanlon and Barnes, 2011).

Perez Gutierrez and Perez (2004) reported following compounds grouped according to structural classes:

- **Alkaloids and nitrogen compounds**
  - Alkaloids and nitrogen compounds found in the roots were pyrrolidine, phenethylamine, nmethylphenethylamine, 1,2′-pyrrolidin-tion-3-il-3-acid-carboxilic-1,2,3,4-tetrahydro-β-carboline, and sinapine (Marquardt, 1976; Wan, 1984; Weilan *et al.*, 1987). Cytokinin (6-benzylamino-9-glucosylpurine) is a major metabolite of 6- benzylaminopurine (6-BAP) in the root radish. A minor metabolite of 6-BAP from radish has been identified as 6-benzylamino-3-β-D-glucopyranosylpurine (Coeley *et al.*, 1975). Total amino acids were 0.5% of dry weight; with proline (0.5%) as the major constituent, however, methionine and cystine were present only in traces (0.02%). Diamines as diaminotoluene (2,4-D), 4,4′-methylenedianiline (4,4-D), and 1,6-hexanediamine (1,6-D) were isolated in the period of germination of young radish seeds. Production of thiamine were recorded higher during germination of radishes (Mal-Nam *et al.*, 2002).
  - Total protein recorded was 6.5% (Shyamala and Singh, 1987). Two chitinases (RRC-A and RRC-B) isolated from radish roots had a molecular weight of 25 kDa (Chen-Tien *et al.*, 2000). N-Bromosuccinimide and di-Etpyrocarbonate inhibited the activities of both chitinases (Perez Gutierrez and Perez, 2004).
  - Arabinogalactan proteins (AGPs) isolated from primary and mature roots contained appreciable amounts of L-fucose (Tsumuraya *et al.*, 1988) and were composed mainly of L-arabinose and D-galactose. Structures of the carbohydrate moieties isolated from the root, seeds and leaves were similar as they consisted of consecutive (1→3)-linked β-D-galactosyl backbone chains having side chains (1→6)-linked β-Dgalactosyl residues, to which α-L-arabinofuranosyl residues were attached in the outer regions (Tsumuraya *et al.*, 1988).
  - Two glycoproteins, containing L-arabino-D-galactan, isolated from the saline extract of mature radish leaves contained L-arabinose, D-galactose, L-fucose-4-O-methyl-D-glucuronic acid and Dglucuronic acid residues. Degradation of the glycoconjugates showed that a large proportion of the polysaccharide chains is conjugated with the polypeptide backbone through a 3-O-D-galactosylserine linkage (Yoichi *et al.*, 1984).

- Arabino-3,6-galactan, associated with a hydroxyproline-rich protein portion, carried a unique sugar residue, α-L-fucopyranosyl-(1-2)-α-L-arabinofuranosyl (Tsumuraya *et al.*, 1984).
- Stigma glycoproteins heritable with S-alleles (S-glycoproteins) were detected in *R. sativus*. Two main glycoproteins with molecular weights of 15,000 and 100,000Da appeared on the SDS-gel electrophoretic pattern. The carbohydrate fraction of the glycoprotein consisted of arabinose 17.3%, galactose 19.1%, xylose 8.1%, mannose 5.4%, glucose 23.7%, and rhamnose or fructose 26.4%. In the stigma surface diffusate of *R. sativus*, the content of protein was established to be 16% and that of carbohydrate was 11% (Yinghua *et al.*, 1983).
- The acanthiformis of *R. sativus* showed two ferredoxin isoproteins indicating that plants have multiple genes for ferredoxin. The relative abundance of the isoproteins varied with leaf stage (Keishiro *et al.*, 1985). The amino acid composition and N-terminal sequence were different in the isoprotein isolated from roots than those of leaf ferredoxin (Perez Gutierrez and Perez, 2004).
- The deduced amino acid sequences of polypeptides RCA1, RCA2 and RCA3 that were purified from seeds of *R. sativus* have agreement with average molecular masses from electrospray mass spectrometry of 4537, 4543, and 4532 kDa, respectively. The only sites for serine phosphorylation are near or at the C terminal and hence adjacent to the sites of proteolytic precursor cleavage (Polya *et al.*, 1993).
- Cysteine-rich peptides (Rs-AFP1 and Rs-AFP2) isolated from *R. sativus* showed peptides 6, 7, 8, and 9 comprising the region from cysteine 27 to cysteine 47 (Terras *et al.*, 1992). Protein AFP1 isolated from radish showed peptide fragments (6-mer, 9-mer, 12-mer, and 15-mer) (Hans *et al.*, 1997).
- Proteins RAP-1 and RAP-2 were isolated from Korean radish seeds and their molecular mass was established to be 6.1 kDa (RAP-1) and 6.2 kDa (RAP-2) by SDS-PAGE and 5.8 kDa (RAP1) and 6.2kDa (RP-2) by gel filtration chromatography (Hans *et al.*, 1997).

- **Coumarins**: Hydroxycoumarins aesculetin and scopoletin were also identified (Stoehr and Herrmann, 1975).
- **Enzymes**
  - A number of enzymes are present in both the cytoplasm and the cell wall and in some cases it has been shown that the cell wall isozymes differ from those of the cytoplasmic (Matile, 1975). When radish

seedlings are grown in the dark, β-fructosidase (βF) first accumulates in the cytoplasm, then slowly increases in the cell wall. Charge heterogeneity of cytoplasmic enzymes resides in the polypeptides while formation of the basic cell wall occurs as a result of post-translational modifications that can be inhibited by tunicamycin (Faye *et al.,* 1986).

- Cysteine synthase (EC 4.2.99.8) was purified to near homogeneity (275-fold) in 11.5% yield from mature roots. It was relatively stable, retaining most of its activity in standing for several days at room temperature (Tamura *et al.,* 1976).
- A basic β-galactosidase (β-Galase) has been purified from imbibed radish. This enzyme consisting of a single polypeptide with an apparent molecular mass of 45 kDa and pI values of 8.6 to 8.8, was maximally active at pH 4.0 on p-nitrophenyl β-D-galactoside and β-1,3-linked galactobiose. Radish seed and leaf arabino-3,6-galactan-proteins were resistant to the β-Galase (Sekimata *et al.,* 1989). β-Amylase (Shigeo and Yuhei, 1976), together with peroxidase c or paraperoxidase (Yuhei, 1973) which is an isoenzyme, were also isolated from Japanese radish roots.
- A hydroxycinnamoyltransferase (EC 2.3.1.-), which catalyzes *in vivo* the formation of 1,2-di-Osinapoyl-β-D-glucose, was isolated from the radish. Cotyledons exhibited activities of 1-O-acyl-glucose– dependent acyltransferases, 1-sinapoyl-glucose:L-malate sinapoyltransferase (SMT), and 1- (hydroxycinnamoyl)-glucose:1-(hydroxycinnamoyl)glucose-hydroxyl cinnamoyl-transferase (CGT), showing contrary developments depending on light conditions. Light-grown seedlings showed high Lmalate sinapoyltransferase and low 1-(hydroxycinnamoyl)glucose-hydroxyl cinnamoyl-transferase activities, while dark-grown seedlings showed low L-malate sinapoyltransferase and high 1- (hydroxycinnamoyl) glucose-hydroxyl cinnamoyl-transferase activities (Dahlbender and Strack, 1986).
- Catalase and glutathione reductase activities increased considerably in the root and leaves after 24h exposure to cadmium, indicating a direct correlation with Cd accumulation. PAGE enzyme activity staining revealed several superoxide dismutase isoenzymes in leaves. The main response may be *via* activation of ascorbate-glutathione cycle for removal of hydrogen peroxide or to ensure availability of glutathione for synthesis of Cd-binding proteins (Vitoria *et al.,* 2001).

- A γ-glutamyl transpeptidase was found. It catalyzed the release of CySH-Gly from glutathione, the release of alanine from γ-glu-Ala, and the formation of γ-glutamyl dipeptides. A dipeptide formed from Smethylcysteine and glutathione or γ-glu-Ala was characterized as γ-glutamyl-S-methylcysteine (Thompson *et al.*, 1964).
- Two cationic isoperoxidases (C1 and C3) and four anionic isoperoxidases (A1, A2, A3n, and A3) were isolated from Korean *R. sativus* root. All the six isoperoxidases are glycoproteins composed of a single polypeptide chain. The molecular weights of C1, C3, A1, and A2 were ca. 44,000, while anionic isoperoxidase A3n and A3 have molecular weights of 31,000 and 50,000, respectively. N-terminal amino acid sequences were determined for A1, A3n, and C3, while A2 was found to have a blocked terminal residue (Lee and Kim, 1994). Analysis of digested products of the two major N-glycans of C3 suggested that corefucosylated trimannosylchitobiose may contain a different linkage from the typical α-1,6 of native Nlinked oligosaccharide (Kim and Kim, 1996).
- Thiamin-binding substances were recorded in radish. There were two kinds of compounds; one was heat labile and Pronase sensitive, and the other was heat stable and Pronase resistant. It would be inferred that the former is protein and the latter is a nonprotein compound (Toshio *et al.*, 1984).
- βF isozyme (glycoprotein) is found in the cytoplasm and cell walls of the radish. The nonglycosylated cytoplasmic and cell wall βF forms have the same relative molecular mass but glycosylated forms have different oligosaccharide side chains with respect to size and susceptibility to αmannosidase and endoglycosidase D digestion (Loic *et al.*, 1986).
- 7-Glucoside de zeatin, isolated from radish cotyledons occurs naturally as glycoside with β-glucose as substituent. A large number of derivatives of purine are glucosylated, but adenine derivatives with alkyl side chains at least three carbon atoms in length at position N6 are preferentially glucosylated (Faye *et al.*, 1986).
- **Gibberellins**: The bolting of *R. sativus* cv. 'Taibyo-sobutori' requires cold treatment and subsequent long-day conditions. It has been suggested that gibberellins (GAs) might be involved in the control of bolting. Eleven gibberellins were identified in extracts of mature seed as 13 hydroxy-GAs [GA1, 3-epi-GA1, GA8, GA17, GA19, GA20, and a new GA, 12α-hydroxyGA20 (GA77)] and four non-13-hydroxy-

GAs [GA9, GA24, 12β-hydroxy-GA24, GA25]. The major GAs were GA8, GA20, and GA77 (Nakayama *et al.*, 1990).

- **Glucosinolates**
  - Glucosinolates are very stable water-soluble precursors of isothiocyanates. The relatively nonreactive glucosinolates are converted to isothiocyanates on wounding of the radish. The tissue damage releases myrosinase (EC 3.2.3.1), a glycoprotein that is physically segregated from its glucosinolate substrates. Large variations in myrosinase-specific activity have been reported in various cruciferous plant sources. Myrosinase, purified to homogeneity from daikon, has a specific activity of 280μMol/min/mg protein with sinigrin as a substrate (Fahey *et al.*, 2001). Glucosinolate contents in radish ranged from 37–87μmol/g of seeds. The 5-vinyl-2-oxazolidinethione, 3-butenyl, 4-pentenyl, and phenethyl isothiocyanate were found in industrially extracted rapeseed oils. The compounds were hydrolysis products from glucosinolates present in the seed (Daun and Hougen, 1977).
  - Desulfoglucosinolates are formed by enzyme desulfation of endogenous glucosilates. The indole glucosinolates, 4-methoxy-3-indolylmethyl glucosinolate and 1-methoxy-3-indolylmethoxy glucosinolate were absent in seed whereas 4-hydroxy-3-indolymethyl glucosinolate was found in highest concentration. The 3-indolymethyl glucosinolate was found in low levels in seeds but indole glucosinolate content dominated in the leaf (Sang *et al*, 1984).

- **Oil Seed Components**
  - Radish seeds contain a high percentage of oil. Chromatographic analysis of these oils showed clearly their complete similarities to cottonseed oil (El-Hinnawy *et al.,* 1975). The steam volatile constituents of fresh radish of Japanese and Kenyan origin have been studied. Overall pattern of compounds in the two materials was similar. Major components are pentyl hexyl, 4-methylpentyl isothiocyanate, dimethyl disulfide, methyl methanethiolsulfinate, and 1-methylthio-3-pentanone (Anders *et al.*, 1978). Oil radish seeds contained 1.21μmol of total alkenylglucosinolates (AG/g), consisting mostly of progoitrin and gluconapin (Jaroslav, 1990).

- **Organic Acids**
  - Four major organic acids are present in the roots of the radish: oxalic, malic, malonic and erythorbic acid. It showed total lipid as

1.23% (Shyamala. and Singh, 1987). Major fatty acids in seed lipids were erucic, oleic, linoleic and linolenic acids. Major fatty acids in the lipids of radish family were linolenic acid (52–55%), followed by erucic acid (30–33%) and palmitic acid (20–22%) (Seiichiro *et al.*, 1984). Through petroleum erther extraction of powdered *R. sativus* seeds also revealed presence of stearic acid though glutamic acid was found in pickled daikon (20–100 mg%) (Changdai, 1984).

- **Phenolic Compounds**
  - The content of phenolic acids in radish roots was much smaller than in leaves. Radishes and horseradishes show caffeic, p-coumaric, ferulic, hydroxycinnamic, p-hydroxybenzoic, vanillic, salicylic and gentisic acid (Stoehr and Herrmann, 1975). Sinapic acid esters (1-sinapolyglucose, sinapoyl-L-malate, and 6,3′- disinapoylsucrose), kaempferol glycosides and free malic acid were isolated from cotyledons of *R. sativus* seedlings (Strack *et al.*, 1985).
  - Among anthocyanins, pelargonidine and cyanidine were responsible for red and violet colours in corollas and roots in all inbred progenies. The absence of pelargonidine and cyanidine resulted in white colour. Flavonoid and quercetine were also found in corolla and root both (Narbut *et al.*, 1972). Anthocyanins extracted from epidermal tissue resulted in juices with fairly low initial °Brix (1.3°), containing 400mg anthocyanin/100ml. This compound provided colour similar to FD&C Red#40. Radish concentrate extract represents a promising natural alternative to the use of FD&C Red#40 (Rodriguez *et al.*, 2001).
  - Other purple root pigment isolated from progeny-radish was an ester of cyanidine triglucoside and three kinds of cinnamic acids. The triglucoside was identified as the 2-diglucoside-5-monoglucoside of cyanidine (rubrobrassicin) (Isikura *et al.*, 1965). Other anthocyanins obtained from red radish are two diacylated pelargonidin 3-O-[2- O-β-glucopyranosyl)-6-O-(trans-p-coumaroyl)-β-glucopyranoside]-5-O-(6-O-malonyl-β-glucopyranoside) and pelargonidin 3-O-[2-O-(β-glucopyranosyl)-6-O-(trans-p-feruloyl)-β-glucopyranoside]-5-O-(6-O malonyl-β-glucopyranoside) and other monoacylated anthocyanins pelargonidin 3-O-[2-O-βglucopyranosyl)-6-O-(trans-p-coumaroyl)-β-D-glucopyranoside]-5-O-(β-glucopyranoside) and pelargonidin 3-O-[2-O-β-glucopyranosidel)-6-O-(trans-p-feruloyl)-β-glucopyranoside]-5-O-(β-glucopyranoside). Pelargonidin-3-diglucosido-5-monoglucoside is known as raphanusin (Guisti *et al.*, 1998).

- The major anthocyanins of radishes are pelargonidin-3-sophoroside-5-glucoside acetylated with malonic acid and either ferulic or p-coumaric acid. Cinnamic acid acylation site for radish anthocyanins was determined to be at position 6 of glucose-1 of the sophorose substituents by one- and two dimensional 1 HNMR-13CNMR (Wrolstad *et al.,* 2001). Also 7-glucoside-pelargonidin has been identified in R. sativus (Harborne, 1963). This compound was stable at 60°C and under light, may be used as a food colorant (Junlian *et al.,* 1992).
- Kaempferol-7-O-rhamnoside, isorhamnetin-7-O-rhamnoside, quercetin-7-O-rhamnoside, kaempferol 3-glucoside-7 rhamnoside, kaempferol-7-glucoside-3-rhamnoside, quercetin-7-O-arabinoside-3-glucoside, and quercetin-7-glucoside-3 rhamnoside were isolated from *R. raphanistrum* (Kamil and Kalina, 1977). Radishes have a high content of flavonoids as quercetin, kaempferol, myricetin, apigenin and luteolin (Lugasi and Hovari, 2000). Malvidin-3,5- diglucoside was produced from the callus of radish via tissue culturing. The callus on dry weight basis contains 16.4% pigments (Kwan *et al.,* 1995).

- **Pigments**
  - Salted radish roots have a characteristic yellow colouration which generates during storage. 4-Methylthio-3- butenyl-glucosinolate (4-MTBG) is the substrate of the main pungent principle of radish and is one of the essential factors for formation of the yellow pigment. The yellow compound 1-(2′-pyrrolidinethion3′-yl)-1,2,3,4-tetrahydro-β-carboline-3-carboxilic acid is presumed to have been the condensation product from the degradation of 4-methylthio-3-butenylisothiocyanate and L-tryptophan. The carboline compound is considered to play an important role in the formation of the yellow pigment in salted radish roots (Ozawa *et al.,* 1990).

- **Polysaccharides**
  - Pectic substances were extracted from the leaves with oxalate buffer of pH 4.25 as weakly acidic pectic polysaccharide (WAP) and pectic acid. WAP was appreciably hydrolyzed by exo- and endopolygalacturonases and the galacturonic acid content (17.3–25.8%) was much lower than the pectic acids, though the neutral sugar components of both pectic substances were almost the same. The arabinose-galactose side chains were very long or highly branched in pectin compared with those in pectic acids. These compounds are probably inherent pectic components of the cell walls of the vegetables (Matsuura and Hatanaka, 1988). Rhamnose,

glucose, and xylose were also isolated. Lipopolysaccharides (LPS) were isolated from radish roots (Genichiro *et al.*, 1991).

- **Proteoglycan**
  - A L-arabino-D-galactan–containing proteoglycan was isolated from hot phosphate-buffered saline extract of radish seeds through ethanol fractionation. The proteoglycan consisted of 86% of a polysaccharide component containing L-arabinose and D-galactose as major sugar constituents, together with small proportions of D-xylose, D-glucose, and uronic acids, and 9% of a hydroxyproline-contained protein. Arabinogalactan from radish seed had a high content (81%) of L-arabinose and its basic structure seemed to be similar to that of the polysaccharide component of the proteoglycan (Yoichi *et al.*, 1987).
- **Sulphur compounds**
  - Radish leaves contain only one of the sulfonium diateroisomers of S-adenosylmethionine (AdoMet), which has a remarkable variety of biochemical functions. It is an allosteric enzyme effector and a precursor of spermine biosynthesis, spermidine, and ethylene. It is also the methyl group donor for most biological transmethylation reactions, wherein transfer of its methyl group converts AdoMet to the homocysteine analog (AdoHcy). Much of the chemistry and biochemistry of AdoMet derives from the fact that it is a sulfonium compound (Creason *et al.*, 1985). 1-(2′-Pyrrolidinethion-3′-yl)-1,2,3,4-tetrahydro-β-carboline-3- carboxilic acid was found in radish root. This carboline compound is considered to play an important role in the formation of the yellow pigment in salted radish roots (Perez Gutierrez and Perez, 2004).
  - Raphanin [(1E)-4-Isothiocyanato-1-(methanesulfinyl)but-1-ene] (also called sulforaphen, sulforaphene and sativin)] is the main sulphur component found in radish seeds of *R. sativus* and is also found in broccoli and red cabbage (Sinha *et al.*, 2010; Michael and Sieghard, 2008). *In vitro*, raphanin inhibits some fungi and various bacteria including *Staphylococcus*, *Streptococcus*, *Pneumococcus* and *Escherichia coli* (Baron, 1950; Florey, 1949; Ivanovics and Horvath, 1947).
- **Other Constituents**
  - β-Carotene was isolated from radish. Vitamin C content in fresh hotbed radishes ranged from 17.95– 27.86 mg (Bulinski and Zhinda, 1962). Also identified was β-sitosterol from R. *sativus* seeds

(Changdai, 1984). The contents of raphanusol A and B in radish increased at the lighted side and decreased in the shaded side. The differential distribution of raphanusol A and B in the hypocotyls is closely correlated with growth suppression at lighted side (Koji and Shigenori, 1989).

Secondary Metabolites with Pharmaceutical Uses

The secondary metabolites having pharmaceutical uses in radishes include glucosinolates, isothiocyanates and polyphenols (Yuan *et al.*, 2010; Salah-Abbes *et al.*, 2015; Nakamura *et al.*, 2008). Glucosinolates (GSL) are secondary metabolites that are exclusively found in cruciferous vegetables (Soundararajan and Kim, 2018). The chemical conformation of GSL was found possessing ß-D-thioglucosides residue bonded to (Z)-N-hydroximinosulfate ester. GSLs are majorly classified into three types based on their precursor amino acids, such as aliphatic glucosinolates (AGSLs), aromatic glucosinolates (ArGSLs) and indolic glucosinolates (IGSLs) (Kliebenstein *et al.*, 2001; Field *et al.*, 2004; Dawson *et al.*, 1993).

List of secondary metabolites with antioxidant properties identified in different parts of radish:

Name	*Class*	*Tissue*	*Reference*
1,2-dihydroxyferuloyl-gentiobiose	Phenolic acid	Leaves	Koley *et al.*, 2020
13Z-ß-Carotene	Carotenoids	Sprouts	Lee *et al.*, 2018
3-Butenyl isothiocyanate	Isothiocyanates	Pod & flower	Pocasap *et al.*, 2013
4-methoxyglucobrassicin	Glucosinolate	Sprouts	Wei *et al.*, 2011
4-OH-glucobrassicin	Glucosinolate	Sprouts	Wei *et al.*, 2011
6-Prenyl-naringenin	Flavonone	Root	Koley *et al.*, 2020
6,7,30,40-Tetrahydroxyisoflavone	Isoflavonoids	Leaves	Koley *et al.*, 2020
9Z-ß–carotene	Carotenoids	Sprouts	Lee *et al.*, 2018
α-Carotene	Carotenoids	Sprouts	Lee *et al.*, 2018
Antheraxanthin	Carotenoids	Sprouts	Lee *et al.*, 2018
Anthocyanin-3-O-(cinnamoyl) sophoroside-5-O-glucoside derivatives	Anthocyanin	Sprouts	Baenas *et al.*, 2015
Anthocyanin 3-O-sophoroside-5-O-(malonyl) glucoside derivatives	Anthocyanin	Sprouts	Baenas *et al.*, 2015
Apigenin	Flavonoids	Sprouts & seeds	Pajak *et al.*, 2014
Apigenin-7-O-neohesperidoside	Flavone	Leaves	Koley *et al.*, 2020
Apigenin-7-O-rutinoside	Flavone	Leaves	Koley *et al.*, 2020
Caffeic acid	Phenolic acid	Sprouts & seeds	Pajak *et al.*, 2014

Name	Class	Tissue	Reference
Caffeoylmalic acid	Polyphenols	Leaves	Luo *et al.*, 2018
Chrysoeriol-7-O-apiosyl-glucoside	Flavone	Leaves	Koley *et al.*, 2020
Cyanidin-3-O-caffeoyl-p-coumaroyl-sophoroside-5-O-glucoside	Anthocyanin	Root	Koley *et al.*, 2020
Cyanidin-3-O-di-p-coumaroyl-sophoroside-5-Omalonylglucoside	Anthocyanin	Root	Koley *et al.*, 2020
Cyanidin-3-O-glucoside	Anthocyanin	Leaves	Koley *et al.*, 2020
Cyanidin-3-O-p-coumaroyl-feruloyl-sophoroside- 5-O-glucoside	Anthocyanin	Root	Koley *et al.*, 2020
Cyanidin-3-O-rhamnoside	Anthocyanin	Leaves	Koley *et al.*, 2020
Cyanidin-3-O-sophoroside-5-O-glucoside	Anthocyanin	Leaves	Koley *et al.*, 2020
Cyanidin-3-O-sophoroside-5-O-malonylglucoside	Anthocyanin	Leaves, root	Koley *et al.*, 2020
Cyanidin-3-O-xylosyl-p-coumaroyl-glucosylgalactoside	Anthocyanin	Leaves	Koley *et al.*, 2020
Delphinidin-3-O-rutinoside	Anthocyanin	Root	Koley *et al.*, 2020
Dihydro-caffeoyl-3-O-glucuronide	Phenolic acid	Root	Koley *et al.*, 2020
Dihydro-kaempherol-3-O-rutinoside	Dihydroflavonol	Leaves	Koley *et al.*, 2020
E- ß –carotene	Carotenoids	Sprouts	Lee *et al.*, 2018
Ferulic acid	Phenolic acid	Aerial parts	Ricardo *et al.*, 2018
Ferulic acid	Phenolic acid	Sprouts & seeds	Pajak *et al.*, 2014
Feruloylmalic acid	Phenolic acid	Leaves	Luo *et al.*, 2018
Gallic acid	Phenolic acid	Sprouts & seeds	Pajak *et al.*, 2014
Genistin	Isoflavonoids	Leaves	Koley *et al.*, 2020
Glucobrassicin	Glucosinolate	Sprouts	Force *et al.*, 2007
Glucodehydroerucin	Glucosinolate	Sprouts	Force *et al.*, 2007
Glucoraphasatin	Glucosinolate	Whole plant, sprouts	Force *et al.*, 2007; Hanlon *et al.*, 2007
Glucoraphenin	Glucosinolate	Sprouts	Li *et al.*, 2017
Indole-3-carbinol	Isothiocyanates	Sprouts	Baenas *et al.*, 2016
Isorhamnetin-3-O-p-coumaroyl-caffeoylsophorotrioside-7-O-malonyl-glucoside	Flavanol	Root	Koley *et al.*, 2020
Isorhamnetin-3-O-p-coumaroyl-sophorotrioside- 7-O-glucoside	Flavanol	Leaves	Koley *et al.*, 2020
Kaemferol	Flavonoids	Sprouts & seeds	Pajak *et al.*, 2014
Kaempferitrin	Polyphenols	Leaves	Luo *et al.*, 2018
Kaempferol-3-O-caffeoyl-sophoroside-7-Oglucoside	Flavanol	Root	Koley *et al.*, 2020

Name	Class	Tissue	Reference
Kaempferol-3-O-feruloyl-sophoroside-7-Oglucoside	Flavanol	Root	Koley *et al.*, 2020
Kaempferol-3-O-glucoside	Flavanol	Leaves	Koley *et al.*, 2020
Kaempferol-3-O-glucosyl-rhamnosyl-glucoside	Flavanol	Leaves	Koley *et al.*, 2020
Kaempferol-3-O-p-coumaroyl-sinapoylsophorotrioside-7-O-malonyl-glucoside	Flavanol	Leaves, root	Koley *et al.*, 2020
Kaempferol-3-O-p-coumaroyl-sophorotrioside- 7-O-glucoside	Flavanol	Leaves	Koley *et al.*, 2020
Kaempferol-3-O-rhamnoside(I)	Flavanol	Leaves	Koley *et al.*, 2020
Kaempferol-3-O-rutinoside	Flavanol	Leaves	Koley *et al.*, 2020
Kaempferol-3-O-xylosyl-rutinoside	Flavanol	Leaves	Koley *et al.*, 2020
Lutein	Carotenoids	Sprouts	Koley *et al.*, 2020
Luteolin-7-O-glucoside	Flavone	Leaves	Koley *et al.*, 2020
m-Coumaric acid	Phenolic acid	Leaves	Koley *et al.*, 2020
Methylgalangin	Flavanol	Leaves	Koley *et al.*, 2020
Naringenin-40-O-glucuronide	Flavonone	Leaves	Koley *et al.*, 2020
Naringenin-7-O-glucuronide	Flavonone	Leaves	Koley *et al.*, 2020
p-Coumaric acid	Phenolic acid	Sprouts & seeds	Pajak *et al.*, 2014
p-Coumarylmalic acid	Polyphenols	Leaves	Luo *et al.*, 2018
Pelargonidin-3-O-caffeoyl-caffeoyl-diglucoside-5-O-malonyl-glucoside	Anthocyanin	Root	Koley *et al.*, 2020
Pelargonidin-3-O-feruloyl-diglucoside-5-Oglucoside	Anthocyanin	Leaves	Koley *et al.*, 2020
Pelargonidin-3-O-p-coumaroyl-diglucoside-5-Oglucoside	Anthocyanin	Leaves	Koley *et al.*, 2020
Pelargonidin-3-O-sambubioside	Anthocyanin	Leaves	Koley *et al.*, 2020
Protocatechuic acid	Phenolic acid	Sprouts & seeds	Pajak *et al.*, 2014
Quercetin	Flavonoids	Sprouts & seeds	Pajak *et al.*, 2014
Quercetin-3-O-p-coumaroyl-sophoroside-7-Oglucoside	Flavanol	Leaves	Koley *et al.*, 2020
Quercetin-3-O-rhamnoside	Flavanol	Leaves	Koley *et al.*, 2020
Quercetin-3-O-rhamnosyl-galactoside	Flavanol	Leaves	Koley i., 2020
Sinapic acid	Phenolic acid	Sprouts & seeds	Pajak *et al.*, 2014
Spinacetin-3-O-(200-p-coumaroyl-glucosyl)(1-6)-(apiosyl(1_2))-glucoside	Flavanol	Root	Koley *et al.*, 2020
ß-Cryptoxanthin	Carotenoids	Sprouts	Lee *et al.*, 2018
Stigmasterol	Phytosterol	Aerial parts	Ricardo *et al.*, 2018

Name	Class	Tissue	Reference
Sulforaphane	Isothiocyanates	Pod & flower	Pocasap *et al*, 2013
Sulforaphene	Isothiocyanates	Pod & flower	Pocasap *et al.*, 2013
Violaxanthin	Carotenoids	Sprouts	Lee *et al.*, 2018
Zeaxanthin	Carotenoids	Sprouts	Lee *et al.*, 2018
β-sitosterol	Phytosterol	Aerial parts	Ricardo *et al.*, 2018
β-sitosterol-3-β-O-D-glucopyranoside	Phytosterol	Aerial parts	Ricardo *et al.*, 2018

Source: Manivannan et al. (2019)

Uses

The entire plant is edible and can be eaten raw or cooked. Radish roots are low in calories and are usually eaten raw in salad but can be cooked or pickled. The young and tender leaves can be eaten in salads, on sandwiches or as a side dish. A typical north Indian salad can be prepared by mixing chopped leaves and roots of radish with tomato slices, 1-2 tablespoon of mustard oil and 1-2 pinches of salt. The young leaves can be cooked like spinach. The seed pods, called 'moongra' or 'mogri' in India are also edible and are often eaten raw, sautėed, in *raita* or can be cooked with potato to prepare *bhujiya*-like side dish. Radish roots can be processed as dried or canned pickles.

The extracts of radishes have been employed to treat stomach disorders, constipation, urinary infections, hepatic inflammation, cardiac disorders and ulcers in folk medicine since the ancient times (Goyeneche *et al.*, 2015). In addition, various reports suggested its antimicrobial (Beevi *et al.*, 2009; Rakhmawati *et al.*, 2009), anticancer (Pocasap *et al.*, 2013), antioxidant (Kim *et al.*, 2017; Baenas *et al.*, 2016) and anxiety reducing properties (Siddiq and Younus, 2018). The radish seed oil is a potential source of biofuel. Wild radish seeds contain up to 48% oil. The daikon grows well in cool climates and, apart from its industrial use, can also be used as a cover crop, grown to increase soil fertility, to scavenge nutrients, to suppress weeds, to help alleviating soil compaction and preventing winter erosion of the soil (en.wikipedia.org/wiki/Radish).

Biological activities

Radish leaves as well as roots show antioxidant activities. Flavonoids show anti-allergic, anti-inflammatory, anti-microbial and anti-cancer activity. Glycosides play an important role in lowering the blood pressure. They are also used in treatment of congestive heart failure and cardiac arrhythmia. The extract contains phytochemicals like phenols and flavonoids that cause inhibition, and this plant property may be important in preventing oxidative-stress-related diseases. Tests for terpenoids and steroids showed variable results (Kumar and Rashmi Patwa, 2018).

Unpleasant Smell

Radishes release a very strong unpleasant odour when cooked or being fermented. Radish consumption causes 'burping & belching' which releases a strong and noxious smell like bad flatulence or rotting cabbage.

Cultivation

Soil

Radishes grow best in a well-drained light and deep sandy-loam soil having 6.5-7 pH range though most ideal range is slightly acidic soil, *i.e.* soil pH of 5.5-6.8 (ecoursesonline.iasri.res.in), however, for late season crop a clayey-loam is ideal (en.wikipedia.org/wiki/Radish).

Growing Season

Radish seeds in warmer climates are usually sown 4 to 6 weeks before the first autumn frost though under cold climates these are sown 4 to 6 weeks before the last spring frost (https://www.almanac.com/plant/radishes). Less pungent varieties that are pleasantly tangy rather than fiery should be grown in the spring or fall but not during summer as the same variety will produce quite pungent roots if weather is hot. Even the use of a garden row-cover to keep leaf-munching flea beetles and root maggot flies off the radishes may raise the soil temperature enough to increase the pungency of radishes (Peterson, 1999).

Though radishes can be grown throughout the year, the main season in North India is from August to January and the best seasons in South India is from October to December and April to June. European varieties in India can be sown from September to March in the plains though on the hills radishes should be sown from March to October.

Throughout the year radish growing schedule in the Indian plains (ecoursesonline.iasri.res.in):

Variety	Sowing time	Harvesting time
Pusa Chetki	Early April to mid-August	Early May- September
Pusa Desi	Mid August to mid-October	September end-early December
Pusa Reshmi	Mid September to mid-November	Last October- early January
Japanese White	Mid October to mid-December	Mid December- early March
Pusa Himani	Mid October to mid-February	Mid February- mid April
White Icicle	Last October to February end	November end- early March

Climate

Due to its rapid growth tendency, it has wider distribution though radish is well suited to a moderately cool climate especially when in vegetative phase. Though for seed to seed cropping it is a biennial crop but for edible root harvesting, strictly it is an annual crop which is harvested before initiation of flowering processs so that roots are tender enough for human consumption. If the soil has enough moisture with temperatures ranging from 18 to 29^{0}C, its seeds germinate quickly with further quicker growth. Good quality roots are obtained under moderate day lengths with air temperatures between 10 and 18°C (en.wikipedia.org/wiki/Radish). For proper plant growth including that of root growth, it requires full sun, especially 6h forenoon sun, however, plants under shaded conditions become lanky with massive leaf growth though root development becomes utterly poor (https://www.almanac.com/plant/radishes). However, too long days force the plants to bolt and seed prematurely on the cost of root development (Peterson, 1999).

Chinese and Indian varieties of Daikon radish tolerate higher temperatures than the Japanese ones and such varieties also grow well at lower elevations in East Africa. If moisture is abundant, it can grow quickly; otherwise, the flesh becomes overly tough and pungent (Grubben, 2004). The seeds of 'Long White Icicle' radish available in Great Britain produce quality roots successfully resembling to those of parsnip by midsummer in Southern England in good garden soil in an average year (en.wikipedia.org/wiki/Daikon).

Certain varieties of Daikon can be grown as a winter cover crop and green manure. These varieties are often named 'tillage radish' due to their huge penetrating root growth that effectively perform deep cultivation by breaking up the hardpan soils or improving tilth in workable but heavy soils. In this case the roots function as the good nutrient scavengers as the nutrients lower in the soil profile are brought up into the higher reaches, proving that they are good partners with legumes instead of grasses. Due to severe winters the roots decompose while still in the soil and release early nitrogen stored in the spring (en.wikipedia.org/wiki/Daikon).

For seed production, a less humid climate is required as long spell of hot and dry period is not suitable for seed production (ecoursesonline.iasri.res.in).

Radish crop being quick-growing requires quite crispy and friable **soil** so the soil should be tilled properly by removing previous crop stubbles, weeds and their rootstocks, rock *cum* brick pieces and any other foreign material which may hinder proper root growth. Root development is badly affected in compact soil as in such cases soil requires to be mixed thoroughly with 10-20 tonnes of well-rotten manure or FYM at the time of field preparation and levelled with proper arrangement of drainage. Looking into the soil fertility and variety, per hectare 60-80kg N (half dose before sowing and half at the stage of root formation) and 50kg each of P and K at the time of field preparation should be applied. Potassium application increases the vitamin C content in radish roots (ecoursesonline.iasri.res.in). In case the soil is deficient in boron, 15-20kg borax/ha should be applied before sowing. In molybdenum deficient soils, 1.2kg sodium molybdate or ammonium molybdate should be applied. Molybdenum deficiency causes stunted growth accompanied with narrow and leathery leaves. In case of those varieties where root growth is quite deep, tilling should be done up to a depth of 20cm (https://www.almanac.com/plant/radishes). For a continuous harvest of radishes in the late spring and early summer, these are planted in small quantities every week or two while weather is still cool, except in July and August (Peterson, 1999).

Seeds are usually directly sown outdoors on flat beds or on the ridges, at the depth of 1cm as recommended for small radishes, 1.5-3cm for semi-long types, 1.0-1.25cm for round cultivars, and up to 4cm for larger varieties (Peterson, 1999) and should be loosely covered with soil. The seed requirement for growing radish in one hectare land is 8-10kg for tropical or Asiatic varieties and 10-12kg for temperate or European types (ecoursesonline.iasri.res.in). The seeds should be sown at a distance of about 2.5cm in rows with inter-row distance of about 25-30cm. After sowing, the seeds must be watered immediately thoroughly down to 15cm deep (https://www.almanac.com/plant/radishes). Afterwards, the radish field is irrigated at 3-4 leaf stage when plants attain the height of 5-7cm. Subsequent irrigation is applied weekly during summer and at 10-12 days during winter. However, irrigation depends entirely on prevalent weather condition. Subsurface drip irrigation is the best option in radish cultivation, especially in sandy-loam soils in which its requirement is about 21-25cm for entire crop season. Moisture fluctuations, *i.e.* heavy application after a dry period results in slow leaf growth, root-cracking *cum* pithiness, *vis-à-vis* hot and bitter roots. After a fortnight of sowing the field is thinned out. Hoeing and weeding should be carried out every fortnight in the temperate types while at every 20 days in the tropical types. To control the weeds even chemical weedicides such as pendimethalin 0.75kg a.i./ha, alachlor at 1.25kg a.i./ha, butachlor at 1.5kg a.i./ ha, fluchloralin (Basalin) @ 0.9kg a.i./ha, Isoproturan @ 1kg a.i./ha, metalachlor @ 1kg a.i./ha or trifluralin @ 0.9kg a.i./ha is quite effective as pre-emergence (ecoursesonline.iasri.res.in; Singh, 2019).

Row-to-row spacing for Asiatic types is 30-45cm and for temperate type 20-30cm. Plant to plant spacing is maintained at 6-8cm for Asiatic type and 2.5-3.5cm for European type, though distance can be adjusted more or less depending on the cultivar and season of cultivation because during summer these may distantly be planted while during winter closely.

Interculture and Cropping System

Radishes can be mulched with compost enriched with wood ashes to help retain moisture in dry condition, to control weeds as well as to keep root maggots at bay. During growing period, the crop needs to be thinned out and weeds controlled. Weeding and thinning must be carried out 15-20 days after sowing keeping a distance of 5-10cm between plants in a row. Earthing up is also necessary to get well developed, elongated and quality roots as generally the growing roots tend to push out of the soil (ecoursesonline.iasri.res.in).

In north-western part of India, the cropping sequence followed is radish-tomato-bittergourd and radish-onion-okra. It is generally intercropped with potato when grown during *rabi* season, though summer radish can be intercropped with amaranth, especially in the eastern part of India. During summer in the northern plains, it can be grown with maize as intercrop or maize can be planted as the border crop so that the crop is protected from summer heat wave.

Companion Plant

Radishes can function as a trap crop, luring insect-pests away from the main crop and are sometimes grown as companion plants for many other crops, probably because their pungent odour deters insect-pests such as aphids, cucumber beetles, tomato hornworms, squash bugs and ants. Cucumbers and radishes seem to thrive when grown in close association with each other, and radishes also grow well with chervil, lettuce, peas and nasturtiums, but they react adversely to growing in close association with hyssop.

You could sprinkle a few seeds alongside a slower-growing crop like cabbage, knowing that the radishes will be harvested before the cabbage needs the room. As the radishes grow, thin them by pulling some as soon as they are the size of dimes or even smaller and add them to salads (Peterson, 1999).

A favourite trick is to mix radish seeds with those of carrots or parsnips to mark the rows and help break the soil for these slow-germinating crops. Parsnips and carrots can take up to three weeks to sprout (old-timers say they go to the devil seven times before they come up) so by this time the radishes will be ready for consumption (Peterson, 1999).

Harvest and Post-harvest

Under average conditions, the crop matures in 3–4 weeks, but in colder weather, 6–7 weeks may be required (Seaman Abby, 2013). European varieties are harvested 25-30 days after sowing. If radishes are not pulled and eaten within a few days of reaching maturity, they become over-mature resulting in pithy or fluffy roots i.e. the cell walls will separate from each other making the interior spongy (ecoursesonline.iasri.res.in). Harvesting periods can be extended by making repeat plantings, spaced a week or two apart. The Asiatic varieties are uprooted 40-45 days after sowing and they remain edible for longer duration compared to European varieties. Before pulling out of roots the crop should be irrigated as this will facilitate easy uprooting.

Radishes remain in prime condition for only a short time, especially in warm weather. Common smaller varieties become ready for consumption within a month while larger daikon varieties take several weeks (ecoursesonline.iasri.res.in). After harvesting, radishes can be stored without loss of quality for 2-3 days at room temperature and about two months at 0°C with a relative humidity of 90–95% (Gopalakrishnan, 2007). Most radishes are marketed in plastic bags after removing the tops. Those sold with tops will lose moisture and quality more rapidly than those without tops.

Yield

Gopalakrishnan (2007) reported that Asiatic cultivars yield 150-200q/ha of roots while European cultivars 150-200 q/ha.

Pests and diseases

Radishes are fast-growing plants therefore generally diseases are not of major importance. Important radish diseases are leaf spot (*Cercospora carotiae*), black rot (*Xanthomonas campestris*) and radish mosaic virus (RaMV). Virus infected plants are immediately removed and burnt.

White rust (*Albugo candida*) survives quite long under dry conditions and its spread is through wind. Its infection is symptomized through leaf rolling and thickening, formation of white pustules on cotyledons and blotting on stems and/or flowers which after coalescing form large patches (https://plantvillage.psu.edu). Chalky-white blisters mainly on leaf undersides, small yellow-green spots or blisters which appear sometimes in circular arrangement on upper leaf surfaces along with possible distortion or galling and stem infection are the symptoms of this rust (https://www.almanac.com/plant/radishes). The disease may be managed through crop rotation, sowing of only disease-free certified seeds and spraying the crop with 0.3% Dithane Z-78 fortnightly (https://plantvillage.psu.edu; https://www.almanac.com/plant/radishes).

Incidence of **Alternaria blight** (*Alternaria brassicae, A. brassicola* and *A. raphanii*) is favoured by warm and wet weather. Its infection causes yellow, dark brown and black circular spots on leaves and concentric rings on leaves, petioles, stems and even on flowers. Lesion centre may dry and drop giving the 'shot-hole' appearance. Later on spots coalesce to form large necrotic patches followed by leaf falling. Control measures include use of certified disease-free seeds, treating of seeds with hot water before sowing, rotation of the crop with non-brassicae crops, irrigation only early in the morning so that by the day plant surface is dry. Its chemical control measure include 0.2% spraying either of Dithane Z-78 or Kavach, or 0.05% spraying of Tricyclazole or Nativo.

Clubroot (***Plasmodiophora brassicae***) damage is akin to root-knot nematode infestation. The fungus can survive in the soil for more than 10 years and it spreads through movement of contaminated soil and irrigation water to uninfected areas. Its infection causes utterly slow growth, stunting of plants, leaf yellowing, showing of leaf wilting during day time but recovering by night, roots show swelling (extensive gall formation) and distortion. Once the pathogen is present in the soil it can survive for many years as its elimination then is economically unfeasible and crop rotation in this case is generally not effective. One can get rid of this pathogen only by planting certified seed and avoiding field-grown transplants unless produced in a fumigated bed. However, lime application in the soil can reduce fungus sporulation (https://plantvillage.psu.edu). Infected plants should be destroyed. Soil should be solarized, soil pH should be maintained around 7.2, tools should be regularly disinfected and crop rotation should be followed (https://www.almanac.com/plant/radishes).

Downy mildew (***Peronospora parasitica***) is favoured by cool *cum* damp weather. In its infection small angular lesions on upper surface of leaves are formed which enlarge into orange or yellow necrotic patches with white fluffy growth on leaf-undersides. To get rid of this problem entire crop debris is removed after harvest, crop rotation with non-brassicae crops should be followed, and more so this may be controlled by applying 0.2% Dithane M-45, Ridomil MZ, Aliette or Matco.

Yellows/*Fusarium* wilt (*Fusarium oxysporum*) spreads through infected transplants, infected tools, soils, water and wind. It survives in the soil for many years. Its infection causes leaf yellowing in one side of plant and defoliation. Its control measures include planting of resistant radish varieties though once it has established it is very difficult to control. Its spread can be prevented by sanitizing all the equipments regularly and solarization of infected soil (https://plantvillage.psu.edu). Spraying the standing crop with 0.1% Bavistin alternate with 0.2% Captafol throughout the crop season each at fortnightly interval is quite effective.

Wirestem/Damping-off (*Rhizoctonia solani*) in radish favours cool *cum* wet soils. Its infection causes death of seedlings after emergence. In its infection the brown-red or black rot girdles (stem constriction) and twist (wirestemming) the plants at seedling stage though seedlings initially may remain upright. To avoid this problem, the seeds or transplants should be pre-treated with G.N. Agrosan and grown in sterilized soil and moreover sowing or planting should be carried out only when weather is warm (https://plantvillage.psu.edu).

Black root fungus (*Aphanomyces raphani*) can survive in the soil for quite long time. Its infection is sumptomized through small black-blue areas appearing on roots which expand and girdle the taproot afterwards, thereby roots become constricted at site of lesions along with black discoloration. One can get rid of this problem by planting resistant varieties and through crop rotation with non-brassicae crops (https://plantvillage.psu.edu). Use of 0.1% benomyl alternate with 0.2% Captafol one by one at fortnightly sprayings will save the crop.

Scabies (*Streptomyces* sp.) may also infect radish similar to those of potato, turnip and rutababga. Brown-yellow, circular, irregular, sunken and cracked lesions form on roots which through coalescing may enlarge. Scab management is not an easy task in radish. However, its incidence can be minimized by crop rotation with non-host crops for four years, maintaining a high level of soil moisture throughout and avoiding the increase of soil pH through soil amendments (https://plantvillage.psu.edu).

***Erwinia* root rot** (*Erwinia rhapontici*) on radish causes flaccid root rot which can be controlled through spraying with 0.2% Mancozeb + Bactrinashak or Bactosan (https://plantvillage.psu.edu).

Ring spot and radish mosaic (RaMV) disease causes small, irregular or circular chlorotic spots between and along the veins which on coalescing later on form larger spots. Foliar mineral oil spray or neem oil at 2%, Dimethoate or

Monocrotophos at 0.05% weekly but no spraying about 20 days of crop harvest is quite effective. ***Turnip mosaic virus*** (TuMV) spreads on radishes by aphids (*Brevicoryne brassicae, Myzus persicae*) causing mosaic mottling of young leaves often with interveinal chlorosis, and finally irregular chlorotic patches through increase in size coupled with dark green patches on light green leaves are its symptoms. Finally the emerging leaves have reduced size and sometimes these necrotic lesions appear even on stems, petioles and midribs. To control further spread of this disease the infected plants are uprooted and destroyed and 2% neem oil or mineral oil is sprayed on the foliage or weekly spraying of Acephate or dimethoate @ 0.05% should be done but spraying should be stopped 20 days ahead of harvesting the crop. **Leaf curl disease** in radishes is a problem only under north Indian conditions which is caused due to radish leaf curl virus (RaLCV). It is spread through whitefly (*Bemicia tabaci*) and situation is aggravated due to beta satellite DNA present in radishes. In its infection the leaves show mottling, vein yellowing, vein clearing, vein twisting, interveinal chlorosis, crinkling, leaf curling and stunted growth. To get rid of this disease, the radishes along with the border crops of bajra, sorghum or maize should be sprayed with Acephate @ 1.5g/l followed by Imadacloprid @ 0.3ml/l or Monocrotophos/Metasystox/ Dimethoate @ 0.05% which is quite effective in controlling the vectors. In rotation with insecticides the sprayings of the chemicals may be alternated with 2% neem seed kernel extract for effective control. **Radish phyllody** appears due to a phytoplasma whose carrier is a leafhopper. Its infection causes light green to dull-gray colouration, especially at flowering as then floral parts appear green, violet and thick though earlier infection causes pale-green colouration, profuse tillering *cum* shortened internodes. At first appearance of such plants, rouging is the best option and field should be kept weed-free. Moreover, one should also opt for insecticide (1.7 ml/l dimethoate or 0.5ml/l of Imidacloprid) spraying along with 2% neem oil alternatively during active vegetative growth weekly so that leafhopper is effectively controlled.

Some insect-pests can be a nuisance in radish crop. The larvae of **flea beetles** (*Phyllotreta striolata*) feed on vegetative parts, especially on seedlings which can be controlled through use of 0.15% Malathion, 0.4% Sevin or 0.5ml/l Lambda cyhalothrin 5EC at 10-15 days intervals. In fact, the larvae live in the soil and this along with its adult feed on the plant parts making shot holes on the leaves. The **swede midge** (*Contarinia nasturtii*) attacks the foliage and growing tip of the plant and causes distortion, multiple or no growing tips, and swollen or crinkled leaves and stems. **Cabbage root fly maggots** sometimes attack the roots. The foliage droops and becomes discoloured, and small, white maggots tunnel through the root, making it unattractive or inedible (en.wikipedia.org/ wiki/Radish). Sticky traps are also quiute effective in controlling the adults. The black larvae of **mustard sawfly** (*Athalia lugens proxima*) feed gregariously on leaves and these can effectively be controlled through spraying with 0.5ml/l of Lambda cyhalothrin at 10-15 days intervals. Black and orange coloured nymphs and adults of **painted bug** (*Bagrada cruciferarum*) have been observed sucking leaf sap where in their severe infestation the leaves wither. Its control is similar to mustard sawfly. Green to black nymphs and adult **aphids** (*Lipaphis erysimi, Myzus persicae*) are found clustered on leaves sucking the sap which causes the plants to die in their severe attack. These can be controlled through fortnightly spraying with Dimethoate 30EC @ 2ml/l of water. **Cabbage worms**/borer (*Hellula undails*) is a whitish-brown caterpillar which mines and skeletonize radish leaves leaving its dark green excrement behind. These can be hand-picked and killed, *vis-à-vis* can be controlled through use of *Bacillus thuringiensis* which feed on its larvae.

Seed Production

Radish is a cross-pollinated entophilous crop (pollinated mainly by wild honey bees and flower flies) which shows considerable inbreeding depression on selfing. Stigma receptivity is maintained up to four days after anthesis. Selfing can be accomplished at bud stage through their own pollen prior to two days of opening whose buds were already bagged before opening. After pollination, the buds are to be protected from foreign pollen by enclosing the particular branch or plant in a muslin cloth bag. In crossing, the same technique is used as in bud-pollination except that in the crossing the buds of the female parent are emasculated a day prior to opening and are pollinated by pollen collected from the flowers of the male parent which were also bagged before opening. Pollination is carried out by shaking the pollen over the stigma directly from the freshly opened but previously bagged buds of male parent.

When a large quantity of crossed seed is required, the roots of female and male parent radishes are planted in alternate rows spaced 60cm apart. About 3-4 days before opening of buds, the plants are covered under an insect-proof wire net or plastic cage of 22-24 mesh. Usually 1-2 plants each of the male and female are covered under small cage. A small honeybee colony is introduced inside the cage 3-4 days before opening of buds. This method is followed when it is possible to rogue out the selfed or sib-mated plants in the seedling or root stage with the help of a dominant marker gene or when the male and female lines are homozygous for self-incompatibility alleles but are cross compatible. This procedure can also be used to produce sib-mated seeds to maintain a variety under insect-proof cages. However, in this case, it will be necessary to place about 20-30 plants under a cage to avoid inbreeding depression. A wire or plastic net cage of 3m×3m×2.5m (length, width and height) with a small door on one side is convenient for this purpose. Honey bee colony is also introduced here.

Mass-selection Breeding Method

This is practiced in landraces/cultivars collected from the farmer's field. Roots are allowed to reach an over-mature stage, then dug-up and gently the leaves are removed but growing points keeping intact. Bare roots after discarding the undesirable types are immersed in a water container from where only large sinking roots are retained for seed production in isolation though floating roots discarded.

Hybrid Breeding Method

This has been a common breeding approach in Japan where self-incompatible inbred lines are produced for 5-6 generations of selfing through bud-pollination. While selfing, only the true to type plants having root shape, size, colour and other quality parameters akin to original type are advanced to the next generation. The F_1 hybrids could be i) single crosses, ii) three-way crosses or iii) double crosses. In a practical breeding programme, the recommended isolation distance is 500-1000m to avoid outcrossing. Usually planting ratio of male and female lines is 1:1.

Seed Production Systems

Both 'root-to-seed' and 'seed-to-seed' systems are used in seed production. The root-to-seed is used for the biennial European types where the roots are lifted in the late autumn, the tops taken off and the radishes are stored usually in clumps during entire winter. Temperate varieties require low temperature (0-40C) for a period of 40-60 days to induce bolting and flowering (Singh, 2019). In such varieties seed production is restricted to temperate hilly regions by sowing seeds during summer and root planting during autumn, i.e. September-October when weather is cooling down. It is also the method used for stock seed production of the annual types (tropical varieties) but in this case the material is replanted immediately after selection but during winter itself. Tropical varieties can produce seeds in tropical as well as in temperate regions of India. Seeds are sown during September-october, lifted after maturity and then after examining the roots their 2/3rd lower portion is removed along with leaves without damaging the top and are immediately planted by maintaining the isolation distance of 1,600m apart to harvest true-to-type seeds. When some 70% pods have started yellowing the crop is harvested, dried and threshed. Now the seeds are cleaned, further dried, graded and packed in standard envelops. Seed yield is 5-6q/ha in case of temperate regions, however, seed yield in case of tropical varieties is little more. In some parts of the globe, especially in Asia, up to half of each steckling's root is removed before replanting. It has been suggested that for seed production of the 'Japanese White' cultivars this results in a higher seed yield.

The seed-to-seed system is used for final multiplication stages where inspections of the mature root are not considered necessary and is normally used only for spring sown seed crops unless the cultivar has a vernalization requirement.

White Radish Seeds

Photo courtesy: By Sanjay Acharya - Own work, CC BY-SA 4.0, https://commons.wikimedia.org/w/index.php?curid=76122410

Roguing Stages (seed-to-seed system)

- Roots at market maturity stage are lifted and those looking true to type, *i.e.* solidity, relative size, shape, colour, proportions of each colour on bi-coloured cultivars are retained for the purpose.
- At stem elongation stage, early bolters and off-types are removed and only true-to-type ones are retained.
- Retain the plants having buds as well flower colours true to the type while the rest are rejected.

Basic Seed Production

The root-to-seed system is used for basic seed production. When selecting plants according to their external morphology, care must be taken to ensure that those with pithy roots are rejected. These are checked for internal solidity through water immersion technique. Without damaging the growing points, the leaves are removed and roots planted. For this purpose 2-3 kg seed is required to raise roots for planting one hectare of area.

Seed Yield

Approximate seed yield is 10-15q/ha where 10g seed may contain some 100 seeds.

References

Aiton, W.T. 1812. *Hortus Kewensis (A Catalogue of the Plants Cultivated in the Royal Botanic Garden at Kew)*, vol. IV (2nd ed.), 129p. Longman, London.

Anders, K., J. Oegaard, M. Yasuhiko, O. Yoshio, and M. Yasushi, 1978. Volatiles in distillates of fresh radishes of Japanese and Kenyan origin. *Agric. Biol. Chem.*, **42**:1715–21.

Baenas, N., F. Ferreres, C. Garcia-Viguera and D.A. Moreno, 2015. Radish sprouts? Characterization and elicitation of novel varieties rich in anthocyanins. *Food Res. Int.*, **69**:305–12. doi: 10.1016/j.foodres.2015.01.009.

Baenas, N., S. Piegholdt, A. Schloesser, D.A. Moreno, C. Garcia-Viguera, G. Rimbach and A.E. Wagner, 2016. Metabolic activity of radish sprouts derived isothiocyanates in *Drosophila melanogaster. Int. J. Mol. Sci.*, **17**:251. doi: 10.3390/ijms17020251.

Baron, A. L. 1950. Handbook of Antibiotics, 215p. Reinhold, England.

Beevi, S.S., L.N. Mangamoori, V. Dhand and D.S. Ramakrishna, 2009. Isothiocyanate profile and selective antibacterial activity of root, stem and leaf extracts derived from *Raphanus sativus* L. *Foodborne Patholog. Dis.*, **6**:129–36. doi: 10.1089/fpd.2008.0166.

Bradley, S., J. Fenton-Smith, Lulu Grimes, Meredith Kirton, Judyth McLeod, Roger Mann and Kim Roney, 2016. Vegetable Gardening: Growing and Harvesting Vegetables, 242p. ISBN 978-1-74045-519-0. Murdoch Books, UK.. Archived from the original on 2016-06-17. Retrieved 2016-01-10.

Brickell, C. (ed.), 1992. The Royal Horticultural Society Encyclopedia of Gardening, pp. 356-7. Dorling Kindersely, London:

Bulinski, R. and M. Zhinda, 1962. Vitamin C in radishes. *Ann. Univ. Mariae Curie-Sklodowska*, **16**: 433–9.

Castro-Torres, Ibrahim Guillermo; Minarda de la O-Arciniega, Janeth Gallegos-Estudillo, Elia Brosla Naranjo-Rodríguez, Mioguel Angel Domínguez-Ortíz, 2013. *Raphanus* sativus L. var *niger* as a source of phytochemicals for the prevention of cholesterol gallstones. *Phytother. Res.*., **28** (2): 167–71. doi:10.1002/ptr.4964. ISSN 0951-418X. PMID 23495001. S2CID 11252934.

Changdai, W. 1984. Studies on chemical constituents in radish (*Raphanus sativus* L) seed II. *Shaanxi Xinyiyao*, **13**: 54–5.

Chen-Tien, C., C. Sheng-Ting, S. Rong-Jen, L. Hsiu-Chiu, C. Li-Ling and S. Hsien-Yi,. 2000. Purification and properties of chitinases from radish roots. *Food Sci. Agric. Chem.*, **2**: 107–14.

Coeley, D.E., I.D. Jenkins, J.K. MacLeod, R.E. Summons, D.S. Lethman and C.W. Parker, 1975. Structure and synthesis of unusual cytokinin metabolites. *Tetrahedron Lett.*, pp. 1015–18.

commons.wikimedia.org/w/index.php?curid=59306701.

commons.wikimedia.org/wiki/Category:Raphanus_sativus_var._oleiformis.

Creason, G.L., J.T. Madison and J.F. Thompson, 1985. Soybeans and radish leaves contain only one of the sulfonium diateroisomers of S-adenosylmethionine. *Phytochem.*, **24**: 1151–5.

Dahlbender, B. and D. Strack, 1986. Purification and properties of 1-(hydroxycinnamoyl)-glucose: 1- (hydroxycinnamoyl)-glucose hydroxycinnamoyl- transferase from radish seedlings. *Phytochem.*, **25**: 1043-6.

Daniel, Z. and H. Maria, 2000. Domestication of Plants in the Old World (3rd ed.), 139p. Oxford University Press, Oxford.

Daun, J.K. and F.W. Hougen, 1977. Identification of sulfur compounds in rapeseed oil. *J. Am. Oil Chem.* Soc., **54**: 351–4.

Dawson, G.W., A.J. Hick, R.N. Bennett, A. Donald, J.A. Pickett and R.M. Wallsgrove, 1993. Synthesis of glucosinolate precursors and investigations into the biosynthesis of phenylalkyl and methylthioalkyl glucosinolates. *J. Biol. Chem.*, **268**:27154–9.

Dixon, G.R. 2007. Vegetable Brassicas and Related Crucifers. *Crop Prod. Sci. Horti.*, **14**: 35 CABI, Wallingford, England.

ecoursesonline.iasri.res.in/mod/page/view.php?id=101592.

El-Hinnawy, I., S. Barakat and K.R. Fouad, 1975. Biochemical studies on some oleaginous seeds of Cruciferae plants. *Grasas Aceites*, **26**: 147–9.

en.wikipedia.org/wiki/Black_radish.

en.wikipedia.org/wiki/Daikon.

en.wikipedia.org/wiki/Gegeol_radish.

en.wikipedia.org/wiki/Korean_radish.

en.wikipedia.org/wiki/Radish.

en.wikipedia.org/wiki/Raphanus_raphanistrum.

en.wikipedia.org/wiki/Root_vegetable.

Fahey, J.W., A,T, Zalcmann and P. Talalay,.2001. The chemical diversity and distribution of glucosinolates and isothiocyanates among plants. *Phytochem.*, **56**: 5–51.

Faust, Joan Lee, 1996. Hail the Speedy Radish in All Its Forms. *The New York Times* (Section WC), p. 13. https://www.nytimes.com/1996/03/03/nyregion/gardening-hail-the-speedy-radish-in-all-its-forms.html.

Faye, L., B. Mouatassim and A. Ghorbel, 1986. Cell wall and cytoplasmic isozymes of radish- β-fructosidase have different N-linked oligosaccharides. *Pl. Physiol.*, **80**: 27–33.

Field, B., G. Cardon, M. Traka, J. Botterman, G. Vancanneyt and R. Mithen, 2004. Glucosinolate and amino acid biosynthesis in *Arabidopsis. Pl. Phys.*, **135**: 828–39. doi: 10.1104/pp.104.039347.

Fitzgerald, J.J. and W.J.M. Black, 1984. Finishing Store Lambs on Green Forage Crops: 1. A Comparison of Rape, Kale and Fodder Radish as Sources of Feed for Finishing Store Lambs in Autumn". *Irish J. agr. Res.*, **23** (2/3): 127–36.

Florey, Howard, 1949. Antibiotics: A Survey of Penicillin, Streptomycin, and other Antimicrobial Substances from Fungi, Actinomycetes, Bacteria and Plants, pp. 625-6. Oxford University Press, England.

Force, L.E., T.J. O'Hare, L.S. Wong and D.E. Irvin, 2007. Impact of cold storage on glucosinolate levels in seed-sprouts of broccoli, rocket, white radish and kohl-rabi. *Postharvest Biol. Technol.*, **44**:175–8. doi: 10.1016/j.postharvbio.2006.11.014.

Genichiro, S., Y. Kiyoshi, T. Daisuke, M. Denichi and O. Haruyuki, 1991. *Eur. Pat. Appl.*, EP. 462,022 (Cl. A61K37/20), 18 dec 1991, JP Appl. 90/155,428, 15 jun 1990.

Gopalakrishnan, T.P. 2007. Vegetable Crops, pp. 244-7. New India Publishing Agency, NBew Delhi. ISBN 978-81-89422-41-7. Archived from the original on 2016-05-20. Retrieved 2016-01-10.

Goyeneche, R., S. Roura, A. Ponce, A. Vega-Galvez, I. Quispe-Fuentes, E. Uribe and K. Di Scala, 2015. Chemical characterization and antioxidant capacity of red radish (*Raphanus sativus* L.) leaves and roots. *J. Funct. Foods.*, **16**: 256–64. doi: 10.1016/j.jff.2015.04.049.

Grubben, G.J.H. 2004. Vegetabkles, PROTA, 444p. ISBN 978-90-5782-147-9.

Guisti, M.M., H. Ghanadan and R.E. Wroslstad, 1998. Elucidation of the structure and conformation of red radish (*Raphanus sativus*) anthocyanins using one-and-two dimensional nuclear magnetic resonance techniques. *J. Agric. Food Chem.*, **46**: 4858–63.

Hanlon, P.R., D.M. Webber and D.M. Barnes, 2007. Aqueous extract from Spanish black radish (Raphanus sativus L. var. niger) induces detoxification enzymes in the HepG2 human hepatoma cell line. *J. Agric. Food Chem.*, **55**:6439–46. doi: 10.1021/jf070530f.

Hanlon, P.R. and D.M. Barnes, 2011. Phytochemical composition and biological activity of 8 varieties of radish (*Raphanus sativus* L.) sprouts and mature taproots. *J. Food Sci.*, **76**(1): C185-92. doi: 10.1111/j.1750-3841.2010.01972.x. PMID: 21535648.

Hans, M.R., P.W. Cornelis, S.W. Martinus, S. Lolke, W.F. Broekaert, R.S. Bronwen and V.W. Martinus, 1997. Antifungal protein fragment-derived peptides and their agricultural, therapeutic or preservative uses. *UK PCT. Appl. Wo*, **97** (21): 815 (cl. Cl2N15/29).

Harborne, J.B. 1963. Flavonoid sophorosides. *Experientia,* 19: 7–8.

https://nhb.gov.in/pdf/vegetable/radish/rad011.pdf.

https://plantvillage.psu.edu/topics/radish/infos.

https://www.almanac.com/plant/radishes.

iivr.icar.gov.in/kashi-mooli-40-vrrad-203.

Isikura, N., T. Hoshi and K. Hayashi, 1965. Anthocyanins. XLV. Crystallization and characterization of the basic triglucoside common to all components in purple pigment of hybrid radish. *Bot. Mag.,* **78**: 8–13.

Ivanovics, G. and S Horvath, 1947. Isolation and Properties of Raphanin, an Antibacterial Substance from Radish Seed. *Experim. Biol. Med.*, **66**(3): 625–30. doi:10.3181/00379727-66-16177. PMID 18900045. S2CID 32292410.

Jaroslav, V. 1990. Content of alkenylglucosinolates in oil-bearing Cruciferae seeds. *Rostl. Vyroba.*, **36**: 227-33.

Junlian, T., J. Qingjun and L. Jiansheng, 1992. Extraction and stability of edible pigment from red heart radish. *Huaxue Shijie*, **33**: 114–7.

Kamil, K. and K. Kalina, 1977. Flavonoid heterosides in the herb of *Raphanus raphanistrum* L. *Herba Pol.*, 23: 291–3.

Keishiro, W., O. Oka and M.H. Hirozo, 1985. Ferredoxin isoproteins and their variation during growth of higher plants. *Physiol. Veg.*, **23**: 679–86.

Kim, S., M. Woo, M. Kim, J.S. Noh and Y.O. Song, 2017. Hot water extracts of pressure-roasted dried radish attenuates hepatic oxidative stress via Nrf2 upregulation in mice fed high-fat diet. *Food Sci. Biotechnol.*, **26**:1063–9. doi: 10.1007/s10068-017-0135-x.

Kim, Haeng-Ran, Ji-Hyun Lee, Kim Yang-Suk and Kyung-Mi Kim, 2007. Chemical characteristics and enzyme activities of Icheon Ge-Geol radish, Gangwha turnip and Korean radish. *Korean J. Food Sci. Technol.*, **39**(3): 255–9. ISSN 0367-6293.

Kim, S.H. and S.S. Kim, 1996. Carbohydrate moieties of three radish peroxidases. *Phytochem.*, **42**: 287–90.

Kliebenstein, D.J., J. Kroymann, P. Brown, A. Figuth, D. Pedersen, J. Gershenzon and T. Mitchell-Olds, 2001. Genetic control of natural variation in Arabidopsis glucosinolate accumulation. *Plant Physiol.*, **126**:811–25. doi: 10.1104/pp.126.2.811.

Koji, H. and T. Shigenori, 1989. Phototropism in hypocotyls of radish. VII. Involvement of the growth inhibitors raphanusol A and R in phototropism of radish hypocotyls. *J. Plant Physiol.*, **135**: 110–13.

Koley, T.K., Z. Khan, D. Oulkar, B.K. Singh, A. Maurya, B. Singh and K. Banerjee, 2020. High resolution LC-MS characterization of phenolic compounds and the evaluation of antioxidant properties of a tropical purple radish genotype. *Arab.. J. Chem.,***13**(1): 1355-66.

Kumar, R. and Roshni Patwa, 2018. Antioxidant activity of *Raphanus sativus* L.of Jhansi district, Uttar Pradesh, India. *Int. Res. J. Pharm.*, **9**(1): 98-101. DOI: 10.7897/2230-8407.09114.

Kwan, H., K. Byong, H.L. Cho and H. Kyu, 1995. Preparation of anthocyanin-type red pigment via tissue cultivation of *Raphanus sativus* L. *Repub. Korea KR*, **9**:504772.

Larkcom, Joy and Elizabeth Douglass, 1994. Oriental Vegetables: The Complete Guide for the Gardening Cook, pp. 114-5. Oxford University Press, USA. ISBN 1-56836-017-7.

Lee, K.B., Y.J. Kim, H.J. Kim, J. Choi and J.K. Kim, 2018. Phytochemical profiles of Brassicaceae vegetables and their multivariate characterization using chemometrics. *Appl. Biol. Chem.*, **61**:131–44. doi: 10.1007/s13765-017-0340-6.

Lee, M.Y. and S.S. Kim, 1994 Characteristics of six isoperoxidases from Korean radish root. *Phytochem.*, **35**: 287–90.

Lewis-Jones, L.J., J.P. Thorpe and G.P. Wallis, 1982. Genetic divergence in four species of the genus *Raphanus:* Implications for the ancestry of the domestic radish R. *sativus. Biol. J. Linn. Soc.*, **18**(1): 35–48. doi:10.1111/j.1095-8312.1982.tb02032.x.

Li, R., D. Song, F. Vriesekoop, L. Cheng, Q. Yuan and H. Liang, 2017. Glucoraphenin, sulforaphene and antiproliferative capacity of radish sprouts in germinating and thermal processes. *Eur. Food Res. Technol.*, **243**: 547–554. doi: 10.1007/s00217-016-2764-3.

Loic, F., M. Bouchaib and G. Abdelwahed, 1986. Cell wall and cytoplasmic isosymes of radish. β-fructosidase have different N-linked oligosaccharides. *Pl. Physiol.*, **80**: 27–33.

Lugasi, A. and J. Hovari, 2000. Flavonoid aglycons in foods of plant origin. I. Vegetables. *Acta Aliment.*, **29**: 345–52.

Luo, X., H. Zhang, Y. Duan and G. Chen, 2018. Protective effects of radish (*Raphanus sativus* L.) leaves extract against hydrogen peroxide-induced oxidative damage in human fetal lung fibroblast (MRC-5) cells. *Biomed. Pharmacother.*,**103**: 406–14. doi: 10.1016/j.biopha.2018.04.049.

Magda, Gamba, Eralda Asllanaj, Peter Francis Raguindin, Marija Glisic, Oscar H. Franco, Beatrice Minder, Weston Bussler, Brandon Metzger, Hua Kern and Taulant Muka, 2021. Nutritional and phytochemical characterization of radish (*Raphanus sativus*): A systematic review. *Trends Food Sci. & Techn.*, **113**: 205-218. https://doi.org/10.1016/j.tifs.2021.04.045.(https://www.sciencedirect.com/science/article/pii/S0924224421003058).

Mal-Nam, K., J. Ji-Chul, L. Ik-Mo, L. Han-Sup and Y. Jin-San, 2002. Toxicity and biodegradation of diamines. *J. environ. Sci. Health*, **B37**: 53–64.

Manivannan, A., J.H. Kim, D.S. Kim, E.S. Lee and H.E. Lee, 2019. Deciphering the Nutraceutical Potential of Raphanus sativus-A Comprehensive Overview. *Nutrients;* **11**(2): 402. doi: 10.3390/nu11020402. PMID: 30769862; PMCID: PMC6412475.

Marquardt, P. 1976. N-methylphenethylamine in vegetables. *Arzneimittelforschung*, **26**: 201–3.

Matile, P. 1975. The Lytic Compartment of Plant Cells, pp. 56-63. Springer-Verlag, New York.

Matsuura, Y. and C. Hatanaka, 1988 Weakly acidic pectic polysaccharides of Japanese radish and cabbage. *Agric. Biol. Chem.*, **52**: 2583–8.

McIntosh, Charles, 1828. The Practical Gardener, and Modern Horticulturist, 288p. Thomas Kelly, London. Retrieved October 2, 2014.

Michael Meyer and T. Adam Sieghard, 2008. Comparison of glucosinolate levels in commercial broccoli and red cabbage from conventional and ecological farming. *Eur. Food Res. & Techn.*, **226**(6): 1429–37. doi:10.1007/s00217-007-0674-0. S2CID 98687281.

missouribotanicalgarden.org1:https://www.missouribotanicalgarden.org/PlantFinder/PlantFinderDetails.aspx?taxonid=263991&isprofile=0&chr=19.

missouribotanicalgarden.org2:https://www.missouribotanicalgarden.org/PlantFinder/PlantFinderDetails.aspx?kempercode=a762.

Nakamura, Y., K. Nakamura, Y. Asai, T. Wada, K. Tanaka, T. Matsuo, S. Okamoto, J. Meijer, Y. Kitamura and A. Nishikawa, 2008. Comparison of the glucosinolate?myrosinase systems among daikon (*Raphanus sativus*, Japanese White radish) varieties. *J. Agric. Food Chem.*, **56**: 2702–7. doi: 10.1021/jf7035774.

Nakayama, M., H. Yamane, T. Yokota, I. Yamaguchi, N. Murofushi, N. Takahashi, T. Nishijima, N. Katsura, M. Nonaka, P. Gaskin, J. MacMillan, L.N. Mander and A. Cu, 1990. Endogenous gibberellins in mature seed of *Raphanus sativus* L. *cv.* Taibyo-sobutori. *Agric. Biol. Chem.*, **54**: 837–40.

Narbut, S.I., G.B. Samorodova and V.S. Fedorov, 1972. Root and corolla color in the plants of varieties and inbred lines of radish. *Vestn. Leningr. Univ. Biol.*, **2**: 128–39.

Nishio, T. 2017. The Radish Genome (eds Takeshi Nishio and Hiroyasu Kitashiba), pp. 3-4. Springer, Holland. ISBN 978-3-319-59253-4.

npgsweb.ars-grin.gov/gringlobal/taxon/taxonomydetail?id=30856.

npgsweb.ars-grin.gov/gringlobal/taxon/taxonomydetail?id=457323.

Ozawa, Y., S. Kawakishi, Y. Uda and Y. Maeda, 1990. Isolation and identification of a novel β-carboline derivative in salted radish roots, *Raphanus sativus* L. *Agric. Biol. Chem.*, **54**: 1241–5.

Pajak, P., R. Socha, D. Gałkowska, J. Roznowski and T. Fortuna, 2014. Phenolic profile and antioxidant activity in selected seeds and sprouts. *Food Chem.*, **143**: 300–06. doi: 10.1016/j.foodchem.2013.07.064.

Perez Gutierrez, R.M. and R.L. Perez, 2004. *Raphanus sativus* (radish): their chemistry and biology. *The ScientificWorld J.*, **4**: 811–37.

Peterson, Cass, 1999. Radishes: Easy to Sprout, Hard to Grow Right. *The New York Times* (May 2, 1999), Section 9, page 8. Archived 2018-02-05 at the Wayback Machine. Retrieved on March 7, 2023 via https://www.nytimes.com/1999/05/02/style/cuttings-radishes-easy-to-sprout-hard-to-grow-right.html.

pfaf.org/user/Plant.aspx?LatinName=Raphanus+sativus+niger..

Pocasap, P., N. Weerapreeyakul and S. Barusrux, 2013. Cancer preventive effect of Thai rat-tailed radish (*Raphanus sativus* L. *var. caudatus* Alef). J. *Funct. Foods,* **5**:1372–81. doi: 10.1016/j.jff.2013.05.005.

Polya, G.M., S. Chandra and R. Condron, 1993. Purification and sequencing of radish seed calmodulin antagonists phosphorylated by calcium-dependent protein kinase. *Pl. Physiol.*, **101**: 545–51.

Price Andrew, J. and Jason K. Norsworthy, 2013. Cover Crops for Weed Management in Southern Reduced-Tillage Vegetable Cropping Systems. *Weed Techn.*, **27**(1): 212–7. doi:10.1614/WT-D-12-00056.1. S2CID 86661605.

Rakhmawati, R., E. Anggarwulan and E. Retnaningtyas, 2009. Potency of Lobak leaves (*Raphanus sativus* L. var. Hortensis Back) as anticancer and antimicrobial candidates. *Biodivers. J. Biol. Divers.*, **10**: 158-62. doi: 10.13057/biodiv/d100310.

Ricardo, L.L., D.I. Bernardi, G.C. Mantovanelli, B.P. Moreno, M.S. Mito, A.A. Silva, R.R. de Oliveira, Jr., E.L. Ishii-Iwamoto, M.H. Sarragiotto and D.C. Baldoqui, 2018. Phytochemical investigation and phytotoxic activity of aerial parts of oilseed radish (*Raphanus sativus* var. *oleifer* Stokes). *Biochem. Syst. Ecol.*, **78**: 52–8. doi: 10.1016/j.bse.2018.03.009.

Rodriguez, S.L., M.M. Giusti, R.W. Durst and R.E. Wrolstad, 2001. Development and process optimization of red radish concentrate extract as potential natural red colorant. *J. Food Process. Preserv.*, **25**: 165–82.

Salah-Abbes, J.B., S. Abbes, H. Zohra, R. Oueslati, 2015. Tunisian radish (*Raphanus sativus*) extract prevents cadmium-induced immunotoxic and biochemical alterations in rats. *J. Immunotox.*, **12**: 40–47. doi: 10.3109/1547691X.2014.880534.

Sang, J.P., I.R. Minchinton, P.K. Johnstone and R.J.W. Truscott, 1984 Glucosinolate profiles in the seed, root and leaf tissue of cabbage, mustard, rapeseed, radish and swede. *Can. J. Plant Sci.*, **64**: 77–93.

Seaman Abby, 2013. Turnips and Radishes (Chapter 28). In: Integrated Crop and Pest Management Guidelines for Commercial Vegetable Production. Cornell Cooperative Extension. Archived from the original on July 23, 2014. Retrieved 2014-07-29. https://web.archive.org/web/20140723053939/http://veg-guidelines.cce.cornell.edu/28frameset.html.

Seiichiro, N., T. Kenjiro, N. Ikuko and M. Kayo, 1984. Lipids in vegetables. III. Variation of lipids and fatty acid composition in seed and growth stage of Chinese vegetable (part 2). *Kenkyu Hokoku-Miyazaki Daigaku Nogakubu,* **31**: 45–56.

Sekimata, M., K. Ogura, Y. Tsumuraya, Y. Hashimoto and S. Yamamoto, 1989. A β-galactosidase from radish (*Raphanus sativus* L.) seeds. *Plant Physiol.*, **90**: 567–74.

Shigeo, A. and M. Yuhei, 1976. β-Amylase of the Japanese radish and soybeans. *Tampakushitsu Kakusan, Koso, Bessatsu*, **6**: 434–6.

Shikita, Mikio, Jed W. Fahey, Tamara R. Golden, W. David Holtzclaw and Paul Talalay, 1999. An unusual case of 'uncompetitive activation' by ascorbic acid: purification and kinetic properties of a myrosinase from *Raphanus sativus* seedlings. *Biochem. J.*, **341**(3): 725–32. doi:10.1042/bj3410725. ISSN 0264-6021. PMC 1220411. PMID 10417337.

Shyamala, G. and P.N. Singh, 1987. An analysis of chemical constituents of *Raphanus sativus. Proc. Natl. Acad. Sci. India, Sect.* B, **57**: 157–9.

Siddiq A. and I. Younus, 2018. The radish, *Raphanus sativus* L. var. caudatus reduces anxiety-like behavior in mice. *Metab. Brain Dis.*, **33**: 1255–60. doi: 10.1007/s11011-018-0240-4.

Singh, B. 2019. Radish. In: Handbook of Horticulture, vol. 1, pp. 604-8. ICAR, New Delhi.

Sinha, Nirmal, K., Y.H. Hui, Siddiq Muhammad and Jasim Ahmed, 2010. Handbook of Vegetables and Vegetable Processing, 156p. John Wiley and Sons, USA. ISBN 978-0-8138-1541-1.

Soundararajan, P. and J. Kim, 2018. Anticarcinogenic glucosinolates in cruciferous vegetables and their antagonistic effects on prevention of cancer. *Molecules,* **23**:2983. doi: 10.3390/molecules23112983.

Stoehr, H. and K. Herrmann, 1975. Phenolic acids of vegetables. III. Hydroxycinnamic and hydroxybenzoic acids of root vegetables. Z. *Lebensm. Unters. Forsch.*, **159**: 219–24.

Strack, D., M. Pieroth, H. Scharf and V. Sharma, 1985. Tissue distribution of phenylpropanoid metabolism in cotyledons of *Raphanus sativus* L. *Planta,* **164**: 507–11.

Tamura, G., T. Iwasawa, M. Masada and K. Fukushima, 1976. Some properties of cysteine synthase from radish roots. *Agric. Biol. Chem.*, **40**: 637–8.

Terras, F..R., H.M. Schoofs, M.F. DeBolle, F. van Leuven, S.B. Rees, J. Vanderleyden, B.P. Cammue and W.F. Broekaert, 1992. Analysis of two novel classes of plant antifungal proteins from Radish (*Raphanus sativus* L.) seeds. *J. Biol. Chem.*, **267**: 15301–9.

Thompson, J.F., D.H. Turner and R.K. Gering, 1964. g-Glutamyl transpeptidase in plants. *Phytochem.*, **3**: 33–46.

Toshio, M., A. Yoshiko, I. Atsuko, S. Mayumi and I. Akio, 1984. Occurrence of thiamin-binding substances in plant foods. *Nippon Eiyo, Shokuryo Gakkaishi,* **37**: 139–43.

tropical.theferns.info/viewtropical.php?id=Raphanus+sativus.

tropicalfruitandveg.com/showdetail.php?srcname=Mogri.

Tsumuraya, Y., K. Nakamura, Y. Hashimoto and S. Yamamoto, 1984. Immunological properties of arabinogalactan proteins from leaves of Cruciferous plants. *Agric. Biol. Chem.*, **48**: 2915–7.

Tsumuraya, Y., K. Ogura, Y. Hashimoto, H. Muroyama and S. Yamamoto, 1988. Arabinogalactan-proteins from primary and mature roots of radish (*Raphanus sativus* L.). *Plant Physiol.*, **86**: 155–60.

Vitoria, A.P., P.J. Lea and R.A.Azevedo, 2001. Antioxidant enzymes response to cadmium in radish tissue. *Phytochem.*, **57**: 701–10.

Wan, C. 1984. Studies on chemical constituents in radish (*Raphanus sativus* L.) seeds. II. *Shaanxi Xinyiyao*, **13**: 54–55.

Warwick, Suzanne, I. and Ardah Francis, 2005. The biology of Canadian weeds. 132. *Raphanus raphanistrum*. L. *Canadian J. Pl. Sci.*, **85** (3): 709–33. doi:10.4141/P04-120.

Wei, J., H. Miao and Q. Wang, 2011. Effect of glucose on glucosinolates, antioxidants and metabolic enzymes in *Brassica sprouts. Sci. Hortic.*, **129**: 535–40. doi: 10.1016/j.scienta.2011.04.026.

Weilan, W., Z. Jin, L. Zhongda and L. Meng, 1987. Hypotensive constituents of Laifuzi (*Semen raphani*). *Zhongcaoyao,* **18**: 101–103.

Wein, K. 1964. Die Geschichte des Rettichs und des Radieschens. *Die Kulturpflanze* (in German), **12**: 31–74. doi:10.1007/BF02095781. S2CID 19391461.

Wrolstad, R.E., M.M. Giusti, S.L. Rodriguez and R. Durst, 2001. Anthocyanins from radishes and red-fleshed potatoes. *Symp. Ser.*, **775**: 66–9.

www.biologydiscussion.com/vegetable-breeding/radish-origin-botany-and-breeding-methods-india/68606.

www.eatingwell.com/article/7631118/what-is-daikon-and-how-can-i-use-it.

www.honestfoodtalks.com/korean-radish-recipe-pickled-mu.

www.nature-and-garden.com/gardening/black-radish.html.

Yinghua, Z., Y. Zhonghan and C. Zongxun, 1983. Studies on stigma pellicle glycoproteins of *Raphanus sativus* L. *Zhiwu Xuebao,* **25**: 544–50.

Yoichi, T., H. Yohichi, Y. Shigeru and S. Naoto, 1984. Structure of L-arabino-D-galactan contained glycoproteins from radish leaves. *Carbohydr. Res.*, **134**: 215–8.

Yoichi, T., H.Yoichi and Y. Shigeru, 1987. An L-arabino-D-galactan and an L-arabino-D-galactan contained proteoglycan from radish (*Raphanus sativus*) seeds. *Carbohydr. Res.*, **161**: 113–26.

Yuan, G., X. Wang, R. Guo and Q. Wang, 2010. Effect of salt stress on phenolic compounds, glucosinolates, myrosinase and antioxidant activity in radish sprouts. *Food Chem.*, **121**: 1014–9. doi: 10.1016/j.foodchem.2010.01.040.

Yuhei, M. 1973. Preliminary crystallographic data for a basic peroxidase from Japanese radish. *Mem. Res. Inst. Food. Sci. Kyoto Univ.*, **36**: 9–11.

Turnip
Brassica rapa var. *rapa* L.

Family: Brassicaceae.

Other common names: White turnip; in **Hindi**: Salgam, Shalgam or Shaljam.

Introduction and Origin

There is a lot of uncertainty about the origin of *Brassica rapa* L. because of its ability to survive as a feral plant worldwide, and distinguishing truly wild species from feral escapees is almost impossible. *B. rapa* probably originated 4000 to 6000 years ago in the Hindu Kush area of Central Asia, and domestication has produced modern vegetables and oil-seed crops (McAlvay, 2021; www.atlasobscura.com) such as Bok choy (*B.r.* subsp. *chinensis*), Bomdong (*B.r.* var. *glabra*), Choy sum (*B.r.* subsp. *parachinensis*), Field mustard (*B.r.* subsp. *oleifera*), Komatsuna (*B.r.* subsp. *perviridis*), Mizuna (*B.r.* var. *nipposinica*), Napa cabbage (*B.r.* subsp. *pekinensis*), Rapini or Broccoli rabe (*B.r.* var. *ruvo*), Tatsoi (*B.r.* subsp. *narinosa*), Turnip (*B.r.* subsp. *rapa*) and Yellow sarson (*B.r.* subsp. *trilocularis*) (en.wikipedia.org/wiki/Turnip).

Wild turnips and its relatives, mustards and radishes, are found over western Asia and Europe. Starting as early as 2000 BC, related oilseed subspecies of *B. rapa* like *oleifera* may have been domesticated several times from the Mediterranean to India, though these are not the same turnips cultivated for their roots (Sanderson,

2005). Although estimates of domestication dates are limited to linguistic analyses of plant names (Zohary *et al.*, 2012), edible turnips were possibly first cultivated in northern Europe and eventually spread east to China, and reached Japan by 700 AD (Sanderson, 2005). Turnip is one of the oldest cultivated vegetables and has been widely consumed in high-altitude localities in Asian countries (Liang *et al.*, 2006).

Turnips (*Brassica rapa*)

By thebittenword.com - https://www.flickr.com/photos/galant/2622027467/, CC BY 2.0, https://commons.wikimedia.org/w/index.php?curid=7462757

Turnip is an important root vegetable worldwide having white and fleshy edible taproot. The common name turnip is combination of 'turn' as in turned/ rounded on a lathe and 'neep', derived from Latin *napus*, the word for the plant. In Northern England, Scotland, Ireland, Cornwall and parts of Canada (Quebec, Newfoundland, Manitoba and the Maritimes), the word turnip (or neep) often refers to rutabaga (*Brassica napus*), also known as swede, a larger, yellow root vegetable in the same genus. In most of England, the smaller white vegetables are called turnips, while the larger yellow ones are referred to as swedes. Its small and tender turnip varieties are grown for human consumption while larger varieties are grown as feed for livestock. In the United States, turnips are almost the same but swedes are usually called rutabagas (en.wikipedia.org/wiki/Turnip). Across Eastern America, Asia and European regions, the turnip is considered as a famous nutritious root vegetable, and in ancient Roman and Greek times, it was cultivated as a staple food (Lo Scalzo *et al.*, 2008). It is grown as a biennial for seed production and as an annual for use as a vegetable (http://ecoursesonline.iasri.res.in). Top turnip producing states in India are Bihar, Punjab, Himachal Pradesh, Tamil Nadu and Haryana (https://www.agrifarming.in). It possesses chromosomes numbers as 2n=2x=20.

Root

Brassica rapa (turnip) is a quick growing biennial, out-crossing and mesopolyploid herbaceous plant generally grown for its white-fleshed and roughly globular taproot that gets developed underneath the flowering stems and leaf petioles. Turnip produces napiform type of taproot that is 5–20cm in diameter, 100-1000g in weight, lacks side roots and has a thin normal root underneath that is around 10cm long and is usually not eaten. Taproot size depends on the turnip variety and the duration for which it has been grown. The most common type of turnip has white-skinned taproot commonly with rosy purple, red or greenish hue on the above 1-6cm part, *i.e.* the root part that protrudes above the ground and is directly hit by the sun. This above ground part of the root gets developed from stem tissue though remains fused with the root (en.wikipedia.org/wiki/Turnip; ecoursesonline.iasri.res.in).

Upper part of the root (hypocotyls) and the lower part of the stem swell by the development of extensive little lignified secondary xylem. The neck is very short and the pith is in the form of an obtuse cone. The root shape in turnip varies from extreme long forms to flat forming types (https://www.slideshare.net).

Aerial part

The plant possesses erect stems and 8–12 leaves forming a crown. The leaves and petioles are hairy, coarse and yellowish-green (https://www.slideshare.net). Light green, thin, sparsely downy and hairy leaves grow directly from the above-ground shoulder of the root, with little or no visible crown or neck. The leaves can reach 30.5–35.5cm in length while the branching flower stems can reach 30.5–91.5cm (https://plantvillage.psu.edu). Leaves are oblong to oval and may be entire, serrate or even pinnate depending on cultivar and time of development because those formed early are less likely to be pinnate. Those forming on inflorescence are alternate, oblong or lanceolate and entire or dentate (https://www.slideshare.net).

Inflorescence

Inflorescence is a corymb with open flowers standing well above the unopened younger buds. Flowers are light-green and very small. They have 4 sepals and 4 petals where outer stamens are shorter with curled filaments (https://www.slideshare.net). Bolting of turnip plants occurs in late winter followed by formation of flower buds. The flowers are attached with short pedicel and are complete, perfect, small, pale yellow in large terminate and/or axillary panicle (https://www.scielo.br; Mishra *et al.*, 2021).

Turnip (flower)

By Brosen - Own work, CC BY 2.5, https://commons.wikimedia.org/w/index.php?curid=324618

Fertilization

Though turnip bears hermaphrodite flowers but varying levels of cross-fertilization takes place due to protogyny and high foraging of insects for nectars. Cross pollination usually occurs due to honey bees, wild honey bees, bumble bees and blow flies. Anthesis starts in the morning at about 7.30 AM and continuous up to 11.00 AM depending on the prevailing weather conditions. Anther dehiscence coincides with the anthesis. The stigma becomes receptive 2-3 days after anthesis (Mishra *et al.,* 2021). Fruit is siliqua. Seeds are very small and highly variable in colour (purple black, brighter red purple, *etc.*) (https://www.slideshare.net).

Classification

There are over 30 varieties of turnips differing in size, colour, flavour and usage. They can be categorized as large sized, small tender ones and the baby turnips or baby bunch turnips. Larger varieties are usually harvested young while they are still small. The baby turnips and other small types are usually grown for human consumption while the larger ones for livestock feed (https://www.scielo.br/j/bjft/a/pH7R3CmhYrcSjxmCRzJmrVN/?lang=en). Baby turnip roots are specialty varieties with mild flavour and very small (about 2.5-5cm in diameter) and can be white, gold, pink or purple-topped having crunchy white, yellow, orange or red flesh (https://www.foodhero.org). They do not keep well and are available only when freshly harvested. Most baby turnips can be eaten whole including their leaves, can be eaten raw in salads like radishes and other vegetables (https://www.foodhero.org). They have a mixed flavour of tart apple and radish and are harvested when they are much smaller than common turnips (https://balconygardenweb.com).

Some varieties of turnip such as *Brassica rapa* var. *ruvo* (rapini or broccoli rabe) native to Europe such as *B.r.* subsp. *chinensis* (bok choy, pak choi, pok choi or Chinese cabbage) and *B.r.* subsp. *pekinensis* (Chinese cabbage or napa cabbage) are grown specifically for their leaves that resemble mustard greens and have small or no storage roots. Similar to raw cabbage or radish, turnip leaves and roots have a pungent flavour but become milder after cooking (en.wikipedia.org/wiki/Chinese_cabbage; en.wikipedia.org/wiki/Bok_choy; en.wikipedia.org/wiki/Rapini; en.wikipedia.org/wiki/Turnip).

Turnips having white skin and a rosy purple top are the most common type. Their crisp white inner flesh and the greens have a mild and slightly sweet flavour which become milder when cooked. Scarlet turnips have a vibrant red colour and look like very large red radishes. They have sweet, crisp, white flesh with occasional splashes of red (https://www.foodhero.org). Scarlet Queen turnips are slightly flat in shape and mature in 40-45 days, with sweet and crispy flavour tasting great when paired with carrots, parsley, lemon and honey (https://balconygardenweb.com).

European/Temperate Type and Asiatic/tropical/subtropical type

There are two distinct groups of turnips: i) biennial, temperate or European type, and ii) annual, tropical or Asiatic type. The seed of former group can be produced in the hills and that of latter in plains. Many good varieties differing in shape and colour are now available for cultivation in both the groups.

Temperate or European Type

Temperate types are quick growing, early and low yielding but excellent in quality. They require temperate climate for seed production, therefore in India these produce seeds only in the hills. These are biennials because they require two seasons for seed production. Some important temperate varieties include 'Purple Top White Globe (PTWG)', 'Snow Ball', 'Golden Ball', 'Pusa Swarnima', 'Pusa Chandrima', 'Early Milan Ted Top' and 'Early White Milan' (http://ecoursesonline.iasri.res.in).

- **Early Milan Red Top** is an extra early maturing variety that becomes ready for harvesting within 45 days after sowing. It is very high yielding variety with deep flat roots having purplish red tops and white underneath, flesh is pure white, crisp and mildly pungent (http://ecoursesonline.iasri.res.in).

- **Golden Ball** was bred at IARI Regional Station, Katrain (Kullu Valley) and is temperate type. Its roots are globe-shaped, smooth, yellow with sweet and yellow flesh having mellow taste if harvested earlier when root size is 7.5-10cm, mildly sweet with almond-like flavour, usual diameter being 12.5-15cm, top is small, erect with cut leaves, maturing in 40-45 days with root yield from 200 to 250 q/ha (http://ecoursesonline.iasri.res.in; http://agropedia.iitk.ac.in; https://balconygardenweb.com).
- **Purple Top White Globe** is the most popular variety of the temperate group (http://agropedia.iitk.ac.in) developed at IARI. It is large-rooted and matures in 60-65 days after sowing with average root yield of 250-300q/ha (http://ecoursesonline.iasri.res.in). Its spicy-flavoured roots are almost round, large (10-15cm in diameter) and smooth where purplish-red skin on upper portion extending above the soil surface while lower portion is creamy, flesh is white, firm and crisp, the top is small, erect with cut leaves and is best for stews, braises and for dehydration (https://balconygardenweb.com).
- **Pusa Chandrima** is an early maturing and high yielding variety developed by hybridization between Japanese White (Asiatic type) x Snow Ball (European type) at IARI Regional Station, Katrain. Its roots are medium to large in size, flattish round, 8-9cm long and 9-10cm in diameter, smooth, white skinned, white flesh, fine grained, sweet and tender (http://ecoursesonline.iasri.res.in). Its top is medium and the leaves not so deeply cut. It is high yielder, *i.e.* 40t/ha. The crop matures in 50-60 days after sowing (http://agropedia.iitk.ac.in).
- **Pusa Swarnima** was developed between Japanese White (Asiatic type) x Golden Ball (European type) at IARI Regional Station, Katrain. Its roots are light creamy yellow, flattish round, 6-7cm in length and 7-8cm in diameter, and the flesh is pale-amber with fine texture and mild flavour (http://ecoursesonline.iasri.res.in). Its medium top has leaf blades not so deeply cut. Crop matures in 65-70 days and is best for sowing during June-July on the hills while November-December in the plains (http://agropedia.iitk.ac.in; https://www.amrafarms.com).
- **Snow Ball** is a medium-sized early variety with round, smooth and white roots. Its flesh is white, fine grained and tender with sweet taste (http://ecoursesonline.iasri.res.in) and the leaves are small and light green. It matures in 55 days (http://agropedia.iitk.ac.in).

Tropical, Subtropical or Asiatic type

These varieties are slow growing, high yielding but poor in quality, however they remain edible for a long period. These varieties do not require chilling for entering into reproductive phase. Thus they can produce seed both in temperate and sub tropical climate but are not economical to produce seeds in hills due to slow growing nature in the temperate climate and thus require more time, land occupation period and labour. Some tropical cultivars include: 'Pusa Kanchan' (A cross between European and Desi type), 'Pusa Swati' and 'Punjab Safed' (http://ecoursesonline.iasri.res.in).

- **Early Milan Red Top** is an extra early and high yielding variety maturing in 45 days. Its roots are deep flat with purplish red tops and white underneath. Flesh is pure white, well-grained, crisp and mildly pungent (http://agropedia.iitk.ac.in).
- **Punjab Safed 4** is an early variety developed at PAU, Ludhiana. Its roots are white, round, medium sized with mild in taste. It matures in about 60 days (http://ecoursesonline.iasri.res.in). Its average root yield is 20t/ha. It is commonly grown in Punjab and Haryana.
- **Pusa Kanchan** is a selection from the cross between Asiatic Red x Golden Ball developed at IARI, New Delhi. Its round and medium sized roots have red skin and light yellow or creamy yellow flesh; excellent in taste and flavour; roots do not develop pithiness even if harvesting is delayed (http://ecoursesonline.iasri.res.in).
- **Pusa Sweti** was developed at IARI, New Delhi. Its pure white attractive roots are flattish round with white, soft, fine-grained and crispy flesh with mild pungency. It matures in 45-60 days after sowing (http://ecoursesonline.iasri.res.in). It is suitable for sowing in September-October in the plains and gives an average yield of 20-30 t/ha (http://agropedia.iitk.ac.in).
- **Turnip L1** bears round, smooth, pure white, rat-tailed and crisp roots which have mild flavour. Its roots are ready for harvesting in 45-60 days with an average yield of 25t/ha. Its dark green leaves are erect and medium-sized.

Varieties Based on The Morphological Characters of Roots and Top

Source: http://ecoursesonline.iasri.res.in; https://www.slideshare.net

White -fleshed

- **Purple-topped**
  - **Flat type:** Purple Top Milan
  - **Globe type**: Purple Top White Globe

- **Green topped**
  - **Globe type:** Green Top White, Green Globe
  - **Long type:** Cowhorn
- **White topped**
  - **Flat type:** White Milan, White Flat Dutch
  - **Globe type:** White Stone, Quick Silver
  - **Half long type**: White Egg, White Gem
  - **Long type:** Lily White

Yellow-fleshed

- **Purple-topped, globe-type:** Aberdeen Purple Top
- **Bronze or green-topped, globe-type**: Aberdeen Green Top, Amber Globe
- **Yellow-topped, globe type**: Yellow Globe, Golden Ball, All Seasons

Foliage Cultivars

- Flat Japan, Shogoin, Seven Top

Uses

Turnip is grown for its edible roots and leaves that have a pungent flavour. The roots can be eaten raw or cooked as vegetable. It can be used as an ingredient in stews and soups. The shoots and leaves can be eaten fresh in salads or the entire plant can be used as forage for livestock. Mature turnips have a stronger taste and their texture becomes firmer and woody as compared to baby turnips, which have a sweeter taste and delicate texture. Turnip leaves resemble their close relative 'mustard greens' in flavour and are sometimes eaten as 'turnip greens'. Smaller leaves are preferred, but the bitter taste of larger leaves can be reduced by pouring off the water from the initial boiling and replacing it with fresh water. Baby turnips, also known as young turnips, are eaten raw in salads.

Nutritional Value

Many essential nutrients are present in the turnip greens which are not present in the turnip roots. Turnip greens are not only abundant in antioxidants such as carotenoids, xanthins, lutein, vitamin A and vitamin C, but are also an excellent source of vitamin K (https://www.scielo.br). It is reported that vitamin B complex, *i.e.* riboflavin; pantothenic acid and thiamine are abundantly present in the top greens of the turnip (Cartea *et al.*, 2010). Likewise, calcium, iron, copper and manganese are important mineral sources present in the fresh turnip greens (https://www.scielo.br).

Turnip greens, cooked, boiled, drained, without salt	
Nutritional value per 100g	
Energy	84 kJ (20 kcal)
Carbohydrates	4.4 g
Sugars	0.5 g
Dietary fiber	3.5 g
Fat	0.2 g
Protein	1.1 g
Vitamins	**Quantity (%DV†)**
Vitamin A equiv. beta-Carotene	381 μg (48%) 4575 μg (42%)
Thiamine (B1)	0.045 mg (4%)
Riboflavin (B2)	0.072 mg (6%)
Niacin (B3)	0.411 mg (3%)
Pantothenic acid (B5)	0.274 mg (5%)
Vitamin B6	0.18 mg (14%)
Folate (B9)	118 μg (30%)
Vitamin C	27.4 mg (33%)
Vitamin E	1.88 mg (13%)
Vitamin K	368 μg (350%)
Minerals	**Quantity (%DV†)**
Calcium	137 mg (14%)
Iron	0.8 mg (6%)
Magnesium	22 mg (6%)
Manganese	0.337 mg (16%)
Phosphorus	29 mg (4%)
Potassium	203 mg (4%)
Sodium	29 mg (2%)
Other constituents	**Quantity**
Water	93.2 g
Lutein	8440 μg
Units: μg = micrograms, mg = milligrams, IU = International units Percentages are roughly approximated using US recommendations for adults. ***Source***: USDA FoodData Central	

Turnips, cooked, boiled, drained, without salt	
Nutritional value per 100g	
Energy	92 kJ (22 kcal)
Carbohydrates	5.1 g
Sugars	3.0
Dietary fiber	2.0 g
Fat	0.1 g
Protein	0.7 g
Vitamins	**Quantity (%DV†)**
Thiamine (B1)	.027 mg (2%)
Riboflavin (B2)	.023 mg (2%)
Niacin (B3)	.299 mg (2%)
Pantothenic acid (B5)	.142 mg (3%)
Vitamin B6	.067 mg (5%)
Folate (B9)	9 μg (2%)
Vitamin C	11.6 mg (14%)
Minerals	**Quantity (%DV†)**
Calcium	33 mg (3%)
Iron	.18 mg (1%)
Magnesium	9 mg (3%)
Manganese	.071 mg (3%)
Phosphorus	26 mg (4%)
Potassium	177 mg (4%)
Sodium	16 mg (1%)
Zinc	.12 mg (1%)
Other constituents	**Quantity**
Water	93.6 g
Units: μg = micrograms, mg = milligrams, IU = International units †Percentages are roughly approximated using US recommendations for adults. ***Source:*** USDA FoodData Central	

Source: (en.wikipedia.org/wiki/Turnip)

Phytochemicals

The main constituents of *Brassica rapa* are glucosinolates, isothiocyanates (Afsharypuor and Tahmasian 2010; Padilla *et al.*, 2007; Pierre *et al.*, 2011), flavonoids (Francisco *et al.*, 2009; Jeong *et al.*, 2013; Romani *et al.*, 2006), indoles (Wu *et al.*, 2012), sulphur compounds, phenolics (Cartea *et al.*, 2010; Fernandes *et al.,* 2007), carbohydrates (Wu *et al.,* 2013a), and volatiles (mainly terpenes, esters, aldehydes, and ketones) (Francisco *et al.,* 2009; Lee *et al.,* 2013; Taveira *et al.,* 2009). The types and concentration of compounds varied a lot with the developmental stage as norisoprenoids, terpenes and aldehydes were found at higher concentration at the germination stage (9-day-old turnip sprouts), sulphur and nitrogenous compounds decreased when the plant reached maturity (Taveira *et al.,* 2009). Furthermore, herbivory pressure and high humidity always increased the abundance of secondary metabolites (Metz *et al.,* 2014).

Like other vegetables of the *Brassica* genus, turnip too is a rich source of glucosinolates which can be hydrolyzed into isothiocyanates *via* plant enzyme myrosinase (Hong and Kim, 2008). During processing, cutting, tissue chewing or when injured, the glucosinolates come into contact with myrosinases in the presence of water and give rise to breakdown products such as isothiocyanates, thiocyanates, nitriles, epithionitriles and oxazolidines which are responsible for the peculiar bitter and pungent taste of turnip leaves (Jones *et al.,* 2007). The persistent bitter taste and aftertaste of turnips are due to the progoitrin and gluconapin present in both turnip greens and tops (Francisco *et al.,* 2011).

In total, 20 glucosinolates and 16 isothiocyanates have been reported from turnip with progoitrin, gluconapoleiferin, gluconapin, 4-hydroxyglucobrassicin, glucobrassicanapin, glucobrassicin, gluconasturtiin and neoglucobrassicin as the main glucosinolates. The highest levels of glucosinolates and N-containing compounds are present in the reproductive organs whereas volatile compounds (terpenes and isoprenoids) are predominant in seedlings which decrease during development. Flavonoids, mainly present as quercetin, kaempferol and isorhamnetin derivatives, have been detected in turnip greens and tops but not in roots yet (https://ift.onlinelibrary.wiley.com/doi/10.1111/1750-3841.14417).

Glucosinolates and Isothiocyanates

Glucosinolates and isothiocyanates, the characteristic constituents of turnip, have shown diverse bioactivities especially for their well-known chemoprotective effect against certain cancers. However, some derivatives of these chemicals are harmful to human and animals because of their goitrogenic potency (Rosa *et al.,* 1997). Structurally, glucosinolates contain a common $C_6H_{11}O_5S$ moiety (1-thio-β-D-glucopyranose) which can be divided into aliphatic (methionine, alanine, valine, leucine, and isoleucine), aromatic (tyrosine and phenylalanine) and indolic (tryptophan) types based upon the amino acids involved in the structures (Clarke, 2010; Fahey *et al.,* 2001). Glucosinolates are unstable and easily converted to the degradation products isothiocyanates after being contacted with myrosinases. 2-Phenylethyl isothiocyanate is the major break down product in the root epidermis followed by 4-pentenyl isothiocyanate and 3-butenyl isothiocyanate that are present in higher content in leaves than other parts (Hong and Kim, 2008). The levels of glucosinolates and bitterness are found to synchronously increase with plant maturity (Jones *et al.,* 2007).

Glucosinolates and isothiocyanates reported in turnip

No.	Compounds	Origin	Detection	References
1	Progoitrin	leaves	LCMS	Francisco *et al.,* 2009; Padilla *et al.,* 2007; Taveira *et al.,* 2009; Yang and Quiros, 2010
2	Sinigrin	leaves	CC	Afsharypuor and Tahmasian, 2010; Yang and Quiros, 2010
3	Epiprogoitrin	leaves	LCMS	Francisco *et al.,* 2009; Padilla *et al.,* 2007
4	Glucoraphanin	leaves	LCMS	Francisco *et al.,* 2009; Padilla *et al.,* 2007
5	Gluconapoleiferin	leaves	LCMS	Francisco *et al.,* 2009; Lee *et al.,* 2013; Padilla *et al.,* 2007
6	Gluconapin	leaves	LCMS	Francisco *et al.,* 2009; Lee *et al.,* 2013; Padilla *et al.,* 2007; Yang and Quiros, 2010
7	4-Hydroxyglucobrassicin	leaves	LCMS	Francisco *et al.,* 2009; Padilla *et al.,* 2007;Yang and Quiros, 2010
8	Dihydrogluconapin	leaves	LCMS	Francisco *et al.,* 2009
9	Glucobrassicanapin	leaves	LCMS	Francisco *et al.,* 2009; Padilla *et al.,* 2007; Taveira *et al.,* 2009; Yang and Quiros, 2010
10	Glucotropaeolin	leaves	LCMS	Francisco *et al.,* 2009
11	Glucobrassicin	leaves	LCMS	Francisco *et al.,* 2009; Padilla *et al.,* 2007; Lee *et al.,* 2013; Yang and Quiros, 2010
12	Gluconasturtiin	leaves	LCMS	Francisco *et al.,* 2009; Lee *et al.,* 2013; Padilla *et al.,* 2007; Yang and Quiros, 2010
13	Neoglucobrassicin	leaves	LCMS	Francisco *et al.,* 2009; Lee *et al.,* 2013; Padilla *et al.,* 2007; Yang and Quiros, 2010,
14	Glucoiberin	leaves	LCMS	Padilla *et al.,* 2007
15	Glucoiberverin	leaves	LCMS	Padilla *et al.,* 2007
16	Glucoalyssin	leaves	LCMS	Padilla *et al.,* 2007; Yang and Quiros, 2010
17	Glucoerucin	leaves	LCMS	Lee *et al.,* 2013; Padilla *et al.,* 2007
18	4-Methoxyglucobrassicin	leaves, roots	LCMS	Lee *et al.,* 2013; Padilla *et al.,* 2007
19	Glucoberteroin	leaves, roots	LCMS	Lee *et al.,* 2013
20	Glucocochlearin	roots	LCMS	Yang and Quiros, 2010
21	2-Phenylethyl isothiocyanate	roots, seeds	GCMS	Afsharypuor and Tahmasian, 2010; Hong and Kim, 2008; Taveira *et al.,* 2009; Robin *et al.,* 2018

No.	Compounds	Origin	Detection	References
22	Butyl isothiocyanate	roots, leaves, seeds	GCMS	Afsharypuor and Tahmasian, 2010
23	4-Pentenyl isothiocyanate	root, leaves, seeds	GCMS	Afsharypuor and Tahmasian, 2010,
24	5-Methylthiopentyl isothiocyanate	roots	GCMS	Afsharypuor and Tahmasian, 2010
25	3-Butenyl isothiocyanate	seeds, leaves	GCMS	Afsharypuor and Tahmasian, 2010, Francisco *et al.*, 2009; Taveira *et al.*, 2009; Yang and Quiros, 2010; Robin *et al.*, 2018
26	Isopropyl isothiocyanate	seeds, sprouts	GCMS	Taveira *et al.*, 2009
27	Allyl thiocyanate	seeds, sprouts	GCMS	Taveira *et al.*, 2009
28	Allyl isothiocyanate	seeds, sprouts	GCMS	Taveira *et al.*, 2009; Robin *et al*; 2018
29	2-butyl isothiocyanate	seeds, sprouts	GCMS	Taveira *et al.*, 2009
30	3-Methylbutyl isothiocyanate	seeds, sprouts	GCMS	Taveira *et al.*, 2009
31	Pentyl isothiocyanate	seeds, sprouts	GCMS	Taveira *et al.*, 2009
32	Hexyl isothiocyanate	seeds, sprouts	GCMS	Taveira *et al.*, 2009
33	4-Methylpentyl isothiocyanate	seed, sprouts	GCMS	Taveira *et al.*, 2009
34	Heptyl isothiocyanate	seeds, sprouts	GCMS	Taveira *et al.*, 2009
35	3-Methylthiopropyl isothiocyanate	seeds	GCMS	Taveira *et al.*, 2009
36	Benzyl isothiocyanate	roots	CC	Robin *et al.*, 2018

GCMS: Gas chromatography–mass spectrometry; LCMS: Liquid chromatography–mass spectrometry; CC: Coupled chromatography.

Flavonoids

Another major type of compounds in turnip are flavonoids which are always present as glycosides and play crucial roles in the protection against UV, pigmentation, stimulation of nitrogen-fixing nodules and disease resistance. Of the 35 flavonoids reported from turnip: kaempferol, quercetin, and isorhamnetin are the most common aglycones (Llorach *et al.*, 2003; Vallejo *et al.*, 2004). The content of phenolic compounds (majority being flavonoids) vary significantly between different parts of turnips as turnip greens contain more phenolic compounds compared to turnip top (Francisco *et al.*, 2009). They are therefore present in high concentration in the epidermis of leaves and fruits (Crozier *et al.*, 2006; Pereira *et al.*, 2009). The flavonoids reported from turnip include 27 flavonols, two flavanones and six chalcones.

Flavonoids reported in *Brassica rapa*

No.	Compounds	Type	Origin	Detection	Ref.
37	kaempferol-3-*O*-triglucoside-7-*O*-glucoside	Flavonols	leaves	LCMS	Francisco *et al.*, 2009
38	kaempferol-3-*O*-sophoroside-7-*O*-glucoside	Flavonols	leaves	LCMS	Francisco *et al.*, 2009
39	kaempferol-3,7-di-*O*-glucoside	Flavonols	leaves	LCMS	Francisco *et al.*, 2009
40	kaempferol-3-*O*-sophoroside	Flavonols	leaves	LCMS	Francisco *et al.*, 2009
41	kaempferol-7-*O*-Glucoside	Flavonols	leaves	LCMS	Francisco *et al.*, 2009
42	kaempferol-3-*O*-(methoxycaffeoyl)sophoroside-7-*O*-glucoside	Flavonols	leaves	LCMS	Francisco *et al.*, 2009
43	kaempferol-3-*O*-(caffeoyl) sophoroside-7-*O*-glucoside	Flavonols	leaves	LCMS	Francisco *et al.*, 2009
44	kaempferol-3-*O*-(sinapoyl) sophoroside-7-*O*-glucoside	Flavonols	leaves	LCMS	Francisco *et al.*, 2009
45	kaempferol-3-*O*-(feruloyl) sophoroside-7-*O*-glucoside	Flavonols	leaves	LCMS	Francisco *et al.*, 2009
46	kaempferol-3-*O*-(*p*-coumaroyl) sophoroside-7-*O*-glucoside	Flavonols	leaves	LCMS	Francisco *et al.*, 2009
47	isomer of 43	Flavonols	leaves	LCMS	Francisco *et al.*, 2009
48	kaempferol-3-*O*-(methoxycaffeoyl sophoroside	Flavonols	leaves	LCMS	Francisco *et al.*, 2009
49	kaempferol-3-*O*-(caffeoyl) sophoroside	Flavonols	leaves	LCMS	Francisco *et al.*, 2009
50	kaempferol-3-*O*-(sinapoyl) sophoroside	Flavonols	leaves	LCMS	Francisco *et al.*, 2009
51	quercetin-3-*O*-(feruloyl) sophoroside	Flavonols	leaves	LCMS	Francisco *et al.*, 2009
52	quercetin-3-*O*-sophoroside-7-*O*-glucoside	Flavonols	leaves	LCMS	Francisco *et al.*, 2009
53	quercetine-3,7-di-*O*-glucoside	Flavonols	leaves	LCMS	Francisco *et al.*, 2009
54	quercetin-3-*O*-sophoroside	Flavonols	leaves	LCMS	Francisco *et al.*, 2009
55	quercetine-7-*O*-Glucoside	Flavonols	leaves	LCMS	Francisco *et al.*, 2009
56	quercetin-3-*O*-(methoxycaffeoyl)sophoroside-7-*O*-glucoside	Flavonols	leaves	LCMS	Francisco *et al.*, 2009

No.	Compounds	Type	Origin	Detection	Ref.
57	quercetin-3-*O*-(caffeoyl) sophoroside-7-*O*-glucoside	Flavonols	leaves	LCMS	Francisco *et al.*, 2009
58	quercetin-3-*O*-(sinapoyl) sophoroside-7-*O*-glucoside	Flavonols	leaves	LCMS	Francisco *et al.*, 2009
59	quercetin-3-*O*-(feruloyl) sophoroside-7-*O*-glucoside	Flavonols	leaves	LCMS	Francisco *et al.*, 2009
60	quercetin-3-*O*-(*p*-coumaroyl) sophoroside-7-*O*-glucoside	Flavonols	leaves	LCMS	Francisco *et al.*, 2009
61	isorhamnetin-3,7-di-*O*-glucoside	Flavonols	leaves	LCMS	Francisco *et al.*, 2009
62	isorhamnetin-3-*O*-glucoside	Flavonols	leaves	LCMS	Igarashi *et al.*, 2008
63	isorhamnetin-7-*O*-glucoside	Flavonols	leaves	LCMS	Francisco *et al.*, 2009
64	Liquiritin	Flavanones	roots	CC	Jeong *et al.*, 2013
65	Liquiritigenin	Flavanones	roots	CC	Jeong *et al.*, 2013
66	4,4′-dihydroxy-3′-methoxychalcone	Chalcones	roots	CC	Jeong *et al.*, 2013
67	Isoliquiritin	Chalcones	roots	CC	Jeong *et al.*, 2013
68	4′-O-*β*-D-glucopyranosyl-4-hydroxy-3′-methoxychalcone	Chalcones	aerial parts	CC	Ninomiya *et al.*, 2010
69	4′-O-*β*-D-glucopyranosyl-3′,4-dimethoxychalcone	Chalcones	aerial parts	CC	Ninomiya *et al.*, 2010
70	4,4′-di-O-*β*-D-glucopyranosyl-3′-methoxychalcone	Chalcones	aerial parts	CC	Ninomiya *et al.*, 2010
71	licochalcone A	Chalcones	roots	CC	Jeong *et al.*, 2013

Other Compounds

Several other types of compounds reported in turnip include 3 sulphur compounds, 5 indoles, 14 phenylpropanoids, 2 lignans, 4 phenyl glycosides, 1 C_{13} sesquiterpene, 7 carbohydrate derivatives, 2 diarylheptanoids and 1 phenanthrene derivative. Carbohydrates have shown various effects to act as enzyme-activated irreversible inhibitors (Wu *et al.,* 2013a), inflammatory modulators, hepatoprotective agents, glycosidase inhibitors and potential therapeutic agents against HIV, diabetes and cancers (Lopez-herrera *et al.,* 1997). Furthermore, carbohydrates are widely used in the food and cosmetic industries as emulsifiers because of their amphiphilic properties (Molinier *et al.,* 2004).

Other compounds reported in *Brassica rapa* are:

No.	Compounds	Type	Origin	Detection	Ref.
72	2,4-dithiapentane,	Sulphur	roots, leaves	GCMS	Pierre *et al.*, 2011; Taveira *et al.*, 2009
73	Dimethyldisulfide	Sulphur	adult plants, sprouts	GCMS	Pierre *et al.*., 2011; Taveira *et al.*., 2009
74	Dimethyltrisulphide	Sulphur	sprouts	GCMS	Taveira *et al.*, 2009
75	2-C-β-D-glucopyranosyl-1-methoxyindole-3-acetonitrile	Indoles	roots	CC	Wu *et al.*, 2012
76	6-hydroxy- 1- methylindole3-acetonitrile	Indoles	roots	CC	Wu *et al.*, 2012
77	caulilexin C	Indoles	roots	CC	Bang *et al.*, 2012; Wu *et al.*, 2012
78	Arvelexin	Indoles	roots	CC	Bang *et al.*, 2012; Wu*et al.*, 2012
79	Indoleacetonitrile	Indoles	roots	CC	Bang *et al.*, 2012
80	Dihydrosyringin	Phenylpropanoids	roots	CC	Wu *et al.*, 2013c
81	caffeic acid	Phenylpropanoids	leaves	LCMS	Francisco *et al.*, 2009
82	sinapic acid	Phenylpropanoids	leaves	LCMS	Francisco *et al.*, 2009
83	Syringing	Phenylpropanoids	roots	CC	Wu *et al.* 2013c
84	Triandrin	Phenylpropanoids	roots	CC	Wu *et al.*, 2013c
85	Sinapoylglucoside	Phenylpropanoids	leaves	LCMS	Francisco *et al.*, 2009
86	1,2-disinapoylgentiobioside	Phenylpropanoids	leaves	LCMS	Francisco *et al.*, 2009
87	1-sinapoyl-2-feruloylgentiobioside	Phenylpropanoids	leaves	LCMS	Francisco *et al.*, 2009
88	1,2,2⊠-trisinapoylgentiobioside	Phenylpropanoids	leaves	LCMS	Francisco *et al.*, 2009
89	1,2⊠-disinapoyl-2-feruloylgentiobioside.	Phenylpropanoids	leaves	LCMS	Francisco *et al.*, 2009
90	3-caffeoylquinic acid	Phenylpropanoids	leaves	LCMS	Francisco *et al.*, 2009
91	3-p-coumaroylquinin acid	Phenylpropanoids	leaves	LCMS	Francisco *et al*, 2009
92	3-phenylpropionitrile	Phenylpropanoids	seeds	GCMS	Robin *et al.*, 2015
93	sinapine thiocyanate	Phenylpropanoids	seeds	UPLC	Fu *et al.*, 2016
94	neoolivil-4-O-β-D-glucopyranoside	Lignans	roots	CC	Wu *et al.*, 2013c

No.	Compounds	Type	Origin	Detection	Ref.
95	Phillyrin	Lignans	roots	CC	Wu *et al.*, 2013c
96	benzyl α-D-fructofuranoside	Phenyl glycosides	roots	CC	Wu *et al.*, 2013c
97	benzyl-β-D-glucopyranoside	Phenyl glycosides	roots	CC	Wu *et al.*, 2013c
98	Picein	Phenyl glycosides	aerial parts	CC	Ninomiya *et al.*, 2010
99	Glucoacetosyringone	Phenyl glycosides	aerial parts	CC	Ninomiya *et al.*, 2010
100	corchoionoside C	C13 sesquiterpene	aerial parts	CC	Ninomiya *et al.*, 2010
101	sucrose 3,3⊠,4⊠-triisovalerate	Carbohydrate derivative	roots	CC	Wu *et al.*, 2013a
102	sucrose 6,3⊠,4⊠-triisovalerate	Carbohydrate derivative	roots	CC	Wu *et al*, 2013a
103	sucrose 2,6,3⊠,4⊠-tetraisovalerate	Carbohydrate derivative	roots	CC	Wu *et al*, 2013a
104	ethanone-1-C-β-D-glucopyranoside	Carbohydrate derivative	roots	CC	Wu *et al*, 2013a
105	ethyl β-D-glucopyranoside	Carbohydrate derivative	roots	CC	Wu *et al.*, 2013a
106	n-butyl β-D-fructofuranoside	Carbohydrate derivative	roots	CC	Wu *et al.*, 2013a
107	n-pentyl β-D-fructofuranoside	Carbohydrate derivative	roots	CC	Wu *et al.*, 2013a
108	6-paradol	Diarylheptanoids	roots	CC	Wu *et al.*, 2013b
109	trans-6-shogaol	Diarylheptanoids	roots	CC	Wu *et al.*, 2013b
110	brassicaphenanthrene A	Phenanthrene derivative	roots	CC	Wu *et al.*, 2013b

Volatiles

Defense chemicals present in plants are of two types: direct that are inherently present in plant and indirect that are herbivore-induced plant volatiles (HIPVs) types released by plants as self-protection when the herbivores attack them (Engelberth *et al.*, 2004). HIPVs are yielded to attract herbivore's natural enemies and are different from the volatile compounds which are released by the plants being mechanically damaged (Heil, 2008). Similar to legumes that release antimicrobial green leaf volatiles (C_6-aldehyde, alcohol and their esters) after being infected by bacterial pathogens (Croft *et al.*, 1993), turnip too is rich in volatiles which act as important defense signals against microbial and herbivore attack as 4-methyltridecane and salicylaldehyde were induced by root herbivory; methyl salicylate was mediated by shoot herbivory whereas hexylacetate (a green leaf volatile) was specifically induced after dual (root and shoot) herbivory in turnip (Pierre *et al.*, 2011).

Volatiles reported in turnip

No.	Compounds	Origin	Ref.
111	1-tetradecanol	leaves, roots	Pierre *et al.*, 2011
112	1-hexadecanol	leaves, roots	Pierre *et al.*, 2011
113	3-hexen-1-ol	leaves, roots	Pierre *et al.*, 2011
114	2-octen-1-ol	leaves, roots	Pierre *et al.*, 2011
115	1-hexanol	leaves, roots	Pierre *et al.*, 2011
116	1-octanol	leaves, roots	Pierre *et al.*, 2011
117	1-dodecanol	leaves, roots	Pierre *et al.*, 2011
118	1-phenyl ethanol	adult plants	Taveira *et al.*, 2009
119	Salicylaldehyde	leaves, roots	Pierre *et al.*, 2011
120	Octadecanal	leaves, roots	Pierre *et al.*, 2011
121	Hexanal	adult plants, sprouts, seeds	Taveira *et al.*, 2009
122	2-hexenal	adult plants, sprouts	Taveira *et al.*, 2009
123	Heptanal	adult plants, sprouts	Taveira *et al.*, 2009
124	2-heptenal	adult plants	Taveira *et al.*, 2009
125	(E,E)-2,4-heptadienal	adult plants	Taveira *et al.*, 2009
126	2,4-nonadienal	adult plants, sprouts	Taveira *et al.*, 2009
127	Octanal	adult plants	Taveira *et al.*, 2009
128	Phenylacetaldehyde	sprouts	Taveira *et al.*, 2009
129	2-octenal	adult plants	Taveira *et al.*, 2009
130	Nonanal	adult plants, sprouts, seeds	Taveira *et al.*, 2009
131	2-nonenal	adult plants	Taveira *et al.*, 2009
132	2-decenal	adult plants, sprouts	Taveira *et al.*, 2009
133	Undecanal	adult plants	Taveira *et al.*, 2009

No.	Compounds	Origin	Ref.
134	10-undecanal	adult plants	Taveira *et al.*, 2009
135	Dodecanal	adult plants	Taveira *et al.*, 2009
136	6-methyldodecane	leaves, roots	Pierre *et al.*, 2011
137	3-methylundecane	leaves, roots	Pierre *et al.*, 2011
138	2,6-dimethylundecane	leaves, roots	Pierre *et al.*, 2011
139	4-methyltridecane	leaves, roots	Pierre *et al.*, 2011
140	Heptadecene	leaves, roots	Pierre *et al.*, 2011
141	hexanoic acid	leaves, roots	Pierre *et al.*, 2011
142	acetic acid	leaves, roots	Pierre *et al.*, 2011
143	methyl salicylate	leaves, roots	Pierre *et al.*, 2011
144	Ethylacetate	leaves, roots	Pierre *et al.*, 2011
145	Hexylacetate	leaves, roots	Pierre *et al.*, 2011
146	(Z)-3-hexen-1-ol-acetate	leaves, roots	Pierre *et al.*, 2011
147	butyl acetate	adult plants, seeds	Taveira *et al.*, 2009
148	ethyl pentanoate	adult plants	Taveira *et al.*, 2009
149	ethyl hexanoate	adult plants	Taveira *et al.*, 2009
150	ethyl benzoate	sprouts	Taveira *et al.*, 2009
151	ethyl octanoate	adult plants, sprouts	Taveira *et al.*, 2009
152	ethyl decanoate	adult plants	Taveira *et al.*, 2009
153	ethyl dodecanoate	sprouts	Taveira *et al.*, 2009
154	ethyl tetradecanoater	sprouts	Taveira *et al.*, 2009
155	ethyl hexadecanoate	sprouts	Taveira *et al.*, 2009
156	ethyl linoleate	sprouts	Taveira *et al.*, 2009
157	(E,E)-4,8,12-trimethyl-1,3,7,11-tridecatetraene	leaves, roots	Pierre *et al.*, 2011
158	(3E)-4,8-dimethyl-1,3,7-nonatriene	leaves, roots	Pierre *et al.*, 2011
159	2-undecanone	leaves, roots	Pierre *et al.*, 2011
160	2,4-pentadione	leaves, roots	Pierre *et al.*, 2011
161	(E)-geranyl acetone	leaves, roots	Pierre *et al.*, 2011
162	2-nonanone	leaves, roots	Pierre *et al.*, 2011
163	Acetone	leaves, roots	Pierre *et al.*, 2011
164	Dihydrojasmone	adult plants	Taveira *et al.*, 2009
165	α-pinene,	adult plants, sprouts	Taveira *et al.*, 2009
166	Limonene	adult plants, seeds	Taveira *et al.*, 2009
167	β-pinene	adult plants	Taveira *et al.*, 2009
168	m-cymene	adult plants, sprouts	Taveira *et al.*, 2009
169	Isomenthone	adult plants	Taveira *et al.*, 2009
170	Menthol	adult plants, sprouts	Taveira *et al.*, 2009
171	Phytol	adult plants	Taveira *et al.*, 2009

No.	Compounds	Origin	Ref.
172	Benzenepropanenitrile	adult plants, sprouts, seeds	Taveira *et al.*, 2009
173	6-methyl-2-heptanone	adult plants, sprouts	Taveira *et al.*, 2009
174	2,2,6-trimethylcyclohexanone	adult plants, sprouts	Taveira *et al.*, 2009
175	Safranal	adult plants, sprouts	Taveira *et al.*, 2009
176	β-cyclocitral	adult plants, sprouts	Taveira *et al.*, 2009
177	β-homocyclocitral	adult plants, sprouts	Taveira *et al.*, 2009
178	Hexahydropseudoionone	adult plants	Taveira *et al.*, 2009
179	α-Ionone	adult plants	Taveira *et al.*, 2009
180	Geranylacetone	adult plants	Taveira *et al.*, 2009
181	β-Ionone	adult plants, sprouts	Taveira *et al.*, 2009
182	epoxy-β-ionone	adult plants, sprouts	Taveira *et al.*, 2009
183	Dihydroactinidiolide	adult plants	Taveira *et al.*, 2009
184	Hexahydrofarnesylacetone	adult plants	Taveira *et al.*, 2009
185	Caryophyllene	leaves, roots	Pierre *et al.*, 2011
186	(E,E)-alpha-Farnesene	leaves, roots	Pierre *et al.*, 2011
187	Longifolene	adult plants	Taveira *et al.*, 2009
188	Dihydroxybenzoquinone	leaves, roots	Pierre *et al.*, 2011

Taste

A bitter taste, particularly sulphurous aroma and pungent flavour are the characteristics of turnips. Likewise, numerous secondary metabolites such as phenolic acid in turnip tops and glucosinolates in turnip greens have been significantly correlating with the flavour and texture. The amount and pattern of the glucosinolates and volatiles in *Brassica* plants vary according to the plant species, cultivar and vegetable part, as well as to the developmental stage of the plant (Kessler and Baldwin, 2002).

Pharmacological Investigation

The major active constituents of turnip are glucosinolates, isothiocyanates, flavonoids and phenylpropanoids which show diverse bioactivities including antimicrobial (Hong and Kim, 2008), antioxidant (Taveira *et al.*, 2009), antitumour (Hong and Kim, 2008), anti-inflammatory (Shin *et al.*, 2011), cardioprotective, hypolipidemic (Bang *et al.*, 2012; Mirzaie *et al.*, 2012; Al-Snafi, 2015) hepatoprotective (Li *et al.*, 2010), antidiabetic (Daryoush *et al.*, 2011; Jung *et al.*, 2008), nephroprotective (Mohajeri *et al.*, 2013), and analgesic (Hosseini *et al.*, 2013) properties together with alleviation of obesity and metabolic syndrome.

Both glucosinolates and its derivatives are known to have a wide range of important biological activities (Padilla *et al.*, 2007); some have shown the chemoprotective effect against certain cancers in humans while some others are detrimental for human and animal consumption (potentially goitrogenic) (Padilla *et al.*, 2007; Taveira *et al.*, 2009). Some of the constituents assessed using crude extracts of turnip and their biological activities are ~ 2-phenylethyl isothiocyanate and isorhamnetin 3-*O*-glucoside which exhibit anti-cancer and hepatoprotective activity; and caulilexin C, indoleacetonitrile and arvelexin showing hypolipidemic and analgesic effects. Among the volatiles, β-ionone exhibits anticancer potency (Duncan *et al.*, 2004); hexahydrofarnesylacetone has bactericidal effects (Yu *et al.*, 2004); menthol has antimicrobial activity and is also useful as flavouring agents and pharmaceutical chemicals (Tholl, 2006); and phytol in having antispasmodic potency (Pongprayoon *et al.*, 1992). Although turnip shows great pharmaceutical potential, the current investigation is mainly concentrated on the bioactivity of crude extracts instead of pure compounds (https://ift.onlinelibrary.wiley.com/doi/10.1111/1750-3841.14417).

Cultivation

Climate

Turnip being a cool weather crop therefore in the plains it is grown during winter while in temperate regions during summer. The ideal temperature for its growing should be 10-15^{0}C for better root growth and good yield though there are varieties for warmer weathers but there roots are quite smaller so root yield is also drastically poor. Turnips require longer nights and shorter days with 6-8 hours of direct sunlight. In India, Nilgiri hills are suitable for turnip cultivation in the south whereas it can be cultivated seasonally at several places in northern parts such as Punjab, Haryana, Himachal, Uttar Pradesh, Bihar, the eastern part of India, and Jammu & Kashmir (https://www.amrafarms.com).

Soil

Turnips require full sun, moist but well-drained soil having high humus and organic matter, and a soil pH of 6-7 (https://plantvillage.psu.edu/). Sandy loam soil is best suited for its cultivation. Heavy soil is not recommended as the roots turn hard and fibrous.

Propagation

Turnips are propagated by seeds. Approximately 1.5kg of seeds is required to cultivate one hectare of land. Turnip seeds are directly sown in the field though antifungal treatment of seeds may be required before sowing. Immediately after sowing, the field should be irrigated (https://www.amrafarms.com).

Sowing Time

Turnip seeds are usually sown in the spring in cold-weather climates, such as the northern US and Canada where growing season is only 3–4 months. In temperate climates with a growing season of 5–6 months, turnips may also be sown during late summer for a second fall crop. In warm-weather climates having 7 or more month growing season, seeds are sown in the fall as 55–60 days is the average time from planting to harvest (https://en.wikipedia.org/wiki/Turnip).

The best time for sowing Asiatic type turnips is July to September and for European type it is October to December in plains while July to september in the Hills (https://krishijagran.com). Depending on the variety, crops can be cultivated once or twice a year. Early varieties can be cultivated twice if planned well. Some varieties can be cultivated in June and July in some areas. Nevertheless, it's important to make sure that the climatic conditions during this period are favourable for its growing (https://www.amrafarms.com).

Land Preparation and Planting

The land needs to be prepared to a fine tilth by deep ploughing, 3-4 harrowings, followed by leveling. Sowing the crop on ridges is usually preferred than flat as ridge sowing as this provides better aeration to roots *cum* drainage so there is better root development (Mishra *et al.,* 2021). Application of manure or compost is imperative. At the time of last ploughing about 20 tonnes of well-rotten farmyard or dung manure per hectare of land should be incorporated in the soil at the time of bed preparation. Rows are made with trenches to remove water. If drip irrigation can be provided that is well and good otherwise these may be irrigated through the channels made for the purpose. On bunds seeds are sown some 4cm deep at a distance of 30 cm in rows with row to row distance being 90cm (https://www.amrafarms.com).

Fertilizers

Turnips require moderate to low content of fertilizers. Initial 20t/ha of FYM should satisfy most of crop requirements but depending on the soil conditions growers may need to add fertilizers that are depleted. Commonly 100kg of nitrogen, 50kg of potash and 50kg of phosphorus are applied during entire course of crop growing. Half of the nitrogen along with whole of potash and phosphorus is applied as a basal application while the remaining 50kg of nitrogen is given in 2 splits ~ once when initial root thickening has initiated and remaining after 20 days when root knobs are in active growth. Micronutrient such as boron, calcium and molybdenum may be required at the rate of 1kg/ha so their 1-2 sprays may be applied depending upon soil condition.

Irrigation

First irrigation should be done immediately after sowing. Thereafter, depending on weather conditions, irrigation may be required every 8 to 12 days. Proper moisture in the soil is required for better development of turnip knobs in terms of quality and quantity as moisture stress reduces the yield drastically (https://www.amrafarms.com). However, over-watering and waterlogging should always be avoided.

Weed Control, Mulching and Intercropping

Depending on the weeds in the field and the growth of the turnip leaves, 2-3 weedings may be required to avoid competition of weeds with main crop. Once the turnip leaves are large enough, these do not permit weeds to grow. The weeding also helps loosening the soil which enables turnip roots developing better. Growers should pull away any weeds as soon as these appear and the soil should initially be cultivated 2.5–5cm deep when the plants are still small but afterwards this should be avoided to save the developing knob from injury as well as the feeder roots as these are very delicate (https://plantvillage.psu.edu/).

Leaving the row of seed sowing, remaining parts of the beds can be mulched with bagasse, dry leaves, straw, grasses and various other organic materials which will conserve moisture, will provide required temperature for root development and will suppress the growth of weeds, vis-à-vis after decomposition will provide organic matter to the next crop. Turnips can be intercropped with beetroots, carrots and radishes.

Harvesting and Yield

Depending on the varieties being grown, turnips are harvested after 40-65 days of sowing when the roots reach 5-10cm in diameter or the fully developed tender roots of turnip are uprooted on attaining the marketable size (http://eagri.org/eagri50). Smaller varieties are harvested when they reach 2.5–3.5cm in diameter (https://plantvillage.psu.edu/). Delayed harvesting makes the roots tough and fibrous. Harvesting is usually done in the evenings so that the roots retain maximum moisture content and freshness. After harvest, the roots should be thoroughly washed and the side roots trimmed. Dipping the roots in hot paraffin will help reduce shrinkage and improve the appearance (http://ecoursesonline.iasri.res.in).

Turnip greens can be harvested from the plant when the leaves are about 10cm tall. If turnips are being grown for greens and roots both, only 2–3 leaves from each plant should be removed but others left for knob development. Small turnips can often be harvested by gently pulling from the soil by hand but larger roots in heavier soils may be gently dug up with a garden fork (https://plantvillage.psu.edu/).

Yield may vary depending on the variety sown and the season when grown. Tropical turnips sown during June-July often provide a much lower yield than those grown during October- December. On an average the root-knob yield is 20-40t/ha (https://www.amrafarms.com).

Post Harvest

Turnips are sent to the market either along with green tops or after cutting them off near the surface of the crown. After removing old and diseased leaves the foliage is made into bunches for marketing. The roots are sorted into different grades according to colour, shape and size to give an attractive appearance in the market. Then these are immediately transported to the market in baskets. The roots can be stored safely for 2–3 days under cool and moist conditions. However, it can be stored for 8–16 weeks at 0°C with 90–95% relative humidity (http://eagri.org/eagri50). Most customers prefer purchasing those knobs which have intact leaves as this indicates freshness, albeit the roots should be quite clean and devoid of yellowed and deformed leaves.

Seed Production

Turnip is a biennial plant, requiring two years from germination to reproduction. The root spends the first year growing and storing nutrients and in the second year it flowers, produces seeds and dies. In areas with less than 7-month growing seasons, temperatures are too cold for the roots to survive the winter. To produce seeds, pulling the turnips and storing them over winter is necessary. During spring, their roots may be set back in the ground to complete their lifecycle (en.wikipedia.org/wiki/Turnip).

Asiatic turnips produce seeds in the plains though European types only on the hills. The techniques of raising seed crop of turnip and its seed production are the same as for radish. Similar to many other biennial crops, in turnip also two methods are used for the purpose of seed production and these are i) seed to seed method (*in-situ*) and, ii) root to seed method (through transplanting of roots). In India, turnip seeds are usually produced through root to seed method but in Japan, USA or European countries *in situ* method is employed for certified seed production (Mishra *et al.*, 2021).

Seed to Seed Method

Seed to seed method is used for certified seed production of those varieties which do not thrive well when transplanted. The source of seed used (foundation seed/certified seed) must be genetically pure and should never be used again for raising certified seed crops. The root crop is grown as for market purpose. The healthy and true-to-type plants should be retained in the field at a spacing of 45x30cm to overwinter and produce seeds in the following spring (Mishra *et al.*, 2021).

This method of seed production is economical and also avoids the extra costs being involved in lifting, storage and replanting. It also helps in early seed maturity thus providing enough time for growing next crop where seasons are short. This method though ensures more yield but the seeds produced have low genetic purity as systematic rouging and selection of true to type is not possible in this method (Mishra *et al.*, 2021).

Root to Seed Method

In this method, the roots produced in the first season are replanted after treatment to get the seeds in the second season.

In the first season stecklings are raised by sowing the good quality certified seeds during last week of August to first week of September. Extra early varieties can be sown by the end of September. Seed rate for raising seed production crop may be slightly higher, *i.e.* 2.5-4kg than for root production, for one hectare of area. The roots produced in one hectare field are sufficient for transplanting in 3 hectares of area.

The field should be prepared to a fine tilth by deep ploughing, three to four harrowing, followed by levelling. Sowing the crop on ridges is usually preferred than on flat beds as ridges provide better drainage and better development of roots, otherwise the roots are likely to become more branched, misshapen and malformed. For tropical varieties, row to row spacing should be 30-45cm and for temperate varieties it should be 40-45cm. The plants are thinned out after 10-15 days of sowing to maintain plant to plant spacing of 10-15cm in tropical and 8-10cm in temperate varieties (Mishra *et al.*, 2021).

It is recommended to apply fertilizer only after the soil test. A light dose of 20t/ha of farmyard manure should be applied at the time of last ploughing.

Before sowing of seeds, 35kg of nitrogen and 50kg each of phosphorus and potash should also be applied in the field as basal dose, and 35kg of nitrogen should be top-dressed when the rootknobbing just start forming. Seeds should be sown thinly (Mishra *et al.*, 2021).

In autumn, about 70-80 days after sowing, the fully developed roots should be uprooted before development of pith. After harvesting, the uprooted roots should be laid bare side by side to facilitate selection of true to type roots. Under-developed, damaged, diseased and off type roots should be discarded (Mishra *et al.*, 2021).

Second Season (Seed Production)

To avoid contamination by crossing with other related crops such as Chinese cabbage, mustard and other turnips, an isolation distance of 1,000-1,600m should be maintained from these crops. The stecklings are prepared from the selected roots by pruning the root tip from the base and leaving one-third of crown after trimming off the top (http://eagri.org/eagri50). The treated stecklings are either transplanted immediately in the field or stored at a temperature of 0^0C and at a relative humidity of 90-95 per cent (Mishra *et al.*, 2021). After treating, the roots are transplanted in well prepared field in the **second season only for seed production** in such a way that only the crown is exposed and rest of root is covered in the soil (Sadhu and Sarkar, 2003). These stecklings are transplanted 60×60cm apart in the hills and 45×35cm apart in the plains (http://eagri.org/eagri50). The soil is then firmed and tapped around the roots. Soon after transplanting the field is irrigated. As later plantings are prone to serious attacks of aphids, planting of stecklings should be finished before second half of December. It is also reported that planting in December is better for the growth and development of turnip seed crops, compared with the January planting as the seed yield usually gets decreased progressively with delayed planting (Singh *et al.*, 1990). For better seed yield, in the second season, about 10t/ha of well rotten FYM should be applied at the time of land preparation along with 50kg/ha of phosphorus as basal dose at the time of root planting whereas 50kg/ha nitrogen should be top-dressed in two equal splits during bolting stage and before starting of flowers (Mishra *et al.*, 2021). When 70% of the pods turn light yellow, the crop is harvested immediately to avoid shattering and damage by birds. On an average, seed yield of 5–6q/ha is obtained (http://eagri.org/eagri50).

Two to three light hoeing and hand weeding are required during the early stages of plant growth. Earthing up the plant once or twice at root formation stage ensures proper development of roots but deep hoeing should be avoided as this may injure the roots. In the second season, one hoeing cum earthing up is necessary during March, so as to give support to the plants which may otherwise lodge during the flowering and fruiting stage. Weeds such as wild mustard and wild turnip should completely be eliminated because it will be extremely difficult to separate them if they are allowed to mature and harvested along with the turnip seed (Mishra *et al.*, 2021).

Pests and Diseases

Fungal Diseases

Several fungi may cause turnip or rutabaga disease on their roots and leaves and these are:

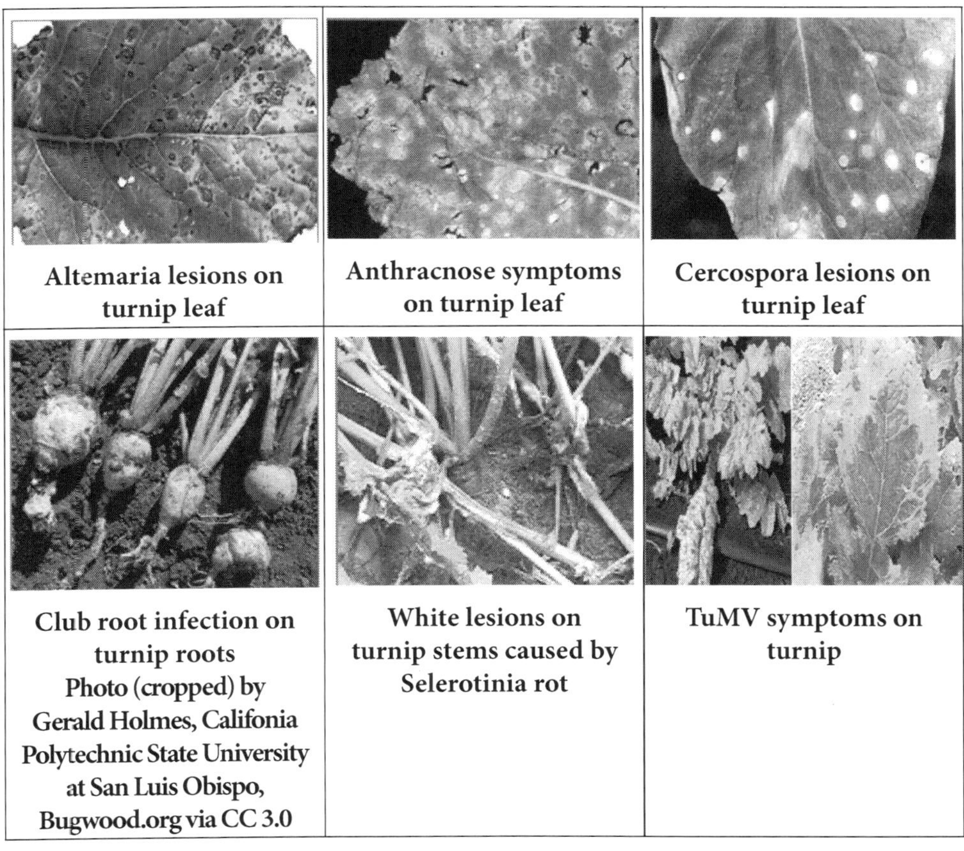

Altemaria lesions on turnip leaf

Anthracnose symptoms on turnip leaf

Cercospora lesions on turnip leaf

Club root infection on turnip roots
Photo (cropped) by Gerald Holmes, Califonia Polytechnic State University at San Luis Obispo, Bugwood.org via CC 3.0

White lesions on turnip stems caused by Selerotinia rot

TuMV symptoms on turnip

- **Alternaria/black/grey leaf spot** caused by *Alternaria brassicicola* and *A. brassicae* may be a problem during cool and wet periods. These on leaves appear as small dark spots (round or angular) ranging from yellow to black which may possess purple-black margins and may form concentric rings with cracking in the centre. Stems and petioles may also exhibit dark brown elongated lesions (https://plantvillage.psu.edu) and these lesions may also extend to flowers and pods (https://gardenerspath.com b). Control of cruciferous weeds, proper crop rotation, use of pathogen-free seeds which are treated before sowing, keeping plant surface dry, spraying with copper hydroxide or synthetic fungicides and use of biofungicides are the options to keep this disease at bay (https://plantvillage.psu.edu; https://gardenerspath.com b).

- **Anthracnose** (*Colletotrichum higginisianum*) in turnips and rutabagas is favoured by moist and warm conditions, and the pathogen overwinters on leaf debris and related weeds. Initially small circular or irregularly-shaped dry gray to straw-coloured spots develop on leaves whose abundance may cause death of leaves. After coalescing, these lesions may have large necrotic patches so leaves turn yellow and wilt. Lesions may also split or crack in dry centers. Dry sunken spots on roots also enlarge and turn gray or brown (https://plantvillage.psu.edu). To avoid this problem, proper sanitary measures should be adopted, only pathogen-free seed should be used and that too treated with HWT before sowing, proper crop rotation should be followed, all the cruciferous weeds should thoroughly be weeded out and there should not be any chance of water stagnation in the field (https://plantvillage.psu.edu).
- **Black root** (*Aphanomyces raphari*) fungus in the soil can survive for prolonged period. In its infection, small black-blue areas on roots are formed which after expanding girdle and constrict the taproots (https://plantvillage.psu.edu). This disease is managed through crop rotation of non-brassica crops.
- **Cercospora or frog-eye leaf spot** (*Cercospora brassicicola*) can infect turnips and rutabaga in cool and wet weather. In its infection, green to gray and angular to circular spots with brown borders develop on leaves and its severity defoliates the leaves (https://plantvillage.psu.edu). To manage this problem only certified seeds should be used, overhead irrigation should be avoided, 3-year crop rotation with non-brassica crop should be followed, and in case of disease occurrence appropriate fungicides should be applied (https://plantvillage.psu.edu).
- **White spot** (*Pseudocercosporella capsellae*, syn. *Cercosporella brassicae*) is another fungal disease that can infect turnip and rutabaga leaves, especially under cool and wet conditions. It spreads through infected seeds, other cruciferous plants, and volunteers. Its infection causes small, brown necrotic spots on leaf tips and margins which mature to light gray or white with the original dark spot in centre. Margins of lesions may

be darker and these on coalescing form large chlorotic areas and cause defoliation (https://plantvillage.psu.edu). In managing this problem at no time the weeds should be allowed to grow in the field, proper crop rotation should be followed, crop canopy requires to be thinned to manage crop wetness, and appropriate fungicide should be used (https://plantvillage.psu.edu).

- **Clubroot** (*Plasmodiophora brassicae*) fungus can survive in soil for up to 10 years and its infection is quite akin to root-knot nematode infestation. It can be spread by movement of contaminated soil and irrigation water to uninfected areas. In its infection, yellowing leaves show wilting during daytime but partly become recovered by the night. Its infection shows swollen *cum* distorted roots, extensive gall formation, slow growth and plant stunting (https://plantvillage.psu.edu). To manage this disease only certified seeds should be used and the trtansplants should be selected for planting only from the fumigated beds and the soil should be applied with lime. Its complete control is impossible as the fungus survives in the soil for several years (https://plantvillage.psu.edu).
- **White mould/Sclerotinia rot** (*Sclerotinia sclerotiorum*) has extremely wide host range and may infect turnips and rutabagas. The disease emergence is favoured by moderate to cool temperatures and high humidity. Its infection causes irregular necrotic grey lesions on leaves, white-grey lesions on stems; reduced pod set and seed shattering. Infections usually begin at the base of the plant where tan water-soaked lesions appear on leaves with cottony-white mycelia growth covering entire infected tissues and from these mycelia seed-like black bodies (sclerotia) appear (https://plantvillage.psu.edu). To get rid of this disease, there should be 3-year rotation with cereals, the plants should have proper spacing and regular weeding and spraying with proper fungicide regularly (https://plantvillage.psu.edu).
- **Wirestem or damping-off** (*Rhizoctonia solani*) is favoured by cool and wet soils and in its infection seedlings start dieing after germination due to brown-red or black rot girdling while still the seedlings may remain erect for sometime before final collapse (https://plantvillage.psu.edu). To manage this problem the seedlings or root-knobs should have been disease-free and produced in sterilized media, before sowing the seeds and before planting the roots should be dressed with fungicides, planting should be delayed until weather becomes a little warm and finally fungicides should have been applied in the field when disease has appeared (https://plantvillage.psu.edu).

- **Oomycete diseases** are caused through water mould fungi. These like wet conditions to react though do not kill their hosts but cause significant damages to the leaves. As with the bacterial diseases, preventing the initial infection is the best way to control oomycete diseases. Destroying cruciferous weeds and minimizing overhead irrigation are also ways to control these diseases (https://gardenerspath.com b). **White rust** (Albugo candida) is usually not a serious problem though rarely this creates systemic infections engulfing the whole plant afterwards (https://gardenerspath.com b). Its infectiuon initially causes yellow spotting on the upper leaf surfaces along with white pustules (blisters) at the leaf bottom (https://gardenerspath.com b). **Downy mildew** (*Peronospora parasitica*) is a serious disease of turnips and rutabaga, especially in cool *cum* wet conditions as occurring during autumn (https://gardenerspath.com b). Its infection initially causes irregular yellow patches on leaves which later on turn light brown with fluffy grey growth on leaf undersides (https://plantvillage.psu.edu). Its infection can kill the young seedlings and seriously affecting the greens. When becomes serious, it can cause cracking starting from the crown into the root (https://gardenerspath.com b). In managing this problem, only sterilized seeds or roots should be sown or planted, the plant debris should be removed and burnt, non-brassica crop rotation should be followed, and crop during infection should be sprayed with appropriate fungicide.

White rust-Photo by Scot Nelson via PD 1.0	**Downy mildew- Photo via Alamy**

Bacterial Diseases

Black Rot
Photo via Alamy

Turnips and rutabagas are infected by a number of bacterial diseases. In case growers are not able to get disease-free seeds they can treat them with 50^{0}C hot water to kill any seed-borne pathogen. Since these bacteria thrive under moist conditions so overhead irrigation should also be avoided and together thinning of the basal spreading leaves should be done to facilitate proper aeration in the cropping-area. Wild crucifer weeds such as groundkeeping radish and shepherd's purse that often harbour these diseases can be removed. Crop rotation to non-*brassica* crops is a critical step, since the bacteria can survive in crop residue in the soil. Growers should grow non-cruciferous crops at least for two years though up to four years, if crop is infected with black rot. However, if the plants develop a bacterial infection, spraying with copper hydroxide (Kocide 3000) might help to keep it from spreading further (https://gardenerspath.com b).

- **Black rot** (*Xanthomonas campestris* pv. *campestris*) is one of the most devastating bacterial diseases infecting turnips and rutabagas whose outbreak is favoured by warm and wet conditions. This bacterium can destroy the leaves and enter the plant's vascular system, causing systemic disease throughout the plant and eventually end up with a black rotten plant (https://gardenerspath.com b). While wilting, the leaves show yellowing to browning, followed by collapse of whole plant. In its infection, V-shaped lesions appear on mature leaf margins and dark ring formation on the cross section of the stem (https://plantvillage.psu.edu) and eventually the veins turn black (https://gardenerspath.com b). To manage this disease, there should be proper field sanitation, disease-free seed and that too chemical treated ones should be used, 2-year crop rotation of non-cruciferous crops should be followed and overhead irrigation should be avoided.
- **Bacterial leaf spot** is caused due to infection of *Xanthomonas campestris* pv. *amoraciae* and *Pseudomonas syringae* pv. *maculicola. P.s.* pv. *maculicola* infects plants at cool temperatures, typically in the early spring and fall, whereas, *X.c.* pv. *amoraciae* infections occur at times with warmer temperatures and is usually a problem during summer (https://gardenerspath.com b). Both these bacteria cause small water-soaked lesions that can spread and coalesce to destroy the greens. These produce yellow haloes though *X.c.* pv. *amoraciae* produces much more noticeable haloes (https://gardenerspath.com b).

Viral

Turnip mosaic [Turnip mosaic virus (TuMV)] can greatly reduce yields. It is transmitted by many species of aphids including those of cabbage and peach aphids. Its symptoms appear as yellow and green mosaic patterns *cum* necrotic areas on leaves coupled with vein clearing *cum* chlorosis on older leaves, black spots and brown necrotic streaks on stems, *vis-a-vis* stunted plant growth and reduced yield (https://plantvillage.psu.edu). Reflective mulches deter aphid to feed on the plants. These can also be controlled by spraying tobacco decoction on affected plants. Mrtasystox 0.2% spraying on affected crop will kill the aphid vectors effectively so virus spread will also be checked.

Whiptail of turnip is the **physiological disorder** caused due to deficiency of Mo, especially in acidic soils. Young leaves become narrow, cupped, showing chlorotic mottling especially around the margins, develop deep patches which ultimately affect the root growth. The affected plants are removed from the field during thinning. Further appearance of this disorder can be controlled by liming the soil and bringing the pH to 6.5. However, it is controlled by the application of 1.2kg/ha of sodium or ammonium molybdate (http://eagri.org/eagri50).

Insect-pests

Various insect-pests infesting turnip at various stages

Cabbage aphid colony | **Crucifer flea beetle** | **Root maggots in soil**

Adult cabbage fly (De lia radicum) | **Cabbage fly root maggots- Photo via Alamy** | **Wireworms**

- **Cabbage aphid** (*Brevicoryne brassicae* and *Lipaphis erysimi*) feed only on cruciferous plants but may survive on related weed species. Apart from these, there are various other aphids feeding on turnip such as **turnip aphid** (*Lipaphis erysimi*), **black bean aphid** (*Aphis fabae*), **green peach aphid** (*Myzus persicae*) and many others. These small, grey-green and soft-bodied insects covered with a white waxy coating may be found clustered on the tender parts of the plants and sucking the sap. Large colonies can cause stunted growth or even plant death (https:// plantvillage.psu.edu; https://gardenerspath.com a). Try to plant the tolerant varieties if available. Reflective mulches such as silver coloured plastic can deter aphids from feeding on plants. Sturdy plants can be sprayed with a strong jet of water to knock down the aphids from leaves. Spraying the insecticidal soaps or oils such as neem or canola oil control the nymphs and adults of aphids (https://plantvillage.psu.edu). In case of very high infestations, 1:800 nicotine sulphate proves quite effective when temperature is about 18.3-21.1^{0}C. Use of insecticides becomes imperative only when infestation is alarmingly high and then 0.02% Oxydemeton or 0.2% Metasystox spraying twice on the plants fortnightly will control the aphids.

- **Cabbage whiteflies** (*Aleyrodes proletella*) nymphs at the bottom of the leaves look like white scales. In its infestation, clouds of tiny white insects can be found flying from the turnip plants when disturbed. Heavy infestation can result in a black sooty mould on top of the leaves (https://gardenerspath.com a). Organic sprays are a good choice because they will kill the whiteflies without affecting beneficial insects such as adult ladybugs. Horticultural oils or insecticidal soaps cause very little harm to beneficial organisms. Insecticidal soaps should be applied when the weather is cloudy and atmospheric temperature is less than 32^{0}C as after use of insecticidal soaps the plants become quite sensitive to sun-rays (https://gardenerspath.com a).
- **Flea beetles** (*Phyllotreta* spp.) may overwinter on nearby weed species, in plant debris or in the soil and may go through a second or third generation in one year. The pest responsible for plant damage is a small (1.5–3mm long) and shining dark beetle which jumps when disturbed (https://plantvillage.psu.edu). Older plants can tolerate infestation but younger ones are quite susceptible to its damage. Their infestation causes small holes or pits on the leaves giving the foliage a characteristic 'shothole' appearance. Young plants and seedlings are particularly very susceptible where its infestation may cause reduction in plant growth. In case of severe damage, the plants may be killed. Where its attack is expected, floating row covers can be used to provide a physical barrier to protect young plants. Enough mature plants are least affected so early sowing is recommended so that plants are well established before menace of beetles starts, Sowing of any other cruciferous plant before the main crop will work as trap crop. Moreover, thick mulching in cropped area will also protect the plants from its attack as this will prevent the beetles reaching the soil surface. Mineral oil or neem oil spray at weekly intervals will control this pest. Weekly spray of 0.2% chemical insecticides such as carbaryl, spinosad, bifenthrin and permethrin can provide adequate

control of beetles (https://plantvillage.psu.edu). Spraying of 0.15% Malathion or 0.4% Sevin at 10-15 days intervals will keep these beetles under check.

- **Mustard saw-fly** (*Athalia* sp.) is a common pest of turnip as well as radish. Its appearance is noted when the crop is either in vegetative phase or in flowering. Its grubs feed on vegetative *cum* floral parts as well as on pods by making holes there. Through spraying with 0.4% Sevin 50WP at weekly intervals this pest is effectively controlled.
- **Root maggots** (*Delia* spp.) are white or white-yellow larvae that are approximately 8-10mm long with tapering towards the head. An adult root maggot (cabbage fly, cabbage root fly, root fly or turnip fly) overwinters as pupae in the soil and attacks all varieties of cruciferous plants at larval stage and on turnip roots these tunnel on root surface (https://gardenerspath.com a). In getting rid of its attack, at least one year gap should be given in case root maggot infestation has occurred there. The field should be kept clean and plant debris destroyed. Use of floating row covers can reduce damage to crops by preventing female flies from laying eggs but in case where root maggots are not already present (https://plantvillage.psu.edu; https://gardenerspath.com a).
- **Wireworms** (*Aeolus*, *Anchastus*, *Melanot*us, *Limonius* species) larvae which are thin *cum* shiny yellow-brown can infest the soil for up to five years. These can kill the seedlings or girdle the stems of the turnip or rutabaga plants (https://plantvillage.psu.edu; https://gardenerspath.com a). In the field infested with wireworm, before sowing or planting of turnip the field should be prepared well by mixing 500g/ha of Carbofuran/Phorate 10G or 5q/ha of neem cake in the soil at the time of crop sowing or planting. There should be crop rotation with non-host crop (https://plantvillage.psu.edu; https://gardenerspath.com a).
- **Slugs:** Slugs prefer cruciferous crops coupled with cool humid weather which turnip also like. These feed voraciously and batter the plants completely. These can be controlled through use of poison baits as well as through the use of parasitic nematodes.

References

Afsharypuor, S. and M. Tahmasian, 2010. Volatile constituents of the tuberous tap-root, leaf and seed of *Brassica rapa* L. ssp. *rapa* cultivated in Isfahan (Iran). *J. Essent. Oil Res.*, **22**: 173–175. https://doi.org/10.1080/10412905.2010.9700295

Al-Snafi, A.E. 2015. The pharmacological importance of Bassica nigra and Brasica rapa grown in Iraq. *J. Pharmaceut. Biol.*, **5**: 240–53. https://ift.onlinelibrary.wiley.com/doi/10.1111/1750-3841.14417

Bang, M.H., D.Y. Lee, Y.J. Oh, M.W. Han Oh, H.J. Yang, H.G. Chung, and N.I. Baek, 2012. Development of biologically active compounds from edible plant sources. XXII. Isolation of indoles from the roots of *Brassica campestris* ssp. *rapa* and their hACAT inhibitory activity. *J. Korean Soc. Appl. Boil. Chem.*, **51**: 65–69.

Cartea, M.E., M. Francisco, P. Soengas and P. Velasco, 2010. Phenolic compounds in Brassica vegetables. *Molec*, **16**(1): 251-80. PMid:21193847. http://dx.doi.org/10.3390/molecules16010251; http://dx.doi.org/10.3390/molecules16010.

Clarke, D.B. 2010. Glucosinolates, structures and analysis in food. *Analyt. Methods*, **2**: 310–25. https://doi.org/10.1039/B9AY00280D.

Croft, K., F. Juttner and A.J. Slusarenko, 1993. Volatile products of the lipoxygenase pathway evolved from *Phaseolus vulgaris* (L.) leaves inoculated with *Pseudomonas syringae* pv. *phaseolicola. Pl. Physiol.*, **101**: 13–24.

Crozier, A., I.B. Jaganath and M.N. Clifford, 2006. Phenols, polyphenols and tannins: An overview. In A. Crozier, M. Clifford, & H. Ashihara (Eds.), Plant Secondary Metabolites: Occurrence, Structure and Role in the Human Diet (eds Crozier, A., M. Clifford and H. Ashihara), pp. 1–24. Blackwell, Oxford, UK.

Daryoush, M., A.T. Bahram, D. Yousef and N. Mehrdad, 2011. Protective effect of turnip root (Brassica rapa L.) ethanolic extract on early hepatic injury in alloxanized diabetic rats. *Aust. J. Basic & Appl. Sci.*, **5**: 748–56.

Duncan, R.E., D. Lau, A. El-Sohemy and M.C. Archer, 2004. Geraniol and beta-ionone inhibit proliferation, cell cycle progression, and cyclin dependent kinase 2 activity in MCF-7 breast cancer cells independent of effects on HMG-CoA reductase activity. *Bioch. Pharmac.*, **68**: 1739–47. https://doi.org/10.1016/j.bcp.2004.06.022.

en.wikipedia.org/wiki/Bok_choy.

en.wikipedia.org/wiki/Chinese_cabbage.

en.wikipedia.org/wiki/Rapini

en.wikipedia.org/wiki/Turnip

Engelberth, J., H.T. Alborn, E.A. Schmelz and J.H. Tumlinson, 2004. Airborne signals prime plants against insect herbivore attack. *Proc. Nat. Acad. Sci.* U.S.A., **101**: 1781–5. https://doi.org/10.1073/pnas.0308037100.

Fahey, J.W., A.T. Zalcmann and P. Talalay, 2001. The chemical diversity and distribution of glucosinolates and isothiocyanates among plants. *Phytochem.*, **56**: 5–51. https://doi.org/10.1016/S0031-9422(00)00316-2.

Fernandes, F., P. Valentao, C. Sousa, J.A. Pereira, R.M. Seabra and P.B. Andrade, 2007. Chemical and antioxidative assessment of dietary turnip (*Brassica rapa* var. *rapa* L). *Food Chem.*, **105**: 1003–10. https://doi.org/10.1016/j.*foodchem*.2007.04.063.

Francisco, M., M.E. Cartea, P. Soengas and P. Velasco, 2011. Effect of genotype and environmental conditions on health-promoting compounds in *Brassica rapa. J. agr. & Food Chem.*, **59**(6): 2421-31. PMid:21341668. https://www.scielo.br/j/bjft/a/pH7R3CmhYrcSjxmCRzJmrVN/?lang=en.

Francisco, M., D.A. Moreno, M.E. Cartea, F. Ferreres, C. García-Viguera and P. Velasco, 2009. Simultaneous identification of glucosinolates and phenolic compounds in a representative collection of vegetable *Brassica rapa. J. Chromat.*, No. 1216: 6611–9. https://doi.org/10.1016/j.chroma.2009.07.055.

Fu, R., Y. Zhan, Y. Gu, T. Pen and F. Chen, 2016. Hepatoprotection using *Brassica rapa* var. *rapa* L.seeds and its bioactive compound, sinapine thiocyanate, for CCl4-induced liver injury. *J. funct Foods*, **22**: 73–81, Retrieved from https://doi.org/10.1016/j.jff.2016.01.013.

gardenerspath.com a. https://gardenerspath.com/how-to/disease-and-pests/turnip-and-rutabaga-pests.

gardenerspath.com b. https://gardenerspath.com/how-to/disease-and-pests/turnip-and-rutabaga-diseases.

Heil, M. 2008. Indirect defence *via* tritrophic interactions. *New Phytol.*, **178**: 41–61. https://doi.org/10.1111/j.1469-8137.2007.02330.x.

Hong E and G.-H. Kim, 2008. Anticancer and antimicrobial activities of BETA-phenylethyl isothiocyanate in *Brassica rapa* L. *Food Sci. technol. Res.*, **14**: 377–82. doi: 10.3136/fstr.14.377. https://ift.onlinelibrary.wiley.com/doi/10.1111/1750-3841.14417.

Hosseini, S. E., S. Zahiri and H. Aqababa, 2013. Effect of alcoholic extract of Brassica rapa root on formalin test pain in adult male rats. *Quart. Horizon Med. Sci.*, **19**: 161–6.

http://agropedia.iitk.ac.in/content/turnip-variety-its-characteristics.

http://eagri.org/eagri50/HORT281/pdf/lec29.pdf.

http://ecoursesonline.iasri.res.in/mod/page/view.php?id=101604.

https://balconygardenweb.com/types-of-turnips-best-varieties-to-grow/.

; https://gardenerspath.com a.

https://gardenerspath.com b.

https://krishijagran.com/agripedia/turnip-know-the-climate-soil-requirements-land-preparation-sowing-spacing-irrigation-harvesting-process/.

https://plantvillage.psu.edu/topics/turnip/infos.

https://www.agrifarming.in/turnip-farming.

https://www.amrafarms.com/turnip-farming.html.

https://www.foodhero.org/healthy-food/types-turnips.

https://www.scielo.br/j/bjft/a/pH7R3CmhYrcSjxmCRzJmrVN/?lang=en.

https://www.slideshare.net/Thejas2015/turnip-250219506.

Igarashi, K., T. Mikami, Y. Takahashi and H. Sato, 2008. Comparison of the preventive activity of isorhamnetin glycosides from atsumi-kabu (red turnip, *Brassica campestris* L.) leaves on carbon tetrachloride-induced liver injury in mice. *Biosci., Biotechn. & Biochem.*, **72**: 856–60. https://doi.org/10.1271/bbb.70558.

Jeong, R.H., Q. Wu, J.G. Cho, D.Y. Lee, S. Shrestha, M.H. Lee and N.I. Baek, 2013. Isolation and identification of flavonoids from the roots of *Brassica rapa* spp. *J. Appl. Boil. Chem.*, **56**: 23–7. https://doi.org/10.3839/jabc.2013.005.

Jones, G., O.G. Sanders and C. Grimm, 2007. Aromatic compounds in three varieties of turnip greens harvested at three maturity levels. *J. Food Qual.*, **30**: 218–27. https://doi.org/10.1111/j.1745-4557.2007.00116.x.

Jung, U.J., N.I. Back, H.G. Chung, M.H. Bang, T.S. Jeong, K.T. Tae-Lee and M.S. Choi, 2008. Effects of the ethanol extract of the roots of *Brassica rapa* on glucose and lipid metabolism in C57BL/KsJ-db/db mice. *Clin. Nutrit.*, **27**: 158–67. https://doi.org/10.1016/j.clnu.2007.09.009.

Kessler, A. and I.T. Baldwin, 2002. Plant responses to insect herbivory: The emerging molecular analysis. *Ann. Rev. Pl. Biol.*, **53**(1): 299-328. PMid:12221978. https://www.scielo.br/j/bjft/a/pH7R3CmhYrcSjxmCRzJmrVN/?lang=en.

Lee, J.G., G. Bonnema, N. Zhang, J.H. Kwak, C. Ric, H. deVos and J. Beekwilder, 2013. Evaluation of glucosinolate variation in a collection of turnip (*Brassica rapa*) germplasm by the analysis of intact and desulfo glucosinolates. *J. agric. Food Chem.*, **61**: 3984–93. https://doi.org/10.1021/jf400890p

Li, L., D.H. Park, Y.C. Li, S.K. Park, Y.L. Lee, H.M. Choi and M.J. Lee, 2010. Anti-hepatofibrogenic effect of turnip water extract on thioacetamide-induced liver fibrosis. *Lab. Anim. Res.*, **26**: 1–6. https://doi.org/10.5625/lar.2010.26.1.1.

Liang, Y.S., H.K. Kim, A.W. Lefeber, C. Erkelens, Y.H. Choi and R. Verpoorte, 2006. Identification of phenylpropanoids in methyl jasmonate treated Brassica rapa leaves using two-dimensional nuclear magnetic resonance spectroscopy. J. *Chromatog.* **A, 1112**:148–55. doi: 10.1016/j.chroma.2005.11.114.

Llorach, R., A. Gil-Izquierdo, F. Ferreres and F.A. Tomas-Barberan, 2003. HPLC-DAD-MS/MS ESI characterization of unusual highly glycosylated acylated flavonoids from cauliflower (*Brassica oleracea* L. var. *botrytis*) agro-industrial byproducts. *J. agric. & Food Chem.*, **51**: 3895–9. https://doi.org/10.1021/jf030077h.

Lo Scalzo, R., A. Genna, F. Branca, M. Chedin and H. Chassaigne, 2008. Anthocyanin composition of cauliflower (*Brassica oleracea* L. var. *botrytis*) and cabbage (*B. oleracea* L. var. *capitata*) and its stability in relation to thermal treatments. *Food Chem.*, **107**(1): 136-44. https://www.scielo.br/j/bjft/a/pH7R3CmhYrcSjxmCRzJmrVN/?lang=en.

Lopez-herrera, F.J., F. Sarabia-Garcia, A. Heras-Lopez and M.S. Pino-Gonzalez, 1997. Stereoselective synthesis of C-amino-substituted-D-mannopyranosides. Easy preparation of novel inhibitors for mannosidases. *J. Org. Chem.*, **62**: 6056–9. https://doi.org/10.1021/jo961875s.

McAlvay, Alex C., P. Ragsdale Aaron, E. Mabry Makenzie, Qi Xinshuai, A. Bird Kevin, Velasco Pablo, An Hong, J. Chris Pires and Eve Emshwiller, 2021. Brassica rapa domestication: untangling wild and feral forms and convergence of crop morphotypes. *Molec. Biol. & Evol.*, **38**(8): 3358–72. doi:10.1093/molbev/msab108. PMC 8321528. PMID 33930151.

Metz, J., K. Ribbers, K. Tielborger and C. Muller, 2014. Long- and medium-term effects of aridity on the chemical defence of a widespread Brassicaceae in the Mediterranean. *Environm. & exper. Bot.*, **105**: 39–45. https://doi.org/10.1016/j.envexpbot.2014.04.007.

Mirzaie, H., H. Johari, M. Najafian and H. Kargar, 2012. Effect of ethanol extract of root turnip (*Brassica rapa*) on changes in blood factors HDL, LDL, triglycerides and total cholesterol in hypercholesterolemic rabbits. *Adv. in envi.* Biol., **6**: 2796–2801.

Mishra, Nityamanjari; Kaushik Kumar and Arghya Mani Panigrahi, 2021. Seed Production in Turnip. In: A Text Book on Vegetable Seed Production, pp.237-45. Pub. by: International Books & Periodical Supply Service, India.. ISBN : 978-93-90425-24-2E-ISBN : 978-93-90425-30-3.

Mohajeri, D., M.N. Gharamaleki, S.S. Hejazi and M. Nazeri, 2013. Preventive effects of turnip (*Brassica rapa* L.) on renal ischemia-reperfusion injury in rats. *Life Sci.J.*, **10**: 1165–70.

Molinier, V., J. Fitremann, A. Bouchu and Y. Queneau, 2004. Sucrose esterification under Mitsunobu conditions: Evidence for the formation of 6-O-acyl-3′,6′-anhydrosucrose besides mono and di-esters of fatty acids. *Tetrahedron Asymmetry*, **15**: 1753–62. https://doi.org/10.1016/j.tetasy.2004.04.021.

Ninomiya, M., M. Efdi, T. Inuzuka and M. Koketsu, 2010. Chalcone glycosides from aerial parts of *Brassica rapa* L. ‘Hidabeni’ turnip. *Phytochem. Letters*, **3**: 96–9. https://doi.org/10.1016/j.phytol.2010.02.004.

Padilla, G., M.E. Cartea, P. Velasco, A. De Haro and A. Ordas, 2007. Variation of glucosinolates in vegetable crops of *Brassica rapa*. *Phytochem.*, **68**(4): 536-45. PMid:17187832. https://www.scielo.br/j/bjft/a/pH7R3CmhYrcSjxmCRzJmrVN/?lang=en.

Pereira, D.M., P. Valentao, J.A. Pereira and P.B. Andrade, 2009. Phenolics: From chemistry to biology. *Molecules*, **14**: 2202–11. https://doi.org/10.3390/molecules14062202.

Pierre, P.S., J.J. Jansen and C.A. Hordijk, 2011. Differences in volatile profiles of turnip plants subjected to single and dual herbivory above and below ground. *J. Chem. Ecol.*, **37**: 368–77. https://doi.org/10.1007/s10886-011-9934-3.

Pongprayoon, U., P. Baeckstrom, U. Jacobsson, M. Lindstrom and L. Bohlin, 1992. Antispasmodic activity of beta-damascenone and E-phytol isolated from *Ipomoea pes-caprae. Planta Medica*, **58**: 19–21. https://doi.org/10.1055/s-2006-961381.

Robin, A.R., S. Arora A.P. Vig, 2018. Development of validated high-temperature reverse-phase UHPLC-PDA analytical method for simultaneous analysis of five natural isothiocyanates in cruciferous vegetables. *Food Chem.*, **239**: 1805–9. https://doi.org/10.1016/j.foodchem.2017.07.059.

Robin, A.S. and A.P. Vig, 2015. Inhibition of DNA oxidative damage and antimutagenic activity by dichloromethane extract of *Brassica rapa* var. *rapa* L. seeds. *Indust. Crops & Prod.*, **74**: 585–91. https://doi.org/10.1016/j.indcrop.2015.05.038.

Romani, A., P. Igolini, L. Isolani, F. Ieri and D. Heimeler, 2006. HPLC-DAD/MS characterization of flavonoids and hydroxycinnamic derivatives in turnip tops (*Brassica rapa* L. subsp. *sylvestris* L.). *J. agric. & Food Chem.*, **54**: 1342–6. https://doi.org/10.1021/jf052629x.

Rosa, E.A.S., R.K. Heaney, G.R. Fenwick and C.A.M. Portas, 1997. Glucosinolates in crop plants. *Hort. Rev.* (ed. Janick, J.), pp. 99–215. John Wiley & Sons, West Lafayette, IN, USA.

Sadhu, M K and K. Sarkar, 2003. Radish. In: Vegetable Crops (eds Bose, T.K., J. Kabir, T.K. Maity, V.A. Parthasarthy and M.G. Som), vol.2 (3rd ed.), pp. 115-26. Pub. by Naya Udyog, 206 Bidhan Sarani, Kolkata 700006.

Sanderson, Helen, 2005. The Cultural History of Plants (eds Prance, Ghillean and Mark Nesbitt), 72p. Routledge, UK. ISBN 0415927463.

Shin, J.S., Y.S. Noh, Y.S. Lee, Y.W. Cho, N.I. Baek, M.S. Choi, and K.T. Lee, 2011. Arvelexin from Brassica rapa suppresses NF-kB-regulated pro-inflammatory gene expression by inhibiting activation of IkB kinase. *Brit. J. Pharmacol.*, **164**: 145–58. https://doi.org/10.1111/j.1476-5381.2011.01351.x.

Singh, C.B., M.L. Pandita and S.C. Khurana,.1990. Effect of planting date, steckling size and spacing on turnip seed production. *Seed Res.*, **18**(2): 106-13.

Taveira, M., F. Fernandes, P.G. Pinho, P.B. Andrade, J.A. Pereira and P. Valentao, 2009. Evolution of *Brassica rapa* var. *rapa* L. volatile composition by HS-SPME and GC/IT-MS. *Microchem. J.*, **93**(2): 140-46. https://www.scielo.br/j/bjft/a/pH7R3CmhYrcSjxmCRzJmrVN/?lang=en.

Tholl, D. 2006. Terpene synthases and the regulation, diversity and biological roles of terpene metabolism. *Curr. Opin. Pl. Biol.*, **9**: 297–304. https://doi.org/10.1016/j.pbi.2006.03.014.

Vallejo, F.A., F.A. Tomás-Barberán and F. Ferreres, 2004. Characterization of flavonols in broccoli (*Brassica oleracea* L. var. *italica*) by liquid chromatography –UV diode array detection-electrospray ionization mass spectrometry. *J. Chromatog. A*, **1054**: 181–93. https://doi.org/10.1016/j.chroma.2004.05.045.

Wu, Q., M.H. Bang, D.Y. Lee, J.G. Cho, R.H. Jeong, S. Shrestra and N.I. Baek,.2012. New indoles from the roots of *Brassica rapa* subsp. *campestris. Chem. Nat. Comp.*, **48**: 281–4. Retrieved from https://doi.org/0009-3130/12/4802-0281.

Wu, Q., J.G. Cho, D.S. Lee, D.Y. Lee, N.Y. Song, Y.C. Kim and N.I. Baek, 2013a. Carbohydrate derivatives from the roots of *Brassica rapa* ssp. *campestris* and their effects on ROS production and glutamate-induced cell death in HT-22 cells. *Carbohyd. Res.*, **372**: 9–14. https://doi.org/10.1016/j.carres.2012.09.015.

Wu, Q., J.G. Cho, K.H. Yoo, T.S. Jeong, J.H. Park, S.Y. Kim and N.I. Baek, 2013b. A new phenanthrene derivative and two diarylheptanoids from the roots of *Brassica rapa* ssp. *campestris* inhibit the growth of cancer cell lines and LDL-oxidation. *Archives pharm. Res.*, **36**: 423–9. https://doi.org/10.1007/s12272-013-0068-8.

Wu, Q., M.H. Bang, J.G. Cho, D.Y. Lee, K.T. Lee, H.G. Chung and N.I. Baek, 2013c. Phenolic compounds from the roots of *Brassica rapa* subsp. *campestris. Chem Nat. Comp.*, **49**: 852–6. Retrieved from https://doi.org/0009-3130/13/4905-0852.

www.atlasobscura.com/articles/brassica-rapa-vegetable-domestication.

Yang, B. and C.F. Quiros, 2010. Survey of glucosinolate variation in leaves of *Brassica rapa* crops. Genetic *Res. & Crop Evol.*, **57**: 1079–89. https://doi.org/10.1007/s10722-010-9549-5.

Yu, J., J. Lei, H. Yu, X. Cai and G. Zou, 2004. Chemical composition and antimicrobial activity of the essential oil of *Scutellaria barbata. Phytochem.*, **65**: 881–4. https://doi.org/10.1016/j.phytochem.2004.02.005.

Zohary, Daniel, Maria Hopf and Ehud Weiss, 2012. Domestication of plants in the Old World: the origin and spread of domesticated plants in Southwest Asia, Europe and the Mediterranean Basin (4th ed.), 139p. Oxford University Press, Oxford, U.K.. 139. ISBN 9780199549061.

5

Rutabaga/Swede (*Brassica napus* L. var. *napobrassica*)

Cultivar Group: Napobrassica Group.

Family: Brassicaceae.

Other names: Swedish turnip (US), Yellow Turnip (US), Tumshie(Scots), Neep (Scots) and Turnip (Scottish and Canadian English, Irish English and Manx English).

Introduction

Rutabaga is a North American English name of a root vegetable that is a form of *Brassica napus*, which also includes rapeseed (https://en.wikipedia.org). Rutabagas probably originated as a cross between turnip (*Brassica rapa* var. *rapa*) and wild cabbage (*Brassica oleracea*) and the resultant hybrid possessed double of its chromosome numbers as 2n=38 so it becomes an allopolyploid. This relationship was first published by Woo Jang-choon in 1935 and is known as the Triangle of U (Dixon, 2007). It is believed to have been first bred in Russia or Scandinavia in the late middle Ages (https://www.britannica.com). Rutabaga looks like a turnip but has ridges on its neck and turns purple towards the crown. It is cultivated for its fleshy roots and edible leaves (https://academics.hamilton.edu).

By pin add - Swede (The Vegetable), CC BY 2.0, https://commons.wikimedia.org/w/index.php?curid=5077916

Longitudinal section of a root

By No machine-readable author provided. Rainer Zenz assumed (based on copyright claims). - No machine-readable source provided. Own work assumed (based on copyright claims)., CC BY-SA 3.0, https://commons.wikimedia.org/w/index.php?curid=1356751

A local farmer's market in the town of Ithaca (New York) is organised with the name of 'International Rutabaga Curling Championship' annually on the last day of the market season. The villages of Askov (Minnesota) and Cumberland (Wisconsin), both hold the annual 'Rutabaga Festival in August (https://en.wikipedia.org/wiki/Rutabaga).

Botany

Rutabaga has 2n=36 chromosome numbers. It is a biennial herbaceous plant which grows quite slow and reaches to the height of 30-46cm bearing thick, lobed, smooth and glaucous leaves with waxy coating and bluish hue. The leaves grow from a stout swollen stem (neck) close to the ground forming the crown of the plant. Taproot of rutabaga is an enlarged, almost perfectly round and bears a distinct neck with well-marked leaf scars. These are purple, white or yellow-skinned with usually yellow or white flesh. The flesh of the root is firm and stores well during winter. White-fleshed varieties have the rough green skin and are of irregular form, while yellow-fleshed varieties are more regularly-shaped and have a smooth skin of a green, purple or bronze colour (https://www.britannica.com; https://plantvillage.psu.edu).

Though it is biennial but is usually grown as an annual, harvested after one growing season. If left to grow a second season the plant bears cross-shaped flowers with four petals that range from pale to bright yellow to pale-orange (https://www.britannica.com). Flowers are clustered at the top of a raceme and unlike turnip the flowers do not extend above the terminal buds (https://plantvillage.psu.edu).

Life Cycle and Seed Production

Rutabaga is a biennial ~ in first season forming the roots though in second season producing the flowers and setting the seeds, similar to those of turnip. Its flowers have the ability for self-pollination as well as cross-pollination with other varieties of rutabagas, some of the turnip varieties and all the varieties of winter rape. For purity of seed production, its male and female plants are caged together with introduction of insects and bees. If grown through seeds it may require 4-7 days if there is sufficient moisture in the bed. First year these are allowed to grow as an usual annual crop but second year these are caged for seed production. Seed sowing is usually done during spring to harvest the roots in the autumn or left in the field through the winter for seed production. However, after full development of the root-knobs, these are harvested and stored in a cool *cum* damp cellar for replanting the second season for seed production. When seeds mature the pods shatter on the stalk if they have not been handpicked. The seeds will remain viable for up to five years if stored properly. If the rutabaga is being grown only for its roots, they are planted in the spring and harvested in the fall. These can be left in the ground over winter except in extremely cold places (https://academics.hamilton.edu).

Classification

Rutabagas can be broadly classified into two types: i) yellow-fleshed, and ii) white-fleshed. Some varieties of both these types are:

- **American Purple Top** or **Champion Purple Top**: An heirloom variety having large, round and bright yellow roots (10-15cm in diameter) with a rich purple crown. Its flesh is creamy-yellow turning to orange when cooked, fine-grained, sweet and of excellent quality. It matures in about 90 days after sowing and its roots store well.
- **Joan**: Round, purple-topped but yellow-fleshed roots which mature in 95 days.
- **Helenor**: It is a hybrid producing round roots with light orange-flesh. It matures in about 90 days after sowing and roots store well.
- **Gilfeather** (Sometimes called a turnip): It produces round- to top-shaped white roots with greenish shoulders above the soil line. Its mild flesh is white and the flavour is quite amicable when picked after frost. It matures in 75-100 days and its roots store well.
- **Laurentian**: It produces smooth roots weighing some 450 grammes each where top is deep purple, base is cream-yellow and the flesh is yellow.It matures in 90–95 days.
- **Burpee's Purple Top**: It produces globe-shaped roots with purple-red shoulders aboveground and white below. It can be sown during spring to mid-summer, matures in 90 days and harvesting of roots is done when roots are 15cm or smaller. Its flesh is fine-grained yellow but on cooking becomes bright orange. Its roots have good winter storage.
- **Thomson Laurentian**: Its globe-shaped smooth roots have purple shoulders and fine-grained yellow flesh. It matures in 120 days and roots store quite well.
- **York**: Its roots are smooth with purple shoulders and creamy-yellow flesh. It matures in 115-120 days and roots store quite well.
- **Doon Major**: It produces large and purple-topped roots. Roots mature in around 87 days.
- **Marian**: It produces short and wide-leaved plants. Its roots are purple-topped with yellow flesh. It is club-root resistant and matures in 85-95 days.
- **Mary**: It produces uniform roots, maturing in 70 days. It has good winter storage.
- **Macomber** or **Westport Macomber Turnip**: Though botanically it is a rutabaga but is usually called a turnip in the Westport MA area where its US popularity originated. It produces large white roots with smooth skin and green or purple shoulders. It requires about 100 days to harvest from seed.

Nutritional Value

Rutabaga, Raw	
Nutritional value per 100 g	
Energy	157 kJ (38 kcal)
Carbohydrates	8.62 g
Sugars	4.46 g
Dietary fiber	2.3 g
Fat	0.16 g
Protein	1.08 g
Vitamins	**Quantity (%DV†)**
Thiamine (B1)	0.09 mg (8%)
Riboflavin (B2)	0.04 mg (3%)
Niacin (B3)	0.7 mg (5%)
Pantothenic acid (B5)	0.16 mg (3%)
Vitamin B6	0.1 mg (8%)
Folate (B9)	21 μg (5%)
Vitamin C	25 mg (30%)
Minerals	**Quantity (%DV†)**
Calcium	43 mg (4%)
Iron	0.44 mg (3%)
Magnesium	20 mg (6%)
Manganese	0.131 mg (6%)
Phosphorus	53 mg (8%)
Potassium	305 mg (6%)
Zinc	0.24 mg (3%)
Units: μg = micrograms, mg = milligrams, IU = International units †Percentages are roughly approximated using US recommendations for adults. ***Source***: USDA FoodData Central ***Table Source***: https://en.wikipedia.org/wiki/Rutabaga	

Phytochemicals

Rutabaga is a rich source of polyphenols and glucosinolates. The bioactive phenolic compounds in rutabaga that show antioxidant activities are 3-caffeoylquinic acid, 5-caffeoylquinic acid, 5-feruloylquinic acid and 4-*p*-coumaroylquinic acid (Jakopic *et al.,* 2021; Stefanucci *et al.,* 2020). Its glucosinolates such as progoitrin, sinigrin and glucoerrucin have antimicrobial properties (Pasko *et al.,* 2018; Dastan *et al.,* 2011).

Rutabaga contains cyanoglucoside and releases cyanide which is subsequently detoxified into thiocyanate. Some other cyanoglucoside-containing foods include sweet potatoes, cassava, maize (corn) and lima beans (https://en.wikipedia.org/wiki/Rutabaga). Glucosinolate and other chemical compounds such as glucocheirolin, glucobrassicanapin, glucoberteroin, gluconapoleiferin and glucoerysolin contribute to flavour and odour of rutabagas (https://en.wikipedia.org/wiki/Rutabaga). Several phytoalexins that aid in defence against plant pathogens have also been isolated from the rutabaga, including three novel phytoalexins (Pedras *et al.*, 2004).

Taste

Rutabaga can best be described as a small turnip-like strong-tasting root. The taste may vary depending on where it is grown. At times, when it is grown along with undesirable companion crops the taste of the vegetable can be greatly affected, making it even more pungent. The outer coat of the vegetable is thick and dark, the inner flesh is white or yellow, and the rutabaga is best harvested when it has reached a diameter of no more than 15cm as beyond that its texture and taste are bound to change (https://ebrary.net).

Glucosinolate and some other compounds cause the bitter taste of rutabaga (Verkerk, 2009). Rutabagas are very nutritious and have a great taste with a delicate sweetness and a hint of both cabbage and turnip flavours. These have a more substantial and robust flavour than the turnip. Young roots are especially tender and pleasant to taste and its flesh colour deepens during cooking (https://academics.hamilton.edu).

Precaution

Over-consumption of cruciferous vegetables may lead to hypothyroidism. Goiters may develop because of dietary imbalances of thiocyanate-containing foods such as rutabaga as this inhibits thyroid iodide transport (Olsson and Jeppsson, 1984; Jones, 1998; Braverman and Utiger, 1991; https://academics.hamilton.edu). However, glucosinolate content in *Brassica* vegetables is around 1% of the dry matter but there have been no reports of ill-effects in humans from the consumption of glucosinolates from normal amounts of *Brassica* vegetables (Verkerk, 2009).

Human Perception of Bitterness in Rutabaga

As with watercress, mustard greens, turnip, broccoli and horseradish, human perception of bitterness in rutabaga is governed by a gene affecting the TAS2R bitter receptor which detects the glucosinolates in rutabaga. Sensitive individuals with the genotype PAV/PAV (supertasters) find rutabaga twice as bitter as insensitive subjects (AVI/AVI). The difference for the mixed type (PAV/AVI) is insignificant for rutabaga (Sandell *et al.*, 2006). As a result, sensitive individuals may find some rutabagas too bitter to eat.

Uses

Rutabaga is a good source of fibre, vitamin C and potassium. Its roots can be eaten raw or pickled and are commonly cooked with other root vegetables or mashed. These are particularly good in soups and stews. The leaves are usually cooked like other mustard greens (https://www.britannica.com; https://academics.hamilton.edu).

Like other crucifers, rutabagas are also high in antioxidant and anticancer compounds. The American Cancer Society recommends an increase in cruciferous vegetables to fight against cancer. In the past, the root has been used as a diuretic and an emollient. The juice from the roots can be used in treatment of bronchial catarrh and for chronic coughs. When added to salt, the powder from rutabaga seeds is folk remedy for cancer. When applied with camphor, the oil from the seeds can be a remedy for stiff joints and rheumatism (https://academics.hamilton.edu).

Cultivation

The rutabagas are extensively cultivated often as a fodder crop in Canada, Great Britain, northern Europe, and to a lesser extent in the United States (https://www.britannica.com). Its culture, insect-pests and diseases are almost similar to turnips.

Propagation

The rutabagas are propagated by seeds and takes four to seven days to germinate. It will germinate in soil temperatures between 10–30°C. Fungicide-dressed seeds are directly sown in the fields. Immediately after sowing, the field should be irrigated lightly for quick and proper germination.

Climate and Season

Rutabaga is a slow-growing, cold-hardy cool-season crop which is even drought-tolerant. It requires a long growing season and is sown only as a main or late season crop sown in late May to early June in an open sunny situation. For seed germination, the soil temperature should be at least 10°C. Its roots grow and develop most during low temperature periods. The sweetness of both turnips and rutabaga is greatly enhanced by exposure to hard frosts (https://academics.hamilton.edu; https://www.britannica.com; https://harvesting-history.com).

Soil

Both turnip and rutabaga prefer well-drained, moderately deep and fertile soils (https://academics.hamilton.edu) rich in potassium and phosphorus though not too high in nitrogen. Addition of bone meal and wood ash to the soil before planting increases the potassium and phosphorus levels (https://harvesting-history.com). Its soil should have around 5.5-7 pH, ideal pH being around 6, *i.e.* a little acidic and this condition locks the boron within the roots. Acidic soil prevents the rutabaga from facing boron deficiency as well as club-root disease. Although it can grow in soil of higher pH but then one will have to compromise the quality of the vegetable, *vis-à-vis* crop becomes quite susceptible to the dreaded club-root disease (https://ebrary.net).

Land Preparation and Planting

The land should be prepared to a fine tilth by deep ploughing and harrowing, followed by levelling. At the time of final ploughing, the soil of the field should be mixed thoroughly with 10t/ha of well-mature farmyard manure or compost. In the prepared beds the seed should be sown thinly, about 1-1½cm deep in rows 30-45cm apart, maintaining plant to plant distance of 10-13cm (https://harvesting-history.com).

Fertilizers and Moisture Requirements

Though in case of rutabaga the soil should be kept moist during growing season but there should be ample drainage facilities. Fertilizers can be applied when seeds are planted as nitrogen and organic materials added to soil will be quite beneficial. If the soil is too acidic, lime can be added, and as much organic material as possible should be added to make the soil moisture-retentive (https://academics.hamilton.edu) apart from supplying the required nutrients for better crop stand.

Companion Plants

Hairy vetch and rutabagas are excellent companions. Aphids will stay off rutabaga as well as turnip leaves since vetch shelters ladybugs which feast on aphids. Wood ashes around rutabagas will control scab (https://academics.hamilton.edu).

Weeds

Herbicides recommended for use on rutabagas will not provide total weed control, therefore, it is important to grow rutabagas in soil where the weed seed population is low. This is best accomplished by a crop rotation programme utilizing crops where weeds can be thoroughly controlled for a year or two prior to rutabagas. Stale seedbed technique can be used to considerable advantage on late seeded rutabagas. Care must be taken to avoid fields where residual herbicides from previous years persist in the soil as crop injury may occur. Cultivation is required to supplement chemical weed control if necessary (https://www.gov.nl.ca).

There are many common weeds in the Cruciferae family including shepherds purse, yellow rocket, wild mustard, wild radish, pennycress, worm seed mustard, hare's ear mustard, false flax, *etc.* Since they are closely related to rutabaga and turnip, they serve as hosts for many insects and pathogens that attack these crops. These weeds often become a serious problem due to their tolerance to herbicides used in these crops. Crop rotation, herbicide rotation, regular hoeing *cum* weeding, and clean culture will help reduce cruciferous weeds in this crop (https://www.gov.nl.ca).

Harvest

Rutabagas should be exposed to at least 2 hard frosts before harvesting (https://harvesting-history.com). The greens can be harvested from the plant when the leaves are about 10-25cm long. If growing for both greens and roots then only 2–3 leaves from each plant should be removed. Roots are ready to harvest when they reach 2.5–7.5cm in diameter. Small rutabagas can often be harvested by hand by pulling the plants gently from the soil but larger roots in heavier soil may require digging up with a garden fork. Unwashed roots are stored in a root cellar or in the basement (https://plantvillage.psu.edu).

The quality and flavour of rutabagas are much improved when the roots are fully matured and are exposed to frosts before harvest. Immature roots have a bitter taste, and if early seeded rutabagas are left in the field until late fall, the roots tend to become fibrous and woody. Certain types of potato diggers can be readily adapted for harvesting rutabagas. Before digging, the tops must be removed with the aid of a topping device. Crops can also be harvested by using a modified sugarbeet or carrot harvester. In preparation for long-term winter storage for 6 to 8 months, the tops should be trimmed to within 1cm of the body (any leaf tissue should not be left as severe rot can occur which may rot the crown and then these will become unmarketable) and the bottom of the taproot and the fibrous side roots should be cut off. Newly harvested rutabagas require protection from sun-burn so these require gentle handling also to avoid bruising and freezing at all times. Growers should not harvest under wet conditions or place wet rutabagas into storage (https://www.gov.nl.ca).

Mechanical harvesters tend to spread bacterial and fungal diseases (particularly if the crop is not surface-dry within a week of harvesting) to crowns and mechanical wounds on roots. To minimize rot in the crown, rutabagas should be topped a week or 10 days before harvesting with a hoe, a knife or through mechanical topping machine.

Turnips are harvested when they are over 5 cm in diameter and are then marketed as quickly as possible maintaining similar storage conditions to rutabagas. Polybags may be used to maintain high quality and as a presentable product.

Storage and Conditioning

Rutabagas infected with diseases should never be stored and the storage area must be sanitized each spring. Rutabagas may be piled up to 2m deep over a slated floor with an air circulation system to keep the temperature uniform. Perforated plastic sheeting placed horizontally at 45 cm intervals helps to reduce moisture loss and excessive accumulation of moisture at the top of the pile or on the roof of the storage area. Bruising should be avoided.

To obtain best results, the storage temperature should be close to 0^0C and the humidity around 95%. Under these conditions rutabagas keep for over 6 months. Rutabaga flavour is improved by conversion of starch to sugar as strong or bitter flavours are reduced. This conversion occurs when rutabagas are held at temperatures of 0-5^0C whether in the field or in storage.

In Atlantic Canada, rutabagas are commonly waxed. In Central Canada they are frequently waxed following storage to prevent loss of moisture during marketing. Waxed rutabagas are considered to have a greater eye appeal and greater consumer demand than unwaxed ones. Temperatures must be kept low after waxing as waxed rutabagas spoil quickly at room temperature. An alternative method is to place rutabagas in perforated plastic containers to prolong shelf life. Rutabagas in plastic bags have been kept at room temperature for 6 weeks without apparent deterioration compared with 3 to 4 days for waxed ones under the same conditions. Peeled and cut rutabagas (diced or sticks) have a storage life of 3 to 4 weeks when packaged in plastic bags and held at 0^0C.

Pests and Diseases

Insects

● Cabbage and Other Root Maggots

The ash-grey adult is a two-winged fly with black stripes on the mid-section which is smaller than housefly. Its legless larvae (maggots) are white, wedge-shaped and with dark feeding hooks at the front end. The pupae are some 6mm long, oval, hard and dark brown. In a year it may have 2-3 generations. Pupae overwinter in the soil near the roots of the host plant. Emergence of the adult flies coincides with flowering of rutabaga. Adult females lay their oval-shaped white eggs close to the ground near host plants at the base of the stem or in nearby crevices in the soil. Eggs hatch in 3-7 days and then maggots enter the roots and feed by rasping plant sap with a pair of hook-like mouth parts and tunnel the roots. These mature in 3-4 weeks and then pupate and from pupa fly emerges in 2-3 weeks. Its feeding may weaken the plants and may cause stunting with reduced seed yield. Severely infested plants wilt similar to those as in case of cutworm infestation. A small amount of tunnelling in rutabaga renders the crop unmarketable. To manage its attack, crop rotation should be followed with non-brassicae crops. Late and early rutabaga and other cruciferous crops should not be grown together as atleast there should be 200 metres distance between these. Moreover, floating mulch is quite useful as it deters the insects from egg laying near rutabaga plants. Insecticide application for root maggots can be timed to coincide with the start of egg-laying activity of each generation of adults. Drenching of soil with 1.6ml/l of profenophos 50EC in early May to mid-June, early July to late August and again in mid-September for all these three generations, respectively is quite useful in controlling cabbage maggot. Actual time and severity of infection fluctuate according to soils, locality and prevailing weather conditions.

● Wireworm (Click beetles)

Early in the spring, adult wireworms lay their eggs around grass roots. The larvae hatch in about a week and depending on the species, live for 1 to 5 years in the ground feeding on roots and seeds. Wireworms require 3 or more years to complete their life cycle. Wireworms of all sizes and ages are present in the soil throughout the year as there is always an overlapping of generations. The wireworms (larvae) are yellow, white or in darker shades of brown. Fully developed larvae may be 1.2-4cm long having a hard and smooth surface. A mature larva pupates usually in the fall and then after becoming adult beetle it starts feeding on rootabaga roots in spring. Wireworms having 3 pairs of legs are found in abundance in land left fallow for several years or in the heavy but poorly drained soil. To manage this pest planting should be avoided in infested soil

- **Aphids**

Aphids are soft-bodied, small and slow-moving insects where nymphs and adults both damage the crops through rasping their sap by clustering at the tender parts of the plant. They are often found in large colonies on the undersurface of leaves. A colony consists of winged and wingless adults and various sizes of nymphs. Aphids may be black, yellow or pink but mostly are in various shades of green. While feeding the saliva is injected to the plants that may carry plant viruses or may be toxic to the host plants. Feeding by large numbers discolours the foliage, curls the leaves and developing buds yellow & die. The plants may be covered by a sticky substance, called honeydew which is excreted by the aphids. Adults and larvae of ladybird beetles predate over aphids. Spraying of NSKE 4% or dimethoate 30EC at 0.2% at 10-days intervals will kill all the stages of aphids.

- **Imported cabbageworm** (*Pieris rapae*)
- **Cabbage looper** (*Trichoplusia ni*)
- **Diamondback moth larvae** (*Plutella xylostella*)
- **Purple-backed cabbageworm** (*Evergestis pallidata*)

Imported cabbageworm adult is a white butterfly which is seen fluttering from plant to plant laying eggs during summer. The eggs hatch into velvety-green larvae with one yellow stripe down the centre of its back and a paler broken stripe along each side and these do not loop while walking. Generally these are the most prevalent larvae on cabbage. Cabbage looper gets its name from the way it loops as it walks. It is a smooth green larvae with white stripes along the back and sides. These are heavy feeders and since they migrate into this region so usually the problem is in late summer. Diamondback moth is much smaller than the other cabbage worms. Three to six generations of 1.1cm long green larvae may develop each year. The larvae squirm actively when disturbed and produce many small holes in leaves of the host plant. Under high population conditions when leaves are severely eaten, then crown is damaged. Hot, dry conditions can cause populations of this insect to explode. Under cold, wet conditions this pest is not much of a problem. Purple-backed cabbageworm is not as commonly seen as the others but it can cause serious damage when high in numbers. The larvae are purple on the back and pale yellow along the sides. There are two generations per year. If larvae are causing severe leaf damage, use one of the recommended insecticides. Neem-seed kernel extract spraying thrice at 10 days interval at 4% after about 4 weeks of sowing or planting is quite effective. Thiocloprid 21.7C @ 0.2% or Spinosad 45SC @ 0.03% spraying are also quite effective to control all these pests.

- **Flea beetles** (*Phylloterata cruciferae*)

Flea beetles prevalent mostly during spring are small shiny black beetles which feed on the leaves of various cruciferous plants including rutabaga. Metalic bluish-green adult beetles are 1.5-2mm long and elongated-oval in shape. Adult beetles feed on the aerial parts, *i.e.* leaves, buds and stems by making holes there though its grubs on tender roots. Seedlings may be seen infested by a large number of adults which while feeding cause small holes. Emerging seedlings are completely destroyed. Damage becomes very serious during hot sunny weather. Per year it has only one generation. Whenever these are first obserbed on the plants, the plants are sprayed with contact insecticides such as lamda cyhalothrin 5EC at 0.05% or Dichlorvos 76EC 0.01%.

- **Cutworms** (*Agrotis ipsilon, A. segetum*)

Black and dark sided cutworms may cause considerable damage to rutabagas and turnips. These during spring as well as late in the season feed on the crown and roots leaving deep scars or burrows into the root. The late season infestations are usually detected only at harvest time. These cutworms are controlled through soil drenching of chlorpyriphos 20EC at 0.25%. However, Lorsban drenching is also very effective.

Diseases

- **Downy mildew** (*Peronospora parasitica*)

It may attack the plants at any stage of growth. Its attack on grown up plants shows as a white fuzzy fungus in patches on underside of leaves while on upper surface purplish and yellow-brown spots are seen, *vis-à-vis* stems become dark brown as well at flowering pods show blackish patches. The fungus overwinters on seed, on cruciferous weeds and perhaps in the soil. Its infection may cause internal discoloration of roots. Moisture and temperature are the important factors for appearance and spread of the disease.

In young seedlings it causes dicolouration but when attack is severe the plant dies. To manage this problem, immediately the field after harvesting should be ploughed and for further planting proper crop roatation should be followed. Before sowing the seed should be treated for 30 minutes at 50^{0}C. Plant density should be proper to allow proper aeration in the field. If available, resistant cultivars should be used. Moreover, as a prophylactic treatment 0.3% maneb or zineb spraying should be effected every week.

- **Powdery mildew** (*Erysiphe polygoni*)

Its infection causes a white powdery superficial patchy growth on the upper surface of leaves which later on may spread even to underside. First the leaves turn light green, then to yellow and finally to tan. Leaves may die and fall off the plant. Disease development is favoured by low relative humidity and water stress. For combating this problem, crop rotation with non-cruciferous crops should be followed, cruciferous weeds should be regularly eradicated and during crop season the rutabaga plants should be sprayed with Karathane or through sulphur dusting.

- **Clubroot (*Plasmodiophora brassicae*)**

It is a soil-borne fungus which infects the roots. It causes wilting and yellowing of aboveground parts. Large spindle shaped galls form on the roots. The fungus causing clubroot in rutabagas and cole crops is usually present in areas where these crops have been grown for many years. Its fungus survives in the soil for up to 7 years or even more. Soil pH less than 7.2 *cum* hot sunny condition favours the development and spread of the disease so to maintain this pH Ca and Mg in the soil should be added. Its infection causes pale-green colouration of foliage. Land will remain infested for 7 years or longer if once disease has occurred. The infection continues year after year due to occurrence of cruciferous weeds so field and nearby areas should be cleaned, waterlogging in any case should not be permitted, summer ploughing and solarization should be done, crop rotation of 7 years with non-cruciferous crops should be followed, if the soil is acidic liming should be done, prior to sowing the seedbed should be treated with 1% formaldehyde, and soil treatment with 0.1% Brassicol or Thiram. While planting the seedlings, the roots should be treated with 4% mercurous chloride. The manure obtained through dung of those animals fed with infected plants also contains its infection so that should also be avoided. 'York' is resistant variety so one should try to grow this instead of any other.

- **Damping off** (*Rhizoctonia solani, Botrytis* spp., *Fusarium* spp., *Phytophthora* spp., *Pythium* spp., *Sclerotinia* spp. and *Sclerotium* spp.)

These all the seven fungi are deeply associated with pre- and post-damping off problem, especially in the nursery. In case of pre-emergence damping off, these fungi kill the growing points even before emergence of sprouts through the soil, though in case of post-emergence, collar rotting of the seedlings occurs. *Rhizoctonia solani* also causes wirestem of other crucifers as well. In rutabagas, the fungus causes a superficial dry rot. Lesions may be sunken, shiny and brown with purplish rims that may develop into large irregular black craters with faint zonation and a scabby appearance. Infection may occur in the field or during storage. Field infections may be more severe where insecticides are not applied to control root maggots. The amount of decay increases rapidly when temperature is raised above 4^0C. Good control can be achieved by storing the roots at 0-2^0C. To avoid this problem, the nursery beds should either be burnt with stubbles or be solarized for 45-60 days during summer by ploughing and covering the soil with 200-400 gauge polythene sheet so that soil temperature is raised up to 60^{0}C or nursery beds should either be treated through 1% formaldehyde drenching or with 0.05^ $CuSO_4$ twice two weeks before sowing. After two weeks of sowing the seeds, the nursery beds should be drenched with 0.2% Thiram, Captan or Aretan. Moreover, when one speculates occurrence of damping off the young seedlings should be sprayed with 0.2% Captan, Difolatan-80 or Fytolan.

- **Blackleg** (*Phoma lingam*)

It is a seed-borne and seedling disease iof cole crops. Early signs of blackleg appear as small spots on leaves of young plants and due to distortion of the vascular bundle *cum* root system infected seedlings collapse. On stems the spots are more linear and often surrounded by purplish borders. Stem lesions at the soil line usually extend to the root system causing dark cankers. Fibrous root system may be destroyed though new roots sent out above the lesion may keep the plants alive. Plants may wilt abruptly and die. Dark cankers may develop in fleshy roots of turnip and rutabaga or may develop as a dry rot on stored roots. To get rid of this problem, before sowing the seeds should be treated for 30 minutes at 50^{0}C together with 0.3% Thiram or Captan, and when standing crop is showing its infection, 0.2% spraying of Captan or Fytolan should be done.

- **Black rot lesions** (*Xanthomonas campestris* pv. *campestris*)

Occurs in the temperate and subtropical zones of the country experiencing 15-22^{0}C temperature during crop duration coupled with copious rainfall or dew. The occurrence of this disease is first seen at margins of leaves. Its infection causes tissues turning yellow and the lesion progressing towards the centre of the leaf, usually in a v-shaped pattern with the base of the v towards the mid-rib. The veins become dark and discolouration frequently extends to the main stem and proceeds upward and downward, and rootabaga roots also show black discolouration.

- **Common scab** (*Streptomyces scabies*)

Though scab is a well known disease of potato but now it is causing severe damage even on rutabaga. Its infection, especially in dry soils causes circular to oval lesions scattered over the root surface which after coalescing form the band around the roots just below the soil line. Affected tissues may consist of a tan-coloured superficial or raised layer or the tissues may be pitted and darken following seccondary decay. During the period of rapid expansion of roots when they are most susceptible there should be sufficient moisture in the soil. To get rid of this problem, crop rotation of non-susceptible crops should be followed. Application of lime against clubroot disease is though effective but aggravates scab problem and so is the case even when wood ashes are added.

- **Water-core or Brown Heart**

This is caused due to boron deficiency in soils where rutabaga is grown. At first appearance the roots are firm but show dark water-soaked lesions in the flesh which eventually turn brownish and spongy inside, especially in storage. It is corrected through foliar application of boron.

References

Braverman, L.E. and R.D. Utiger, 1991. Werner and Ingbar's The Thyroid: A Fundamental and Clinical Text (6th ed.), pp. 371-2. J.B. Lippincott Company, Philadelphia, Pennsylvania.

Dastan, K., A. Tehranifard and M. Vazirian, 2011. Investigation of anti-pseudomanal activity of *Brassica napus* L. *Intern. Conf. Environ. & Computer Sci. (IPCBEE)*, pp. 144-6. IACSIT Press, Singapore.

Dixon, G.R. 2007. Vegetable Brassicas and Related Crucifers, pp. 6-36.. CABI, Oxfordshire, UK.

https://academics.hamilton.edu/foodforthought/our_research_files/rutabaga.pdf.

https://ebrary.net/28093/environment/rutabaga.

https://en.wikipedia.org/wiki/Rutabaga.

https://harvesting-history.com/the-rutabaga-and-the-turnip.

https://plantvillage.psu.edu/topics/rutabaga/infos.

https://www.britannica.com/plant/rutabaga.

https://www.gov.nl.ca/ffa/files/agrifoods-plants-pdf-turnip.pdf.

Jakopic, J., R. Veberic, A. Slatnar, 2021. Changes in quality parameters in rutabaga (*Brassica napus* var. *napobrassica*) roots during long term storage. *LWT*, **147**:111587.

Jones, D.A. 1998. Why are so many food plants cyanogenic? *Phytochem.*, **47**(2): 155–62.

Olsson, K. and L. Jeppsson, 1984. Undesirable glucosinolates in *Brassica* vegetables. *Acta Hort.*, **163**: 83–4. doi:10.17660/ActaHortic.1984.163.9.

Pasko, P., K. Okon, M. Krosniak, E. Prochownik, P. Zmudzki and J. Kryczyk-Kozioł, 2018. Interaction between iodine and glucosinolates in rutabaga sprouts and selected biomarkers of thyroid function in male rats. *J. Trace Elem. in Medic.& Biol.*, **46**:110-6.

Pedras, M.S.C., S. Montaut and M. Suchy, 2004. Phytoalexins from the crucifer rutabaga: structures, syntheses, biosyntheses and antifungal activity. *J. Org. Chem.*, **69**(13): 4471–6. doi:10.1021/jo049648a. PMID 15202903.

Sandell, Mari A. and Paul A.S. Breslin, 2006. Variability in a taste-receptor gene determines whether we taste toxins in food. *Curr. Biol.*, **16**(18): R792-4. doi:10.1016/j.cub.2006.08.049. PMID 16979544. S2CID 17133799.

Stefanucci, A., G. Zengin, E.J. Llorent-Martinez, M.P. Dimmito, A. Della Valle and S. Pieretti, 2020. Chemical characterization, antioxidant properties and enzyme inhibition of rutabaga root's pulp and peel (*Brassica napus* L.). *Arab. J. Chem.*, ***13***(9): 7078-86.

Verkerk, R., M. Schreiner, A. Krumbein, E. Ciska, B. Holst, I. Rowland, R. De Schrijver, M. Hansen, C. Gerhäuser, R. Mithen and M. Dekker, 2009. Glucosinolates in Brassica vegetables: The influence of the food supply chain on intake, bioavailability and human health. *Mol. Nutr. Food Res.*, **53**:S219–S265.

6

Horseradish/Pepper Rod/Red Cole (*Armoracia rusticana* G.Gaertn., B.Mey. & Scherb.)

Family: Brassicaceae.

Synonyms: *Armoracia armoracia* Cockerell ex Daniels; *A. lapathifolia* Gilib.; *A. rustica* Schur; *A. sativa* Bernh.; *Cardamine armoracia* (L.) Kuntze; *Cochlearia armoracia* L.; *C. lancifolia* Stokes; *C. lapathifolia* Gilib.; *C. rusticana* Lam.; *C. variifolia* Salisb.; *Crucifera armoracia* E.H.L.Krause; *Nasturtium armoracia* (L.) Fr.; *Raphanis magna* Moench; *Raphanus rusticanus* Garsault; *Rorippa armorac ia* (L.) Hitchc.; *R. rusticana* (G. Gaertn., B. Mey. & Scherb.) Godr.

Other names: Red cole or Pepper rod (plantvillage.psu.edu); in Central and Eastern Europe horseradish is called *khren, hren* and *ren* (in various spellings like *kren*) in many Slavic languages, in Austria, in parts of Germany (where the other German name *Meerrettich* is not used), in North-East Italy, and in Yiddish (transliterated as *khreyn*).

Introduction and Origin

Armoracia is a genus of 3 species of erect perennials that grow at a range of habitats at low altitudes ~ the wasteland, streamsides and roadsides in Eurasia and E. USA. Their leaves are simple or pinnatifid, dock-like, toothed and coarse. Basal roots emerge from the deep and fleshy or woody taproots. Their cruciferous, small, white and 4-petalled flowers are borne on leafy *cum* branched stems in large terminal racemes or panicles from late spring to late summer, and the fruits are oblong to obovate.

Armoracia rusticana

Horseradish (*Armoracia rusticana* syn. *Cochlearia armoracia*), a perennial plant native to SE Europe and possibly W. Asia (Misra *et al.*, 2023; en.wikipedia.org/wiki/Horseradish), has been grown for its large roots throughout the temperate zones for over 2,000 years (foodprint.org), typically as a spice and condiment but also for medicinal purposes (www.healthline.com). The word horseradish is combination of the word 'horse' (used in a figurative sense to mean strong or coarse) and the word 'radish' (www.theplantlist.org). Horseradish is called 'seiyowasabi' or 'Western wasabi' in Japan (en.wikipedia.org/wiki/Horseradish) where its root is commonly used as a substitute for true wasabi (*Eutrema japonicum*) in Japanese cuisine (www.britannica.com). Horseradish may become invasive unless carefully managed. Its deep roots can be difficult to eradicate so it is best planted in a container to limit its spread. Horseradish is mainly sold as long pieces of root (known as thongs), available in spring (www.rhs.org.uk).

Botany

Its chromosome number is 2n=4x=32. Horseradish is a vigorous and fully hardy clump-forming glabrous perennial herb that goes dormant over winter, the leaves dying down in autumn then re-sprouting from the roots every spring (www.rhs.org.uk). The plant grows up to 1.5m tall with more than 45cm spread and is cultivated primarily for its long, fleshy, pungently aromatic, fusiform or cylindrical taproot that can grow to about 60cm in loose soils (Misra *et al.,* 2023; Lim, 2013; Shehata *et al.,* 2009) and can be in a variety of shades of beige and white. Around the main root and near the collar of the crown, thin lateral roots are also formed. If left undisturbed for years, the root can penetrate 3-4m deep into the soil with a lateral spread of about 1m having many secondary roots and rootlets in a tangled mass (a-z-animals.com; Shehata *et al.,* 2009).

Sections of Roots of the Horseradish Plant

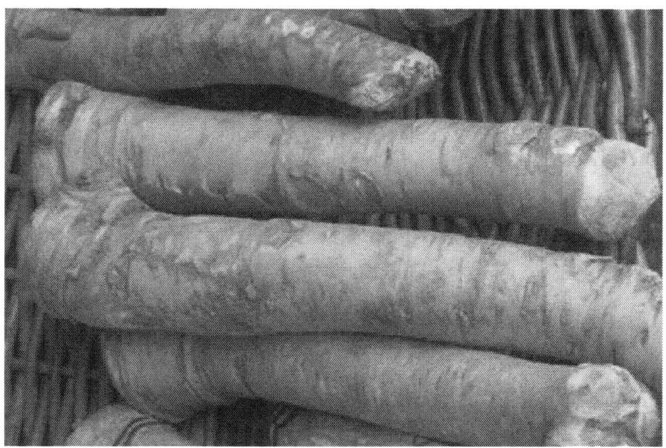

By Anna reg - Own work, CC BY-SA 3.0, https://commons.wikimedia.org/w/index.php?curid=9740114

The large glossy bluish-green and hairless basal leaves that grow to a length of 30 to 100cm and may be mistaken for docks (*Rumex*) are long-petioled with elliptic-ovate or oblong-lanceolate, cordate at the base and unevenly crenate lamina. The lower cauline leaves have shorter petioles and may be lobed or pinnately comb-shaped, having linear-oblong, entire-margined or serrate sections. The upper cauline leaves have narrow bases and are mostly sessile, oblong or lanceolate, unevenly crenate to serrate and obtuse at the apex. The uppermost leaves of the plant are linear or almost entire-margined (Lim, 2013; Shehata *et al.,* 2009; Stace, 2019). The leaf morphology varies through the season. In the summer horseradish produces laminate leaves while pinnate during autumn (Agneta *et al.,* 2013).

Horseradish produces numerous fragrant flowers with 5 to 7cm long pedicels borne on axillary racemes that have four 2.5-3mm long sepals that are broadly ovate and with a membranous white margin; four 5-7mm long white petals that are broadly obovate; and six 2.5mm long tetradynamous stamens where outer ones are 1.5mm long. The stigma is broad, round and gently 2-lobed. Small fruits (4-6mm long) are globose to obovate siliqua pods that have persistent styles on 2cm long upright-spreading stems (Lim, 2013; Shehata *et al.,* 2009; www.britannica.com). Only a few seeds (0-6) per siliqua are produced and when mature they are smooth and brown (Agneta *et al.,* 2013).

Foliage of the Horseradish Plant

By Raffi Kojian - http://Gardenology.org, CC BY-SA 3.0, https://commons.wikimedia.org/w/index.php?curid=12734857

Varieties

Horseradish is generally divided into two types (horticulture.oregonstate.edu):

- **Common types**: They have broad crinkled leaves and are considered to have superior quality but are more susceptible to diseases.
- **Bohemian types**: They have smooth, narrow and poorer leaves though with better disease resistance. 'Maliner Kren' ~ a Bohemian type horseradish is a German variety from which many local selections have been made. It has most vigorous growth and largest roots. Bohemian and improved Bohemian form the basis of the current industry.

Nutritional Value

Source:
USDA: U.S. DEPARTMENT OF AGRICULTURE via https://fdc.nal.usda.gov/fdc-app.html

Per 100 g

Name	Amount	Unit
Water	85.08	g
Energy	48	kcal
Protein	1.18	g
Total lipid (fat)	0.69	g
Carbohydrate, by difference	11.29	g
Fiber, total dietary	3.3	g
Sugars, total including NLEA	7.99	g
Calcium, Ca	56	mg
Iron, Fe	0.42	mg
Magnesium, Mg	27	mg
Phosphorus, P	31	mg
Potassium, K	246	mg
Sodium, Na	420	mg
Zinc, Zn	0.83	mg
Copper, Cu	0.058	mg
Selenium, Se	2.8	µg
Vitamin C, total ascorbic acid	24.9	mg
Thiamin	0.008	mg
Riboflavin	0.024	mg
Niacin	0.386	mg
Vitamin B-6	0.073	mg
Folate, total	57	µg
Folate, food	57	µg
Folate, DFE	57	µg
Choline, total	6.5	mg
Carotene, beta	1 µg	
Lutein + zeaxanthin	10	µg
Vitamin E (alpha-tocopherol)	0.01	mg
Vitamin K (phylloquinone)	1.3	µg
Fatty acids, total saturated	0.09	g
Fatty acids, total monounsaturated	0.13	g
Fatty acids, total polyunsaturated	0.339	g

1 Tablespoon (15 g)

Name	Amount	Unit
Water	12.8	g
Energy	7.2	kcal
Protein	0.177	g
Total lipid (fat)	0.103	g
Carbohydrate, by difference	1.69	g
Fiber, total dietary	0.495	g
Sugars, total including NLEA	1.2	g
Calcium, Ca	8.4	mg
Iron, Fe	0.063	mg
Magnesium, Mg	4.05	mg
Phosphorus, P	4.65	mg
Potassium, K	36.9	mg
Sodium, Na	63	mg
Zinc, Zn	0.124	mg
Copper, Cu	0.009	mg
Selenium, Se	0.42	µg
Vitamin C, total ascorbic acid	3.74	mg
Thiamin	0.001	mg
Riboflavin	0.004	mg
Niacin	0.058	mg
Vitamin B-6	0.011	mg
Folate, total	8.55	µg
Folate, food	8.55	µg
Folate, DFE	8.55	µg
Choline, total	0.975	mg
Carotene, beta	0.15	µg
Lutein + zeaxanthin	1.5	µg
Vitamin E (alpha-tocopherol)	0.002	mg
Vitamin K (phylloquinone)	0.195	µg
Fatty acids, total saturated	0.013	g
Fatty acids, total monounsaturated	0.019	g
Fatty acids, total polyunsaturated	0.051	g

Though horseradish root is high in vitamin C and fibre, *vis-à-vis* contains decent amounts of folate, potassium, calcium and manganese, it is unlikely that an individual ingests enough horseradish in one sitting to make much of a nutritional dent. According to the agriculture folks at North Carolina State University, the consumption of a large amount of either the root or leaves of the horseradish plant could cause profuse sweating, irritation of the stomach and intestines, and loss of strength (foodprint.org).

Phytochemicals

The taste and aroma of horseradish is due to presence of sulphur glucosides known as glucosinolates (Ciska *et al.*, 2017) that are secondary metabolites and play a major role in the defensive system of plants against insects, certain food bacteria and fungi (Singh, 2017). Their structure consists of β-thioglucoside N-hydroxysulfates with a side chain and a sulphur-linked β-D-glucopyranose moiety (Nguyen *et al.*, 2013). Glucosinolates are usually classified according to the structure of the side chain in aliphatic, aromatic and indole glucosinolates that result from different amino acid precursors (Agneta *et al.*, 2013). Though glucosinolates are stable and soluble in water, they degrade at temperatures above 100°C. Indole glucosinolates are more unstable than aliphatic glucosinolates at temperatures below 110°C. Distribution of glucosinolates ~ both quantitative and qualitative, varies with the plants part, age, environmental factors, etc. (Nguyen *et al.*, 2013; Agneta *et al.*, 2013).

Using reversed-phase liquid chromatography coupled with electrospray ionization and a hybrid quadrupole linear ion trap and fourier transform ion cyclotron resonance mass spectrometry (LC-ESI-FTICR MS), the glucosinolates isolated from sprouts and root were sinigrin, 4-hydroxyglucobrassicin, glucobrassicin, gluconasturtin, 4-methoxyglucobrassicin, glucoiberin, gluconapin, glucocochlearin, glucoconringianin, glucosativin, glucoibarin, 5-hydroxyglucobrassicin, glucocapparilinearisin or glucobrassicanapin, glucotropaeolin, glucoarabishirsutain and methylsulfonyl-oxo-ethyl-glucosinolate (Agneta *et al.*, 2012), 2(S)-hydroxy-2-phenylethyl-glucosinolate (glucobarbarin) and/or 2(R)-hydroxy-2-phenylethyl-glucosinolate (epiglucobarbarin) (Agneta *et al.*, 2014), glucoraphanin, glucoraphenin and napoleiferin in horseradish (Ciska *et al.*, 2017). Out of several glucosinolates in horseradish, sinigrin represents 64-86% followed by gluconasturtin which represents 4-26% (Lim, 2013).

When horseradish is cut or grated, glucosinolates hydrolysis is triggered, and is mediated by membrane bound myrosinase enzyme existing in vacuoles until damage is caused to the cells (Nguyen *et al.*, 2013; Ciska *et al.*, 2017). An unstable thiohydroximate-*O*-sulphonate gets formed when myrosinase and glucosinolates come in contact, which then rearranges to a variety of compounds including isothiocyanates, thiocyanates, nitriles, epithionitrile and oxazolidine thione, depending on the reaction conditions ~ isothiocyanates are the most common products (Bones and Rossiter, 1996).

In 2004, volatile organic compounds found using SPME–GC–MS (solid phase microextraction coupled with gas chromatography mass spectrometry) in fresh horseradish samples were identified as allyl isothiocyanate (AITC), 4-isothiocyanato-1- butene, and 2-phenylethyl isothiocyanate (PEITC). Allyl isothiocyanate represented 81% of the overall compounds found; and when horseradish has been cut and kept at 5°C for 12 hours the profile of volatile compounds changed and new compounds were found. Some of them decreased their concentrations and 2-phenylethyl isothiocyanate became the most abundant compound (Agneta *et al.*, 2013).

Isothiocyanates have been reported as unstable compounds due to their electrophilic properties. It is assumed that the side chain influences the decomposition rate and route of different isothiocyanates. Even though around 200 isothiocyanates have been identified though merely stability of allyl isothiocyanate only have been studied probably due to the fact that allyl isothiocyanate is the main occurring isothiocyanate found in plants and the simplicity of its structure (repositorio.ul.pt).

Detailed phytochemical analysis additionally revealed a variety of compounds, not described until now for horseradish roots: the flavonol glycosides kaempferol-3-O-β-d-xylopyranosyl-(1‴→ 2″)-β-d-galactopyranoside, kaempferol-3-O-α-l-rhamnopyranosyl-(1‴→ 6″)-β-d-glucopyranoside, kaempferol-3-O-β-d-glucopyranoside, Kaempferol-3-O-β-d-xylopyranosyl-7-O-β-d-glucopyranoside, Kaempferol-3-O-β-d-xylopyranosyl-(1⁗→ 2‴)-β-d-galactopyranoside-7-O-β-d-glucopyranoside, the oxo-indole derivative spirobrassinin, the phenylthiazole 2-methylsulfanyl-4-phenyl-4,5-dihydro-1,3-thiazole, a series of lysophophatidylethanolamine and 13 different N-phenylpropenoyl-L-amino acids (Negro *et al.*, 2022).

The enzyme horseradish peroxidase (HRP) that is found in horseradish plant has the ability to amplify a weak signal and increase detectability of a target molecule. For this reason, it is used in molecular biology and biochemistry (Bladha *et al.*, 2011).

Flavour

Horseradish has a slightly sweet and slightly bitter but sharp spicy flavour that is similar to but not as intense as wasabi (a-z-animals.com). It is more pungent than bitter (roundglass.com). Intact horseradish root has little aroma. The distinctive pungent taste of horseradish is from the compound allyl isothiocyanate. Upon crushing the flesh of horseradish, the enzyme myrosinase is released and acts on the glucosinolates sinigrin and gluconasturtiin which are precursors to the allyl isothiocyanate. When cut or grated the enzymes from within the plant cells digest sinigrin (a glucosinolate) to produce allyl isothiocyanate (mustard oil) which irritates the mucous membranes of the sinuses and eyes. Once exposed to air or heat, horseradish loses its pungency, darkens in colour and develops a bitter flavour (en.wikipedia.org/wiki/Horseradish).

The allyl isothiocyanate serves the plant as a natural defense against herbivores. Since allyl isothiocyanate is harmful to the plant itself, it is stored in the harmless form of the glucosinolate separate from the enzyme myrosinase. When an animal chews the plant, the allyl isothiocyanate is released repelling the animal (Cole, 1976). Allyl isothiocyanate is an unstable compound, degrading over the course of days at 37°C (Ohta *et al.*, 1995). Because of this instability the horseradish sauces lack the pungency of the freshly crushed roots.

Culinary Uses

Horseradish root is generally not cooked but grated or mashed and served as a condiment or spice. Cooking grated horseradish greatly diminishes the flavour and pungency of the root so it should be added at the end of cooking. Grated mash should be used immediately or preserved in vinegar for best flavour. Once exposed to air or heat it darkens in colour and starts losing its pungency so in taste over time it becomes unpleasantly bitter. Because of this instability horseradish sauces lack the pungency of the freshly crushed roots. The grated root is commonly mixed with dairy products such as cream or sour cream to add its peppery bite (foodprint.org; a-z-animals.com; Misra *et al.*, 2023).

Horseradish sauce is made from grated horseradish root and vinegar while in some other cases the vinegar may be substituted with certain other products such as lemon juice or citric acid. Horseradish sauce is normally white or creamy beige but will darken as it loses flavour. Refrigerated horseradish sauce lasts about three months (foodprint.org; a-z-animals.com; en.wikipedia.org/wiki/Horseradish; Misra *et al.*, 2023). The grated root is commonly mixed with dairy products such as cream or sour cream to add its peppery bite.

Horseradish sauce can be served with roasted beef and can be used in a number of other dishes including sandwiches or salads. In England, a blend of mustard and grated horseradish called 'Tewkesbury mustard' is very popular. In Russia, horseradish root is usually mixed with grated garlic and small amount of tomatoes for colour. In USA, the term 'horseradish sauce' refers to grated horseradish combined with mayonnaise or salad dressing. The Japanese condiment 'wasabi' although traditionally prepared from the true wasabi plant (*Wasabia japonica*) is now usually made with horseradish due to scarcity of wasabi plant (en.wikipedia.org/wiki/Horseradish; Misra *et al.*, 2023). Though not commonly eaten but horseradish greens are edible and have a flavour similar to those of the roots.

Potential Health Benefits

Like its mustard-family relatives, horseradish contains a powerful anti-inflammatory phytochemical called sinigrin. It is also rich in antioxidants and contains minerals such as calcium, potassium, magnesium and zinc. Evidently horseradish can impede growth of certain cancer cells and can also help with respiratory infections.

Horseradish has been used in traditional medicine since early times for its stimulant, rubefacient, stomachic, diaphoretic, diuretic, expectorant, antiseptic and vermifuge properties (Lim, 2013). Syrup prepared from horseradish was reported to be a reliever for hoarseness and whooping cough and to remove freckles when applied on the skin. The root is high in vitamin C and since a long it is being used as a cure for rheumatism, dropsy and scurvy (Shehata *et al.*, 2009). In herbal medicine, horseradish is used for sinus remedies and other mucusy ailments such as cold and flu because its pungency helps to relieve nasal discharge and lung congestion. It is also used as a natural remedy for urinary infections and for intestinal worms (foodprint.org).

Glucosinolates and isothiocyanates in horseradish may protect against cancer by inhibiting the growth of cancer cells, as well as promoting their death (Yuan *et al.*, 1999; Auborn *et al.*, 2003). Some horseradish compounds such as sinigrin may also act as antioxidants and fight cell damage caused by free radicals that may increase the risk of diseases including cancer when levels become too high in the body (Meng *et al.*, 2000; Lampe and Peterson, 2002). When horseradish root is cut, allyl isothiocyanate oil is released, which may have powerful antibacterial and antifungal properties and may fight a range of dangerous bacteria including *E. coli*, *H. pylori*, and *Salmonella* (Kojima *et al.*, 1994; Sakao and Singh, 2012), and may prevent the growth of fungi that may lead to chronic nail infections (Lee *et al.*, 2013).

Bio-medical Uses

Horseradish enzyme peroxidase (HRP) is isolated from horseradish plant which is used extensively in molecular biology and biochemistry primarily for its ability to amplify a weak signal and increase detectability of a target molecule (Bladha *et al.*, 2011). HRP has been used in decades of research to visualize under microscopy and assess non-quantitatively the permeability of capillaries, particularly those of the brain (Lossinsky and Shivers, 2004).

Cultivation

Horseradish is very easy to grow and can become hard to control if grown in the open ground as its deep roots form a vigorous spreading clump that becomes difficult to eradicate. It will re-grow from any piece of root left in the ground. Therefore, it is recommended to grow this in large containers or in the area where it can be grown under control (www.rhs.org.uk).

Propagation

Though horseradish seeds are occasionally available but growing through seeds is much slower to reach harvesting size (www.rhs.org.uk). Therefore, horseradish is usually vegetatively propagated by dividing crowns or from root cuttings. If crown division is being used, all but the terminal shoots should be removed to prevent later formation of multiple crowns and the side roots should also be sheared off to minimize branching and crooked formation (www.britannica.com).

It can be grown from supermarket bought root as it is not treated with any chemicals that affect growth. Root pieces are small, pencil size lateral roots called sets which grow from the main root. Since horseradish root exhibits polarity, it is important to make a note of which end of the set is the top and which is the bottom in order to plant the piece in the correct orientation. A good way of doing this is to use a straight cut across the top end of the set and an oblique cut at the bottom. If planting by dividing an established crown, the plant should be carefully dug from the soil and split into four equal pieces each with some leaf and root (plantvillage.psu.edu).

Climate and Soil

Horseradish thrives in the rich moist loamy (a-z-animals.com) and cool soils of temperate climates (roundglass.com) in full sun or partial shade (www.rhs.org.uk) at temperatures 15.5-28°C. If grown as an annual, the plant requires a long growing season with a warm summer and cooler temperatures in late summer and fall to allow the flavour to develop in the root (plantvillage.psu.edu).

Soil

Horseradish can be grown in most soils but that should be well-drained, deep, fertile with a soil pH between 6 to 7.5 (plantvillage.psu.edu). Shallow and stony soil should be avoided (a-z-animals.com).

Field Preparation and Planting

Horseradish sets should be planted in spring or early summer, ideally in a large container filled with multi-purpose compost. It can also be planted in a small raised bed or similarly enclosed site (www.rhs.org.uk). The soil should be incorporated with composted organic matter at least 15q/h with the final preparation of soil, along with all purpose fertilizer. Cultivators usually propagate horseradish as soon as the soil is workable in the spring by placing pieces of pencil-sized roots in the soil at a slight angle (45° angle) (www.britannica.com) so that the top of the set is higher in the soil than the bottom end. They should be planted 30-45cm apart and at a depth of about 5 cm (www.rhs.org.uk; a-z-animals.com).

Irrigation and Fertilizers

Care should be taken not to overwater the plants or allow them to become too dry as this will affect the flavour of the harvested root. When grown in a container, horseradish needs regular watering throughout the growing season, especially in warm weather. Horseradish on raised beds should be provided with approximately 2.5-5cm of water each week. Additional irrigation may be required during dry spells. Once established, horseradish in raised beds should only need watering in long dry spells. The leaves may flop if plants get too dry. A thick layer of mulch, such as garden compost around horseradish clumps can be used to help hold moisture in the soil (www.rhs.org.uk).

A balanced liquid fertilizer can be applied in summer especially to horseradish plants in containers (www.rhs.org.uk). An additional dose of fertilizer high in potassium and phosphorous but low in nitrogen should be applied 4-8 weeks after planting. Excess nitrogen should be avoided as it will encourage too much top growth on the plants and will also cause the roots to branch (plantvillage.psu.edu).

Interculture

Adding organic mulch around horseradish plant will help to suppress weed growth and retain moisture. If no mulch is added then any weed coming up around young horseradish should be gently removed. Many growers make use of the practice of lifting and suckering their horseradish plants. Once the plant has begun to grow, it is gently lifted to expose the crown and any branching roots are removed so that the plant can focus its energy on growing the main root (plantvillage.psu.edu). The faded leaves should be cut back once they die down in late autumn or before start of new growth in spring (www.rhs.org.uk).

Harvest

Horseradish is harvested after one growing season once the leaves die back in autumn or winter, ideally after frost, which enhances its flavour. A horseradish growing in container can be tipped out of it, taking the roots from the clump and replanting the remaining roots back in the container in fresh potting compost (www.rhs.org.uk). If the horseradish clump is growing in a raised bed, this can simply be dug up with a garden fork and divided, the main root is harvested and one or more large offshoots of the main root are replanted to produce next year's crop. Harvesting some of the roots annually helps to control the spread of this very vigorous plant. Older roots left in the ground become woody after which these are no longer culinarily useful though older plants can be dug and re-divided to start new plants (en.wikipedia.org/wiki/Horseradish; www.rhs.org.uk).

Post Harvest

After harvesting, roots are washed and trimmed for sale (www.britannica.com). Uncut horseradish roots can be kept for several weeks in the crisper drawer of the fridge. Cut horseradish should be used right away. Grated fresh horseradish preserved in vinegar will keep for several months in the fridge (foodprint.org).

Pests and Diseases

Horseradish is a hardy and vigorous plant that is rarely troubled by any pest or disease though there are some causing damage to the plants. Chemical sprays should be avoided unless very necessary or selected carefully as it may also damage populations of natural enemies of insect-pests.

Insect-Pests

- **Cabbageworms** (*Pieris rapae*): These are common larval pests of small white butterfly attacking horseradish. Butterfly having black spots on the forewings are commonly seen flying around plants during daytime. There are three to five overlapping generations a year. Its adults start appearing in gardens after the last frost and are a problem through the remainder of the growing season. Its caterpillars are velvety green with faint yellow stripes, 2.5cm in length when fully grown, these move sluggishly when prodded and these overwinter in green pupal cases. Mature caterpillars chew and create ragged holes in the leaves leaving the large veins intact. Handpicking is an effective control strategy in home gardens (en.wikipedia.org; www.extension.umn.edu).

- **Cabbage looper** (*Trichoplusia ni*): These looper-caterpillars have a wide host range and overwinter as pupae under crop debris in soil. Adult insect is a dark coloured moth. Caterpillars are pale-green with a white line running down either side of their body. These are easily distinguished by the way they arch their body when moving. Female butterfly lays white or pale-green single egg usually on the lower leaf surface close to the leaf margin. The larvae cause extensive damage by feeding on leaves and other parts of plants and making holes in leaves. Its population is usually controlled by natural enemies. Loopers can be hand-picked or through application of *Bacillus thuringiensis* which effectively kills younger larvae (plantvillage.psu.edu).
- **Horseradish flea beetle** (*Phyllotreta armoraciae*): It is a small (1.5–3mm) dark coloured beetle that jumps when disturbed. The pest overwinters on nearby weed species, in plant debris or in the soil and may go through a second or third generation in one year. Young plants and seedlings are particularly susceptible. While feeding the beetles make small holes in leaves giving the foliage a characteristic 'shothole' appearance so plant growth is reduced and in its severe attack the plant may be killed also. A thick layer of mulch may help prevent beetles reaching the soil surface. Application of diamotecoeus earth or neem oil is also effective in controlling this. Insecticides such as carbaryl, spinosad, bifenthrin and permethrin can provide adequate control of beetles for up to a week but will need to be reapplied (plantvillage.psu.edu). Whenever their infestation is first observed, the plants are sprayed with contact insecticides such as lamda cyhalothrin 5EC at 0.05% or Dichlorvos 76EC 0.01%.

Diseases

- **Brittle root** (*Spiroplasma citri*): This is a fungal disease being transmitted through beet leafhoppers. It is one of the most destructive diseases of horseradish. In its attack plants show poor growth with chlorotic leaves which collapse and dry out. Roots become brittle and discoloured dark brown with a dark ring. It is controlled by controlling the insect vector so appropriate insecticide such as lamda cyhalothrin 5EC at 0.05% or

Dichlorvos 76EC 0.01% spraying can be used to kill the leafhopper vector (plantvillage.psu.edu)

- **Cercospora leaf spot** (*Cercospora amoraciae*): Emergence of this fungal disease is favoured by periods of cool and wet weather and is spread by splashing water. Its infection causes round or angular tan spots with lighter centres on leaves, so leaves die and fall so plants become defoliated. Infected plants along with weeds should be removed and destroyed. One should avoid working in the field when the plants are wet. Hot water treatment at 50^{0}C for 30 minutes to the seeds will eliminate the fungus prior to planting (plantvillage.psu.edu). Under field condition, 0.1% thiophanate methyl or Baycor; or 0.05% Difenaconazole/Tricyclazole/ native spraying will be quite effective.
- **Ramularia leaf spot** (*Ramularia cynarae*): Fungus overwinters on plant debris. Its initial symptoms appear as yellow-green circular patches between leaf veins which become distinct lesions with irregular margins where centres of lesions may dry out and drop from plant producing a shothole appearance and finally entire leaf may dry out. Proper field sanitation should be followed including weeding, sprinkler irrigation should be avoided, infected plants should immediately be removed and destroyed, crop rotation with non-host crops for a period of 3 years should be followed and use of appropriate fungicides if required (plantvillage.psu.edu). Sprayiung of plants with 0.2% maneb or zineb may prove effective.
- **White rust** (*Albugo candida*): Fungus overwinters in plant debris or on perennial plants in the vicinity and its spores spread usually through wind. The disease emergence is favoured by cool weather and periods of extended rainfall. Its infection causes pale-yellow areas on upper leaf surface followed by creamy white pustules on leaf undersides, *vis-a-vis* leaf curling *cum* distortion with pustules covering entire leaf surface. Proper field sanitation should be maintained, no weed should be allowed to grow in between, overhead irrigation should be avoided, infected plants should be removed and destroyed, and crop rotation with non-host crop for a period of 3 years should be followed. Spraying of the fungicides such as mancozeb, bupirimate with triforine or penconazole at 0.2% may prove quite effective.
- **Bacterial leaf spot** (*Phytomonas campestre armoraciae*): This bacterium overwinters on plant debris and its emergence is favoured by wet weather. Its infection causes small translucent spots on leaves in between the veins, which later on become black. These spots may enlarge after periods of

rainfall. Its infection causes leaf curling and drying up. To get rid of this disease proper field sanitation should be maintained, infected leaves should be removed and burnt and spraying with Bordeaux mixture or any other copper-based fungicide should be used. Bactrinashak + Blitox at 300 ppm should be sprayed

- **Turnip mosaic virus** (TuMV): This virus causes black streaks on petioles and green and yellow mottling on leaves which make the leaves underdeveloped and crinkled. Though horseradish plants can tolerate low levels of infection, weeds and after-harvest plant debris should be removed and destroyed. As soon as first symptom on any plant appears that should be uprooted and destroyed and the crop should be sprayed with 0.05% Dimethoate, Metasystox or Monocrotophos to control the aphid vector. However, before opting for chemical control of vector aphids, mineral oil or neem oil @ 2% at 10 days interval may prove beneficial.

References

Agneta, R., A.R. Rivelli, E. Ventrella, F. Lelario, G. Sarli and S.A. Bufo, 2012. Investigation of glucosinolate profile and qualitative aspects in sprouts and roots of horseradish (*Armoracia rusticana*) using LC-ESI-hybrid linear ion trap with fourier transform ion cyclotron resonance mass spectrometry and infrared multiphoton dissocia. *J. Agric. Food Chem.*, **60**(30): 7474–82.

Agneta, R., C. Mollers and A.R. Rivelli, 2013. Horseradish (*Armoracia rusticana*), a neglected medical and condiment species with a relevant glucosinolate profile: A review. *Genet. Resour. Crop Evol.*, **60**(7): 1923–43.

Agneta, R., F. Lelario, S. De Maria, C. Mollers, S.A. Bufo and A.R. Rivelli, 2014. Glucosinolate profile and distribution among plant tissues and phenological stages of field-grown horseradish. *Phytochem.*, **106**: 178–87.

Auborn, K.J., S. Fan, E.M. Rosen, L. Goodwin, A. Chandraskaren, D.E. Williams and T.H. Carter, 2003. Indole-3-carbinol is a negative regulator of estrogen. *J. Nutrit.*, **133**(7): 2470S-2475S.

a-z-animals.com/blog/wasabi-vs-horseradish/.

Bladha, K. Wedelsbäck and K.M. Olssonb, 2011. Introduction and use of horseradish (*Armoracia rusticana*) as food and medicine from antiquity to the present: Emphasis on the nordic countries. *J. Herbs, Spices & Med. Pl.*, **17**(3): 197–213. doi:10.1080/10496475.2011.595055. S2CID 84556980.

Bones, A.M. and J.T. Rossiter, 1996. The myrosinase-glucosinolate system, its organization and biochemistry. *Physiol. Plant.*, **97**: 194–208.

Ciska, E., M. Horbowicz, M. Rogowska, R. Kosson, N. Drabińska and J. Honke, 2017. Evaluation of seasonal variations in the glucosinolate content in leaves and roots of four European horseradish (*Armoracia rusticana*) landraces. *Polish J. Food Nutr. Sci.*, **67** (4):117-23.

Cole, R.A. 1976. Isothiocyanates, nitriles and thiocyanates as products of autolysis of glucosinolates in Cruciferae. *Phytochem.*, **15**(5): 759–62. doi:10.1016/S0031-9422(00)94437-6.

en.wikipedia.org/wiki/Horseradish.

foodprint.org/real-food/horseradish/.

horticulture.oregonstate.edu/oregon-vegetables/horseradish-0.

https://fdc.nal.usda.gov/fdc-app.html.

Kojima, T., T. Tanaka and H. Mori, 1994. Chemoprevention of spontaneous endometrial cancer in female Donryu rats by dietary indole-3-carbinol. *Cancer Res.*, **54**(6): 1446-9.

Lampe, J.W. and S. Peterson, 2002. Brassica, biotransformation and cancer risk: genetic polymorphisms alter the preventive effects of cruciferous vegetables. *J. Nutrit.*, **132**(10): 2991-4.

Lee, J.H., T.O. Khor, L. Shu, Z.Y. Su, F. Fuentes and A.N.T. Kong, 2013. Dietary phytochemicals and cancer prevention: Nrf2 signaling, epigenetics and cell death mechanisms in blocking cancer initiation and progression. *Pharmacol. Ttherap.*, **137**(2): 153-71.

Lim, T.K. 2013.. Edible Medicinal and Non-Medicinal Plants (vol. 9), pp. 746-60.. Springer Science, Holland.

Lossinsky, A.S. and R.R. Shivers, 2004. Structural pathways for macromolecular and cellular transport across the blood-brain barrier during inflammatory conditions: A Review. *Histol. & Histopath.*, **19**(2): 535–64. doi:10.14670/HH-19.535. PMID 15024715.

Meng, Q., F. Yuan, I..D. Goldberg, E.M. Rosen, K. Auborn and S. Fan, 2000. Indole-3-carbinol is a negative regulator of estrogen receptor-α signaling in human tumor cells. *J. Nutrit.*, **130**(12): 2927-31.

Negro, E.J., J. Sendker, T. Stark, B. Lipowicz and A. Hensel, 2022. Phytochemical and functional analysis of horseradish (*Armoracia rusticana*) fermented and non-fermented root extracts. *Fitoterapia*, **162**:105282.

Nguyen, N.M., S. Gonda and G. Vasas, 2013. A review on the phytochemical composition and potential medicinal uses of horseradish (*Armoracia rusticana*) root. *Food Rev. Int.*, **29**(3): 261–75.

Ohta, Y., K. Takatani and S. Kawakishi, 1995. Decomposition rate of allyl isothiocyanate in aqueous solution. *Biosc., Biotechn.& Biochem.*, **59**: 102–3. doi:10.1271/bbb.59.102.

plantvillage.psu.edu/topics/horseradish/infos.

repositorio.ul.pt/bitstream/10451/35955/1/MICF_Leonardo_Cordeiro_Lourenco.pdf.

roundglass.com/food/ingredients/horseradish.

Sakao, K. and S.V. Singh, 2012. D, L-sulforaphane-induced apoptosis in human breast cancer cells is regulated by the adapter protein p66Shc. *J. Cellul. Biochem.*, **113**(2): 599-610.

Misra, Sanyat, S. Mishra, R.L. Misra, P. Ranjan and J.K. Ranjan, 2023. Herbs, Spices and Medicinal plants (vol. 1), pp. 122-3. Daya Publishing House, Darya Ganj, New Delhi.

Shehata, A., R.M.S. Mulwa, M. Babadoost, M. Uchanski, M.A. Norton and R. Skirvin, 2009. Horseradish: Botany, Horticulture, Breeding. In: Horticultural Reviews (ed. Janick, J.), vol. 35, pp. 221-61. John Wiley & Sons, Inc., USA.

Singh, A. 2017. Glucosinolates and Plant Defense. In: Glucosinolates (eds Mérillon J-M. and K.G. Ramawat), pp. 237-46. Springer International Publishing, Cham, Holland.

Stace, C.A. 2019. New Flora of the British Isles (4th ed.), pp. 420-25. C&M Floristics, Middlewood Green, Suffolk, U.K. ISBN 978-1-5272-2630-2.

www.britannica.com/plant/chard-plant.

www.extension.umn.edu/yardandgarden/ygbriefs/e253caterpillarpests-cole.html.

www.healthline.com/nutrition/horseradish.

www.rhs.org.uk/herbs/horseradish/grow-your-own.

www.theplantlist.org/tpl1.1/record/kew-2652952.

Yuan, F., D.Z. Chen, K. Liu, D.W. Sepkovic, H.L. Bradlow and K. Auborn, 1999. Anti-estrogenic activities of indole-3-carbinol in cervical cells: implication for prevention of cervical cancer. *Anticanc. Res.*, **19**(3A):1673-7.

7

Parsnip (*Pastinaca sativa* L.)

Family: Apiaceae.

Other common names: Bird's nest, Hart's-eye, Heeltrot, Hockweed, Madnip and Queen weed (Averill and DiTommaso, 2007); in Persian: Zardak and Wild carrot; in French: Cujtive and Panipainais; in Arabic: Jazar; and in Indian: Kajer (Emami *et al.*, 2010).

Introduction and Origin

Parsnip, a lesser-known root vegetable closely related to carrot and parsley, is native to Eurasia and has been cultivated since ancient times for its edible, distinct-flavoured root. It was used as a sweetener in Europe before the arrival of sugarbeets and cane sugar (Venema, 2015). The root is found on roadsides and in open places in Great Britain and throughout Europe and temperate Asia. It has become extensively naturalized in North America after it was introduced there early in the 17th century (www.britannica.com).

The wild parsnip from which the modern cultivated varieties were derived is a plant of dry *cum* rough grassland and waste places particularly on chalk and limestone soils (McKlintock and Fitter, 1956). Some authorities in Eurasia, distinguish cultivated parsnips with wild parsnips by using subspecies *P. s. sylvestris* for the latter or even elevating it to species status as *Pastinaca sylvestris*. In Europe, various subspecies have been named based on characteristics such as the hairiness of the leaves, the extent to which the stems are angled or rounded, and the size and shape of the terminal umbel (Cain *et al.*, 2010).

Parsnip

By Goldlocki - Own work, CC BY-SA 3.0, https://commons.wikimedia.org/w/index.php?curid=2914134

The generic name *Pastinaca* is probably derived from either the Latin word *pastino* meaning 'to prepare the ground for planting of the vine' or *pastus* meaning 'food'. The specific epithet *sativa* means 'sown' (Averill and DiTommaso, 2007).

Botany

Its diploid chromosome number is 2n=22 (Kalloo, 1993). Although usually grown as an annual, parsnip is a biennial plant whose apical meristem produces a rosette of roughly hairy, 1- or 2-pinnate, mid-green and up to 40cm long aromatic leaves in its first growing season, and if left unharvested, it produces a more than 1.5m tall, hairy, grooved, hollow (except at the nodes) and sparsely branched flowering stem topped by a loose and compound umbel (10-20cm in diameter) of small yellow flowers (3.5mm) in its second growing season, later producing pale brown, flat and winged seeds. By this time, the stem would become woody and the taproot inedible (Rubatsky *et al.*, 1999; Cain *et al.*, 2010).

The long taproot with creamy skin and flesh is generally smooth but sometimes the lateral roots get formed. The taproot is generally narrowly conical, although some cultivars have a more bulbous shape and are more resistant to breakage. If the root is left in the ground to mature, it becomes sweeter in flavour after winter frosts (Rubatsky *et al.*, 1999).

Each leaf has several pairs of broad, ovate and sometimes lobed leaflets with toothed margins. The petioles are grooved and have sheathed bases. The floral stem bears a few 5-10cm long, single-lobed and stalkless leaves arranged in opposite pairs and when crushed these emit a pungent odour (Rubatsky *et al.*, 1999).

The compound umbel comprises 2-5cm long 6-25 straight pedicels that support the umbellets (secondary umbels). Both umbels and umbellets usually have no upper or lower bracts. The flowers, with or without tiny sepals, consist of 5 yellow petals that are curled inward, 5 stamens and one pistil. The straw to light brown and 4-8 mm long fruits (or schizocarps) are oval and flat with narrow wings and short but spreading styles (Cain *et al.*, 2010).

Flowering parsnip, second year

By Skogkatten at English Wikipedia, CC BY-SA 4.0, https://commons.wikimedia.org/w/index.php?curid=56154151

The flowers are usually hermaphrodite. The stamen matures before the pistil. Self-fertilization does not therefore occur within the same flower. Yet since the flowers do not bloom at the same time, self-fertilization is possible within the same umbel or between two umbels on the same plant. Fertilization also occurs between the umbels of different plants. The parsnip is an allogamous plant primarily pollinated by insects, particularly from *Diptera* and *Lepidoptera* species as well as by ladybugs, which are attracted to umbels that are often covered with aphids (www.diyseeds.org).

Classification and Varieties

Based on shapes, the parsnips can be classified into three main types (www.producebluebook.com):

- Bulbous parsnips: These are rounder, squatter and generally used consistently in food processing.
- Wedge parsnips: These are similar in shape to bulbous but are longer resembling a carrot.
- Bayonet parsnips: These are long and narrow.

Varieties

There are several varieties of parsnips for commercial and home garden use, ranging from 'Albion', 'All American', 'Archer' and 'Cobham Marrow' to 'Harris Model', 'Viking', 'Warrior' and 'White Spear'. Cultivars are differentiated whether one is heirloom or hybrid. Heirloom cultivars such as 'Hollow Crown' or 'Student' are open-pollinated, allowing more genetic diversity in parsnip growth and letting growers save seeds from year to year. Hybrid cultivars such as the popular 'Gladiator' or 'Javelin' are blended genetically. Hybrids tend to be more resistant to some diseases, such as canker. Newer varieties include 'Andover' and 'Lancer' (www.producebluebook.com).

Some of the best parsnip varieties for UK gardeners include (plantura.garden):

- **Albion F_1** is a smooth-skinned variety producing very sweet, long and tapered roots. This variety is very resistant to many diseases including canker.
- **Countess F_1** is a high-yielding variety with uniform long smooth-skinned roots.
- **Gladiator F_1** has smooth white skin, long roots and a sweet taste. It has a good resistance to disease particularly to canker (*Itersonilla perplexans*).
- **Hollow Crown** is one of the oldest parsnip types, dating back to the 1800s. It is a heirloom variety and its name may be due to its sunken crown-like structure at the base of the leaves. It matures in 100-120 days and has long white tapered roots growing to around 35cm long that can be harvested from mid-fall through winter. It can take up to three weeks to germinate but can be forced if the soil is warmed with a covering of clear plastic (plantura.garden; www.weekand.com; gardenerspath.com).
- **Javelin F_1** is a British-bred variety favoured by commercial growers for its consistent high yields of smooth-skinned long and thin roots.

- **Panorama F_1** is a vigorous variety that produces long roots with a great sweet flavour.
- **Tender and True**, a very popular variety, having long tapered roots and pale skin. It is an old and consistent type that offers a very sweet flavour.
- **White Gem** is shorter and broader than other varieties. It is high-yielding, reliable, smooth-skinned and has good resistance to canker.

Some varieties suitable for cultivation in U.S. (www.weekand.com):

- **All American** has tender flesh with a sweet and nutty flavour. This fast-growing and disease- tolerant variety can be harvested in about 95 days, best during late fall.
- **Cobham Marrow** develops roots up to 20cm long and is one of the sweetest parsnip varieties used most often in desserts or glazed with brown sugar. It is highly resistant to canker disease. Although it can be harvested in about 120 days, the flavour will further develop if allowed to remain in the beds past the first couple of hard frosts.
- **Harris Model** has smooth white skin, straight roots and a more finely grained flesh than other cultivars. This sweet-flavoured parsnip grows to about 30-38cm long in about 130 days.
- **Hollow Crown** is an heirloom variety from England.
- **The Student** is an American heirloom parsnip dating to the mid-1850s. Its roots can grow up to 76cm long if the plant beds are deeply tilled. Its mild and sweet flavour increases if left in the ground beyond the first couple of frosts. It takes 95-125 days to mature.

Nutritional Value

Parsnip is rich in vitamins including various B-group vitamins but levels of vitamin C are reduced in cooking. It is also rich in minerals with particular reference to potassium which is 375mg per 100g. Since most of the vitamins and minerals are found close to the skin, many will be lost unless the root is finely peeled or cooked whole. During frosty weather, part of the starch is converted into sugar and the root tastes sweeter.

Parsnip, raw	
Nutritional value per 100g	
Energy	314 kJ (75 kcal)
Carbohydrates	18 g
Sugars	4.8
Dietary fiber	4.9 g

Parsnip, raw	
Fat	0.2 g
Protein	1.2 g
Vitamins	**Quantity (%DV†)**
Thiamine (B1)	0.09 mg (8%)
Riboflavin (B2)	0.05 mg (4%)
Niacin (B3)	0.7 mg (5%)
Pantothenic acid (B5)	0.6 mg (12%)
Vitamin B6	0.09 mg (7%)
Folate (B9)	67 μg (17%)
Vitamin C	17 mg (20%)
Vitamin E	1.49 mg (10%)
Vitamin K	22.5 μg (21%)
Minerals	**Quantity (%DV†)**
Calcium	36 mg (4%)
Iron	0.59 mg (5%)
Magnesium	29 mg (8%)
Manganese	0.56 mg (27%)
Phosphorus	71 mg (10%)
Potassium	375 mg (8%)
Sodium	10 mg (1%)
Zinc	(6%)
Other constituents	**Quantity**
Water	79.53 g
Units: μg = micrograms; mg = milligrams; IU = International units †Percentages are roughly approximated using US recommendations for adults. ***Source***: USDA FoodData Central	
Table source: https://en.wikipedia.org/wiki/Parsnip.	

The dietary fibre in parsnips is partly of the soluble and partly the insoluble type and comprises cellulose, hemicellulose and lignin. The high fibre content of parsnips may help prevent constipation and reduce blood cholesterol levels (Siddiqui, 1989).

Phytochemicals

The phytochemicals found in parsnips are alpha-thujene, beta-carotene, beta-pinene, isobergapten, isorhamnetin, kaempferol, limonene, quercetin and rutin. Kaempferol, quercetin and rutin are flavonoids that have anti-inflammatory, cancer-preventing, anti-oxidative, cardioprotective and immune system-enhancing properties. Isobergapten is a furanocoumarin found in grapefruit as well and can block the liver enzyme cytochrome P450, having interactions with some medications (www.caasn.com).

Parsnips contain antioxidants such as falcarinol, falcarindiol (www.caasn.com), panaxydiol, and methyl-falcarindiol which may potentially have anticancer, anti-inflammatory and antifungal properties (Christensen, 2011). Falcarinol is one of the allergy-inducing substances in the Apiaceae but falcarindiol is not an allergen (Paulsen *et al.*, 2014; Czyzewska *et al.*, 2014; Christensen and Brandt, 2006; Bogucka-Kocka *et al.*, 2008).

Kenari *et al.* (2021) stated various active ingredients of parsnip as obtained from the websites of www.sciencedirect.com and www.researchgate.net ~ the most important active ingredients in parsnip include coumarins, furanocoumarins, polyacetylenes, essential oils, terpenes and flavonoids (Walling and Walling, 2018; Harper *et al.*, 2015; Seavy and Rizzolo, 2013; Heywood, 2014; Jensen and Hansen, 1939; Amri, 2014). Al-Barwani and Eltayeb (2004) reported that xanthotoxin, bergapten, isopimpinellin, angelicin, psoralen, sphondin and imperatorin are the most important furanocoumarins found in parsnip. All aerial parts of parsnip contain xanthotoxin, bergapten and imperatorin though angelicin and sphondin are present only in germinal parts of some species (Emami *et al.*, 2010; Conforti *et al.*, 2009).

As per various workers (Matejic *et al.*, 2014; Emami *et al.*, 2010; Chizzola, 2010; Kurkcuoglu *et al.*, 2006; Kviesis *et al.*, 2019; Jamshidi *et al.*, 2020; Janeczko and Skoczowski, 2005; Soine *et al.*, 1956; Madari and Jacobs, 2004; Fahmy *et al.*, 1956; Nemych *et al.*, 2018; Mongeau *et al.*, 1994; Picardo *et al.*, 1986; Terreaux *et al.*, 2003), the active ingredients in various parts of parsnip are listed in the below mentioned table:

Parts of parsnip	***Active ingredients***
Aerial parts	Furanocoumarins (in particular: angelicin, bergaptene, xanthotoxin, imperatorin, psoralen) Essential oils (chief components: E-β-ocimene, Z-β-ocimene, α-terpinolene, E-β-farnesene, γ-palmitolactone, hexyl butanoate, lavandulyl acetate, sesquiterpene) Polysaccharides Organic acids Hydroxycinnamic acids Amino acids Polyphenols
Root	Furanocoumarins (in particular: angelicin, bergaptene, xanthotoxin, imperatorin, psoralen) Essential oils (chief components: aliphatic ester, in particular octylbutyrate, Myristicin, β-ocimene, β-farnesene, octyl butyrate, octyl hexanoate, α-terpinolene) Sex hormone (5a-androstenone) Polysaccharides Organic acids Hydroxycinnamic acids Amino acids Polyphenols

Parts of parsnip	Active ingredients
Seeds	Furanocoumarins (bergapten, pimpinellin, methoxsalen, isopimpinellin, imperatorin, phellopterin, byakangelicol, heraclenin, isobergapten, byakangelicin, heraclenol, psoralen, and isobyakangelicin, xanthotoxin) Essential oils {myristicin, β-ocimene, β-farnesene, octyl butyrate, hexanoate, Butyl butyrate, (Z)-b-Ocimene, Octanal, Hexyl butyrate, Octyl acetate, Decanal, Octanol, a-Zingiberene, Decanol, b-Sesquiphellandrene, ar-Curcumene, Decyl butyrate, Benzyl butyrate, Phenylethyl butyrate, (E)-Nerolidol, Phenylethyl hexanoate, butyric acid, acetyl ester) Coumarins (osthol, umbelliferone, 4-hydroxycoumarin) Hydrocarbons (e.g. methanol, ether, chloroform, ethanol, chlorophyll) Resinous substance Sulphuric acid

Table source: Kenari *et al.* (2021) via www.sciencedirect.com; www.researchgate.net.

Pharmacological properties of the chemical components of parsnip

Pharmacological properties			Chemical components
CNS	Neurologic	Treatment of multiple sclerosis	Furanocoumarins
		Anticonvulsant	Xanthotoxin, Imperatorin
		Improving cognitive function and memory	Xanthotoxin, Imperatorin
		Sedative	Furanocoumarins
		Interacting with GABA receptors	Falcarinol
	Psychiatric	Antidepressant	Furanocoumarins
		MAO inhibitor	Xanthotoxin
Respiratory	Eliminating bronchospasm		Furanocoumarins
Gastrointestinal	Eliminating abdominal smooth muscle spasms		Furanocoumarins
Liver and gallbladder	Eliminating spasms		Furanocoumarins
Skin	Treatment of vitiligo		Bergapten + Xanthotoxin
	Treatment of psoriasis		Psoralen, Bergapten + Xanthotoxin
	Treatment of mycosis fungoides		Psoralen
	Stimulating melanogenesis		Oxypeucedanin hydrate (dose-dependent) Heraclenol (dose-dependent)
	Inhibiting melanogenesis		Oxypeucedanin hydrate (dose-dependent), Heraclenol (dose-dependent), Byakangelicin (dose-dependent)
	Changing in the size and number of melanocytes		Psoralen
	Increasing hair growth		Bergapten + Xanthotoxin

Pharmacological properties		Chemical components
Blood	Antiplatelet aggregation	Falcarinol, Falcarindiol
	Treatment of ALL	
Cardiovascular	Peripheral vasodilator Coronary artery expander	Furanocoumarins
Urogenital	Eliminating spasms	Furanocoumarins
	Impact on sexual activity	Sex hormone (5a-androstenone)
	Diuretic	-------
Microorganisms	Antibacterial	Xanthotoxin, Angelicin + Psoralen, Falcarinol, Essential oils
	Antifungal	Bergapten, Xanthotoxin, Angelicin, Falcarinol, Falcarindiol, Essential oils
Cancers	Improving the effect of chemotherapy drugs on cancer cells	Byakangelicin
	Cytotoxic effect on human KB cell line	Angelicin
	Anti-proliferative effect on HeLaS3 cells	Bergapten, Isopimpinellin, Xanthotoxin, Xanthotoxol, Imperatorin, Coumarins
	Cytotoxic effect on human gastric adenocarcinoma	Falcarinol
	Inhibiting the growth of pre-neoplastic cells in the rat colon	Falcarinol
	Anti-mutagenic and cytotoxic	Falcarindiol
	Inhibiting the growth of cancer cells in the breast, colon, lung and prostate	Coumarins
	Preventing angiogenesis of cancer tissue and direct invasion of cancer cells into the body	Coumarins
	Apoptotic effect on Leukemic cell lines	-------
Others	Antioxidant	Xanthotoxin
	Anti-fever	Angelicin
	Anti-inflammatory	Angelicin, Sphondin, Falcarinol, Falcarindiol

Source: Kenari et al. (2021) via www.sciencedirect.com; www.researchgate.net

Flavour

Parsnip taste is somewhere between carrot and potato and has a unique aroma somewhat similar to fresh parsley (timesofindia.indiatimes.com).

Uses

Although parsnips can be used similar to carrots and can be substituted for carrots in recipes, they are slightly sweeter and nuttier in taste especially when cooked. They can be eaten raw in salads or baked, boiled, pureed, roasted, fried, grilled, steamed or cooked like vegetables. They give a rich flavour when used in salads, soups, stews, casseroles, pies and puddings and even dried ones for seasoning soups and processed for canning. Parsnips can also be used for making jams, marmalades and converted into sweet flour for making cakes. In Britain and Ireland, parsnips are used for making wine and beer. Apart from the roots, the leaves and young shoots of this plant are cooked with other greens as a vegetable or added to soups (www.netmeds.com).

Warning

Use protective gloves and clothing when handling parsnip plants as the shoots and leaves contain furanocoumarin ~ a chemical that can cause serious skin burns. Burning, tingling, redness and streak-like blisters may develop within 24 to 48 hours of exposure. Expert medical attention is required in case if a serious rash or blister develops (www.weekand.com).

Cultivation

Propagation

Parsnips are usually grown from freshly harvested seeds that are directly sown in the bed in early spring. For this the ground is worked to a fine tilth by mixing some 10t/ha FYM or compost. The bed should be watered well after sowing. Its seeds remain viable only for one year or maximally for two years. It may take up to 4 weeks for seedlings to appear (Brickell, 1992; plantura.garden) and up to 4 months to develop fully (www.producebluebook.com).

Climate

Parsnip is a cool season crop and can be sown in early spring and harvested in fall before the last frost. The germination is promoted by moderately cool temperatures, sun and adequate moisture. Unlike carrots, parsnips do not tolerate high heat as warmer temperatures will cause bolting (www.producebluebook.com). Its seeds require near to 8°C temperature for germination (plantura.garden). The root-solids consist largely of starch but at the end of summer when it encounters a period of low temperature then it changes much of its starch into sugar. The root is hardy and not damaged by hard freezing of the soil (www.britannica.com).

Soil

Parsnips prefer well-drained sandy-loam soils with a pH of 6–6.8. Silt, clay and stony ground produce short and forked roots (Brickell, 1992).

Planting

Prior to sowing, field should be deeply ploughed 2-3 times and disked properly to eliminate polythene shreds, rootstocks of previous season crop and perennial weeds, brick and stone pieces and any other material which in any way may hinder proper root development. If the soil is not rich in nutrients, FYM, vermicompost or poultry manure should be mixed well while preparing the field after the last ploughing.

The seeds are sown thinly into drills 1cm deep (www.producebluebook.com) in rows about a half metre apart, and the plants are thinned to stand 5 to 7cm apart in the rows (www.britannica.com).

Irrigation

Parsnips are fairly drought-tolerant and once established there is no need to continue watering regularly. However, extra watering would be required in a very warm weather when soil dries out completely. Watering can be done once every two weeks if the foliage starts to wilt (plantura.garden).

Fertilizers

Since parsnip plants grow slowly, these require only a small amount of fertilizer that too only if the planting site was not amended with compost or well-rotted organic matter prior to sowing. When the seedlings are around 10-15cm tall, a small dose of fertilizer high in potassium and magnesium can be applied (plantura.garden).

Cultural Practices

The growing plants are thinned and kept weed-free. Harvesting begins in late fall after the first frost and continues through winter. The rows can be covered with straw to enable the crop to be lifted during frosty weather (Brickell, 1992).

Harvest

Once seeded, parsnips can take up to 4 months to fully develop, depending on the time of sowing and climatic conditions (www.producebluebook.com). The root can be lifted with a garden fork and will be ready to harvest from September onwards. These hardy plants can be left in the ground all winter and harvested as needed. They taste better in cold weather because starch in the parsnip root is converted to sugar when exposed to frost, thus enhancing the sweetness and flavour of the root (plantura.garden). At harvest, the roots should have attained 2.5-5cm diameter and 17-20cm length though this also depends on the variety planted, *vis-a-vis* growing conditions. Parsnips taste sweetest if tops have frozen and roots have been exposed to temperatures under 5°C (www.producebluebook.com).

Postharvest

After harvest, tops should be removed and the roots thoroughly washed before storage. Freezing is generally preferred for storing parsnips. These can be stored for up to six months at 0 to 4°C temperatures and 90-95% relative humidity. Parsnips can be kept in pallet bins with other crops such as tomatoes, cabbage, parsley and carrots. They should not be stored with crops such as cantaloupes or apples which generate ethylene which causes bitterness. At home, they can be put in a plastic bag and placed in the coolest part of the refrigerator or can be placed unbagged in the vegetable crisper of the refrigerator. If the parsnips are washed before putting into moist storage, excess moisture should be removed to prevent from rotting. The average storage life for parsnips, using this method is two to six months (www.canr.msu.edu; www.producebluebook.com).

Pests and Diseases

Insect-pests

- **Celery fly** (*Euleia heraclei*): The larvae of this insect sometimes form tunnels on parsnip leaves that can be seen as irregular pale-brown passages between the upper and lower surfaces of the leaves. If plant is young, whole leaves may shrivel and die. Treatment is by removing affected leaflets, whole leaves or by chemical means (Brickell, 1992). This can be controlled by spraying the leaves with 0.2% malathion or primiphos methyl.
- **Carrot fly** (*Chamaepsila rosae*): Carrot fly maggots feed on the outer layers of the root, burrowing its way inside later in the season. Larger roots get spoilt and the seedlings may get killed. The damage may become entry for fungal rots and canker and the smell of bruised tissue attracts the fly (en.wikipedia.org/wiki/Parsnip). These maggots can be controlled by spraying with 0.15% Triazophos 40EC or 0.2% Dimethoate 30EC whenever maggots have been observed feeding on the crop.

- **Parsnip moth** (*Depressaria radiella*): It is native to Europe which was accidentally introduced to North America in the mid-1800s. Their larvae construct webs on the umbels and continue feeding on flowers and partially developed seeds (Cain *et al.*, 2010). In this case, the egg masses can be collected and destroyed and plants should be sprayed with 0.2% Chlorpyrifos 20EC or 0.03% Spinosad 45SC.

Though only seldom but infestation of swallowtail (*Papilio polyxenes*), the common swift moth (*Korscheltellus lupulina*), the garden dart moth (*Euxoa nigricans*), and the ghost moth (*Hepialus humuli*) has been recorded (Robinson *et al.*, 2023) and all these can be controlled as to those of parsnip moth as defined above.

Diseases

- **Parsnip canker** (*Phoma complanata, Ilyonectria radicicola, Itersonilia pastinaceae, I. perplexans*) is a serious disease of this crop caused by various fungi. Its infection can be identified by black or orange-brown patches forming around the root's crown and shoulders, accompanied by cracking and hardening of the flesh. It usually occurs when the roots have already been damaged by carrot fly larvae or when the seed is sown into cold and wet soil and/or when soil pH is too low (www.which.co.uk). Root shoulder is most commonly affected though lesions may appear anywhere. Initially when symptoms are first seen, the affected areas should be sheared off and destroyed. The pathogen may enter the soil through leaf spots and usually infects through injured fine roots or through the injury made by carrot fly. Since there is no control so sowing resistant cultivars such as 'Avonresister' is only the option. Moreover, the field should be well-drained. Late sown crop is less affected due to thin and smaller roots.
- **Black rot** (*Mycocentrospora acerina*) has been observed in Europe killing parsnip plants early (Cain *et al.*, 2010) but in India as this crop is rarely grown anywhere so its prevalence has yet not been noted. However, in any case its occurrence can be controlled through spraying with 0.2% Captaf alternate with 0.1% Bavistin at fortnightly intervals.
- **Watery soft rot** (*Sclerotinia minor, S. sclerotiorum*) occurs when taproot surface starts showing with a white or buff-coloured mould so taproots become soft and watery. It is most common in temperate and subtropical regions with a cool and wet season (Snowdon, 2010). Fortnightly spraying of Blitox at 0.2% alternate with 0.1% Bavistin will control this disease to some extent.

- **Violet root rot** (*Helicobasidium purpureum*) sometimes affects the roots, covering them with a purplish mat to which soil particles adhere. The leaves become distorted and discoloured, and the mycelium can spread through the soil between plants. Some weeds can harbour this fungus, and it is more prevalent in wet acid conditions (Brickell, 1992). Field sanitation, following proper crop rotation and spraying the crops with 0.2% Blitox alternate with Bavistin may be effective.
- **Powdery mildew** (*Erysiphe heraclei*) can cause significant crop loss. Its infection results in yellowing *cum* loss of foliage. Moderate temperatures *cum* high humidity favour the development of the disease (Koike *et al.*, 2007). Its initial infection causes white tufts of mould on upper leaf surface but afterwards engulfs the entire leaf. Karathane spraying at 0.2% or sulphur dusting is quite effective.
- Several **viruses** are known to infect parsnips such as 'seed-borne strawberry latent ringspot virus', 'parsnip yellow fleck virus', 'parsnip leaf curl virus', 'parsnip mosaic potyvirus', and 'potyvirus celery mosaic virus'. The latter causes clearing or yellowing of the areas of the leaf immediately beside the veins, the appearance of ochre mosaic spots, and the crinkling of the leaves in infected plants (Cain *et al.*, 2010). Viruses cause stunting and distortion of foliage and retardation of growth. Parsley yellow fleck virus spreads through aphids. To keep plants safe from viruses, infected plants should immediately be removed and destroyed. Aphids should also be controlled through the use of nicotinic acid.

Seed Production

(Source:www.diyseeds.org.)

Parsnips are grown for seed production in the same way as to those for consumption but they will produce the seeds only in the second year of the growth cycle. It may take a long time for seeds to germinate at the start of the season ~ around three weeks. After maturity, different methods exist for winter storing of parsnips intended for seed production. It is possible to dig up the roots in autumn. They should be selected for characteristics specific to the variety, *i.e.* colour, shape and vigour of the roots. The roots are cleaned without water and the leaves are cut above the collar. They are then left to dry a short while in the open air and stored in the cellar. During winter the parsnips should be regularly checked and rotting roots should be removed.

The easiest way is to leave them in the ground in the garden. The parsnip is one of the hardiest plants and can stay in the ground in winter even in very cold regions. Its flavour is even improved by frost. After the worst of the cold has passed, they can be uprooted to select those intended for seed production. At the beginning of second season, *i.e.* spring, the selected roots are replanted at a distance of 60-70cm between each plant either way by burying them so that the collar is either at ground level or slightly below. Seed crop of parsnip can grow up to 2m or even more so sometimes these require to be staked. Parsnip plants produce several umbels that bloom at different times. Primary umbel is formed at the tip of the main stem, secondary umbels developing from the main stem and the tertiary umbels forming on the secondary stems.

By and large, the parsnip is pollinated by insects especially from the *Diptera* and *Lepidoptera* species as well as by ladybugs. Parsnips can cross with other varieties or with wild parsnips. To avoid cross-pollination, two varieties of parsnip should be grown 300 metres apart. This distance can be reduced to 100 metres if a natural barrier such as a hedge exists between them. To isolate varieties growers can cover two varieties with different nets, net of one can be kept open while the other closed on one day and alternate the next or covering one variety with an insect net and putting a bumble beehive inside.

Since primary umbels produce the best seeds so it is preferable to harvest these first on maturity as these produce the best seeds. The secondary umbels should be harvested only if necessary. The umbels are cut along with the top of the stalk when the first mature seeds begin to fall. They can also be cut earlier as they tend to fall very easily to the ground. In any case, they should continue to dry in a well ventilated and dry place. The seeds are removed from the umbels by hand. The person who collects the seeds should wear gloves because the seeds release essential oils that may cause blisters in direct sunlight. To clean the seeds, coarse mesh sieves should be used to retain the chaff inside while exiting the seeds while a finer sieve is used that will retain the seeds but let the dust pass through. Labels are attached with the name of the variety and species, as well as the year, as writing on the outside may rub off. Storing the seeds in the freezer for several days kills larvae parasites. Parsnip seeds remain viable only for one year or at the most for two years when stored in freezer. One gramme of seed contains around 220 individual seeds.

References

Al-Barwani, F.M. and E.A. Eltayeb, 2004. Xanthotoxin and other furanocoumarins as phytoalexins in *Pastinaca sativa* L. roots. *SQU J. Sci.*, **9**:7–17.

Amri, E. 2014. The role of selected plant families with dietary ethnomedicinal species used as anticancer. *J. Med. Plants Stud.*, **2**(1):28–39.

Averill, Kristine M. and Antonio DiTommaso, 2005. Wild parsnip (*Pastinaca sativa*): A troublesome species of increasing concern. *Weed Techn.*, **21**(1): 279–87. doi:10.1614/WT-05-186.1. S2CID 86774319.

Bogucka-Kocka, A., H.D. Smolarz and J. Kocki, 2008. Apoptotic activities of ethanol extracts from some Apiaceae on human leukaemia cell lines. *Fitoterapia,* **79**(7-8):487–97. doi: 10.1016/j.fitote.2008.07.002.

Brickell, C. (ed.), 1992. The Royal Horticultural Society Encyclopedia of Gardening. Dorling Kindersely, London:

Cain, N., S.J. Darbyshire, A. Francis, R.E. Nurse and M.-J. Simard, 2010. The Biology of Canadian Weeds. 144. *Pastinaca sativa* L. *Can. J. Pl. Sci.*, **90** (2): 217–40. doi:10.4141/CJPS09110.

Chizzola, R. 2010. Essential oil composition of wild growing Apiaceae from Europe and the Mediterranean. *Nat. Prod. Commun.*, **5**(9):1477–92.

Christensen, L.P. 2011. Aliphatic C(17)-polyacetylenes of the falcarinol type as potential health-promoting compounds in food plants of the Apiaceae family. *Rec. Patents Food, Nutrit. & Agric.*, **3**(1): 64–77.

Christensen, L.P. and K. Brandt, 2006. Bioactive polyacetylenes in food plants of the Apiaceae family: occurrence, bioactivity and analysis. *J. Pharm. Biomed. Anal.*, **41**(3):683–93. doi: 10.1016/j.jpba.2006.01.057.

Conforti, F., M. Marrelli, F. Menichini, M. Bonesi, G. Statti, E. Provenzano and F. Menichini, 2009. Natural and synthetic furanocoumarins as treatment for vitiligo and psoriasis. Curr. *Drug Ther.*, **4**(1), 38-58.

Czyzewska, M.M., L. Chrobok, A. Kania, M. Jatczak, F. Pollastro and G. Appendino, 2014. Dietary acetylenic oxylipin falcarinol differentially modulates GABAA receptors. *J. Nat. Prod.*, **77**(12):2671–7.

Emami, A., S. Fasihi and I. Mehregan, 2010. Reference Book on Medicinal Plants, 1002p. Islamic and Complementary Medicine at Iran University of Medical Sciences, Tehran.

en.wikipedia.org/wiki/Parsnip.

Fahmy, I.R., E.A. Kadir and A.H. Saber, 1956. A pharmacognostical study of the fruit of *Pastinaca sativa* L. cultivated in Egypt. *J. Pharm. Pharmacol.*, **8**(9): 653–60. doi: 10.1111/j.2042-7158.1956.tb12197.x.

gardenerspath.com/plants/vegetables/grow-parsnips/.

Harper, N., S. Orpin and A. Mehta, 2015. The root cause of a blistering eruption. *Arch. Dis. Child.*, **100**(1): 72.

Heywood, V.H. 2014. The socio-economic importance of the Apiales. *J. Fac. Pharm. Istanbul*, **44**(2): 113–30.

https://commons.wikipedia.org/w/index.php?curid=2914134.

https://commons.wikimedia.org/w/index.php?curid=56154151.

https://enwikipedia.org/wiki/Parsnip.

Jamshidi, A.H., F. Eghbalian, S. Mahroozade, H.M. Kenari, A. Ghobadi and B.S. Yousefsani, 2020. Recommended natural products in Alzheimer's disease based on traditional Persian medicine. *J. Med. Plants*, **19**(75): 17–29.

Janeczko, A. and A. Skoczowski, 2005. Mammalian sex hormones in plants. *Folia Histochem. Cytobiol.*, **43**(2): 71–9.

Jensen, T. and K.G. Hansen, 1939. Active spectral range for phytogenic photodermatosis produced by *Pastinaca sativa* (dermatitis bullosa striata pratensis, Oppenheim). *Arch. Derm. Syphilol.*, **40**(4): 566–77.

Kalloo, G. 1993. Genetic Improvement of Vegetable Crops (eds Kaloo, G. and B.O. Bergh), pp. 485-6. Permagon Press, USA. doi:10.1016/B978-0-08-040826-2.50038-2. ISBN 978-0-08-040826-2.

Kenari, H.M., G. Kordafshari, M. Moghimi, F. Eghbalian and D. Taher Khani, 2021. Review of Pharmacological Properties and Chemical Constituents of *Pastinaca sativa. J. Pharmacopuncture*, **24**(1): 14-23.

Koike, Steven, T., Peter Gladders and Albert O. Paulus, 2007. Vegetable Diseases: A Color Handbook, 118p. Gulf Professional Publishing, Canada. ISBN 978-0-12-373675-8.

Kurkcuoglu, M., K.H.C. Baser and M. Vural, 2006. Composition of the essential oil of Pastinaca sativa L. subsp. urens (Req. ex Godron) Celak. *Chem. Nat. Comp.*, **42**(1): 114–5.

Kviesis, J., I. Kļimenkovs, I. Arbidans, A. Podjava, M. Kļaviņs and E. Liepiņs, 2019. Evaluation of furanocoumarins from seeds of the wild parsnip (*Pastinaca sativa* L. s.l.). *J. Chromatogr. B. Analyt. Technol. Biomed. Life Sci.*, **1105**: 54–66. doi: 10.1016/j.jchromb.2018.12.012.

Madari, H. and R.S. Jacobs, 2004. An analysis of cytotoxic botanical formulations used in the traditional medicine of ancient Persia as abortifacients. *J. Nat. Prod.*, **67**(8):1204–10. doi: 10.1021/np049953r.

Matejic, J.S., A.M. Dzamic, T. Mihajilov-Krstev, V.N. Randelovic, Z.D. Krivosej and P.D. Marin, 2014. Antimicrobial potential of essential oil from *Pastinaca sativa* L. *Biol. Nyssana.*, **5**(1):31–5.

McKlintock, David and R.S.R. Fitter, 1956. The Pocket Guide to Wild Flowers, 102p. Collins, U.K.

Mongeau, R., R. Brassard, R. Cerkauskas, M. Chiba, E. Lok, E.A. Nera, P. Jee, E. McMullen and D.B. Clayson, 1994. Effect of addition of dried healthy or diseased parsnip root tissue to a modified AIN-76A diet on cell proliferation and histopathology in the liver, oesophagus and forestomach of male Swiss Webster mice. *Food Chem. Toxicol.*, **32**(3): 265–71. doi: 10.1016/0278-6915(94)90199-6.

Nemych, V.A., N.S. Donakanian, O.M. Novosel and L.M. Horyacha (eds), 2018. Phytochemical analysis of the leaves and root crops of *Pastinaca sativa*. Abstracts of XXV International Scientific and Practical Conference of Young Scientists and Student; Proceedings of Topical Issues of New Drugs Development.

Paulsen, E., T.H. Petersen, X.C. Frette, K.E. Andersen and L.P. Christensen, 2014. Systemic allergic dermatitis caused by Apiaceae root vegetables. *Contact Dermatitis*, **70**(2): 98–103. doi: 10.1111/cod.12122.

Picardo, M., A. Cristaudo, C. de Luca and B. Santucci, 1986. Contact dermatitis to *Pastinaca sativa*. *Contact Dermatitis*, **15**(2): 98–9. doi: 10.1111/j.1600-0536.1986.tb01291.x.

plantura.garden/uk/vegetables/parsnips/parsnip-overview.

Robinson, G.S., P.R. Ackery, I. Kitching, G.W. Beccaloni and L.M. Hernández, 2023. HOSTS (from HOSTS - a Database of the World's Lepidopteran Hostplants) [Data set resource]. Natural History Museum. https://data.nhm.ac.uk/dataset/hosts/resource/877f387a-36a3-486c-a0c1-b8d5fb69f85a.

Rubatsky, V.E., C.F. Quiros and P.W. Siman, 1999. Carrots and Related Vegetables ofUmbelliferae, pp. 30, 31, 225.. CABI Publishing, U.K.. ISBN 978-0-85199-129-0.

Seavy, L. and D. Rizzolo, 2013. Blistering rash and erythema on the chest and arms. JAAPA, **26**(2):12.

Siddiqui, I.R. 1989. Studies on vegetables: fiber content and chemical composition of ethanol-insoluble and -soluble residues. *J. agric. Food Chem.*, **37** (3): 647–50. doi:10.1021/jf00087a015.

Snowdon, Anna L. 2010. Post-Harvest Diseases and Disorders of Fruits and Vegetables. Vol. 2: Vegetables, 290p. Manson Publishing.. ISBN 978-1-84076-598-4.

Soine, T.O., H. Abu-Shady and F.E. Digangi, 1956. A note on the isolation of bergapten and imperatorin from the fruits of *Pastinaca sativa* L. *J. Am. Pharm. Assoc.*, **45**(6): 426–7. doi: 10.1002/jps.3030450623.

Terreaux, C., J. Polasek and K. Hostettmann, 2003. Plant constituents with hormonal effects. *Curr. Org. Chem.*, **7**(12): 1151–61.

timesofindia.indiatimes.com/life-style/food-news/what-is-parsnip-and-why-you-should-eat-it-daily/photostory/85463861.cms?from=mdr.

Venema, Christine, 2015. Parsnips: A vegetable from antiquity. Michigan State University, USA. https://www.canr.msu.edu/news/parsnips_a_vegetable_from_antiquity.

Walling, A.L. and H.W. Walling, 2018. Phytophotodermatitis induced by wild parsnip. *Dermatol. Online J.*, **24**(2):19.

www.britannica.com/plant/parsnip.

www.caasn.com/parsnip.html.

www.canr.msu.edu/news/parsnips_a_vegetable_from_antiquity.

www.diyseeds.org/en/film/parsnip/.

www.netmeds.com/health-library/post/parsnip-health-benefits-nutrition-ayurvedic-uses-recipes-side-effects.

www.producebluebook.com/know-your-commodity/parsnips.

www.researchgate.net/publication/350526467_Review_of_Pharmacological_Properties_and_Chemical_Constituents_of_Pastinaca_sativa.

www.sciencedirect.com/science/article/abs/pii/0278691594901996?via%3Dihub.

www.weekand.com/home-garden/article/parsnip-varieties-18051072.php.

www.weekand.com:gardenerspath.com.

www.which.co.uk/documents/pdf/parsnip-canker-153659.pdf.

Parsley (*Petroselinum crispum* (Mill.) Fuss)

Family: Apiaceae.

Subfamily: Apioideae.

Origin and Introduction

Parsley is native to Greece, Morocco and the former Yugoslavia (POWO, 2023) from where it was introduced elsewhere. *Petroselinum* is the classical Latin name for parsley which is derived from the Greek *petroselinon* for 'parsley'. *Petroselinon* comprises of 2 words, *petros* meaning 'rock or stone' and *selinon* standing for 'celery' or 'wild parsley' as in Greece people feel only a little distinction between the two. The species name *crispum* may be due to the crispate shape of leaves and this tendency in some cultivars is more pronounced as in 'curly parsley' (GernotKatzer'sSpicePages, 2023). The countries having no traditional use of parsley mostly think it a variant of coriander due to similar shaped leaves and they use it so. Coriander is often termed Indian parsley in Western countries as the name French parsley is sometimes used for chervil (GernotKatzer'sSpicePages, 2023).

Botany

The chromosome number in garden parsley is 2n=2x=22 (Kalloo, 1993). In temperate regions it is a bright green biennial plant while remains an annual herb in subtropical and tropical regions. In temperate climates it forms a rosette of tripinnate leaves 10–25cm long with numerous 1–3cm leaflets and a taproot used as a food store over the winter. In the second year it produces a flowering stem up to 75cm tall with sparser leaves and flat-topped umbels having 3–10cm diameter which comprise of numerous yellow to yellowish-green flowers with 2mm diameter. Ovoid seeds are 2–3mm long with prominent style remnants at the apex. One of the compounds of the essential oil is apiol. The plant usually dies after seed maturation (Huxley, 1992).

'Root parsley' (var. *tuberosum*) has an aromatic, thick, tender and edible taproot whereas in 'leaf parsley' the taproot is quite small and tough with woody texture and it is cultivated exclusively for its leaves but not the roots. The leaves of 'leaf parsley' are chopped and used as a garnish in many European countries whereas roots of 'root parsley' are used as vegetable similar to carrots (GernotKatzer'sSpicePages, 2023).

Cultivated Types

Crispum Group parsley	Crispum Group parsley
By Donovan Govan. - Image taken using a Canon PowerShot G3 (reference 7828)., CC BY-SA 3.0, https://commons.wikimedia.org/w/ index.php?curid=122217	**By Tharish - Own work, CC BY-SA 3.0, https://commons.wikimedia.org/w/ index.php?curid=18213364**

In cultivation, parsley can be broadly subdivided into three cultivated selections related to their end use and depending upon the form of the plant. Though these selections are often treated as botanical varieties but actually they are not of natural botanical origin. Of the three main groups of parsley, French or curly leaf (*P. crispum* 'Crispum group', syn. *P. crispum* var. *crispum*) and Italian, broad or flat leaf (*P. crispum* 'Neapolitanum group', syn. *P. crispum* var. *neapolitanum*) parsleys are herbs being cultivated for their leaves, while the third is a root vegetable which is called the Tuberous or Hamburg root parsley (*P. crispum* 'Radicosum group', syn. *P. crispum* var. *tuberosum*). Hamburg root parsley produces much thicker roots than types cultivated for their leaves. Although seldom used in Britain and the United States, root parsley is common in central and eastern European cuisine where it is used in soups and stews or simply eaten raw as a snack (similar to carrots). Root parsley looks similar to the parsnip which is among its closest relatives in the family Apiaceae though its taste is quite different (en.wikipedia.org, 2023).

Leaves and flowers of flat parsley (*Petroselinum crispum*)

By Jonathunder - Own work, GFDL 1.2, https://commons.wikimedia.org/w/index.php?curid=29637295

Chemical Constituents

The essential oils of leaf and root show approximately the same composition. The main components (10–30%) are myristicin, limonene and 1,3,8-p-menthatriene while minor components are mono- and sesquiterpenes. The curly leaf types tend to be richer in myristicin but contain much less essential oil than flat leaf types (0.01 and 0.04%, respectively). In contrast, the essential oil from the fruits (3–6%) is either dominated by myristicin (60 to 80%; mostly var. *tuberosum* and var. *crispum*) or by apiole (70%; mostly var. *latifolium*). A third chemical race shows allyl tetramethoxy benzene (55 to 75%), which can also appear in apiol-dominated oils (up to 20%) (GernotKatzer'sSpicePages, 2023).

Dietary flavonoids (aglycones) present in parsley, thyme, celery or hot peppers are apigenin, luteolin, baicalein and chrysin (Delage, 2015). Parsley is a source of flavonoids and antioxidants, especially luteolin, apigenin (Meyer *et al.*, 2006), folate and vitamins K, C & A.

In a very low concentration, toxic poly-ynes have also been found in parsley. Another matter of concern is that the photosensitizing furano-coumarins bergaptene and isoimperatorin have been found in the root (GernotKatzer'sSpicePages, 2023).

Nutritional Value

One gram of dried parsley contains about 6μg of lycopene, 10.7μg of alpha carotene as well as 82.9μg of lutein+zeaxanthin and 80.7μg of beta carotene. Dried parsley can contain about 45 mg/gramme apigenin (en.wikipedia.org, 2023; Shankar *et al.*, 2017). The apigenin content of fresh parsley is reportedly 215.5mg/100g which is much higher than the next highest food source, green celery hearts providing 19.1mg/100g (Delage, 2015).

Parsley, fresh	
Nutritional value per 100g	
Energy	151 kJ (36 kcal)
Carbohydrates	6.33 g
Sugars	0.85 g
Dietary fiber	3.3 g
Fat	0.79 g
Protein	2.97 g
Vitamins	Quantity (%DV†)
Vitamin A equiv. beta-Carotene lutein zeaxanthin	421 μg (53%) 5054 μg (47%) 5561 μg
Thiamine (B1)	0.086 mg (7%)
Riboflavin (B2)	0.09 mg (8%)
Niacin (B3)	1.313 mg (9%)
Pantothenic acid (B5)	0.4 mg (8%)
Vitamin B6	0.09 mg (7%)
Folate (B9)	152 μg (38%)
Vitamin C	133 mg (160%)
Vitamin E	0.75 mg (5%)
Vitamin K	1640 μg (1562%)
Minerals	Quantity (%DV†)
Calcium	138 mg (14%)
Iron	6.2 mg (48%)

Parsley, fresh	
Magnesium	50 mg (14%)
Manganese	0.16 mg (8%)
Phosphorus	58 mg (8%)
Potassium	554 mg (12%)
Sodium	56 mg (4%)
Zinc	1.07 mg (11%)
Units: µg = micrograms, mg = milligrams, IU = International units †Percentages are roughly approximated using US recommendations for adults. ***Source***: USDA FoodData Central. https://fdc.nal.usda.gov/fdc-app	

Precautions

Excessive consumption of parsley should be avoided by pregnant women. Normal food quantities are safe for pregnant women but consuming excessively large amounts may have uterotonic effects (en.wikipedia.org, 2023).

Uses

Chopped fresh green curly leaf parsley is often used as a garnish by CE&S Europe as well as in W. Asia. Flat leaf parsley is similar but with slightly stronger flavour and easier to cultivate. Root parsley is very common in CE&S European cuisines where it is used as a snack or a vegetable in many soups, stews and casseroles (en.wikipedia.org, 2023).

Cultivation

Parsley is grown from seeds that are usually sown directly in the garden after the last frosts (TerrePromise, 2023). From seeds it requires 4-6 weeks as germination is very slow due to presence of furanocoumarins in the seed coat. Parsley grows best in moist but well-drained fertile soil in full sun or partial shade in autumn or spring (ShootGardening, 2023) at 22-30^{0}C temperatures (Huxley, 1992). The seeds are sown at a depth of 5mm (TerrePromise, 2023). Typically, plants grown for the leaf crop are spaced 10cm apart while those for root crops are spaced 15-20cm apart to allow for root development. Other growing requirements, and pests and diseases are almost similar to carrots or parsnips.

Harvest

The plant matures in about 180 days (TerrePromise, 2023). Leaves can be picked throughout the growing season. Growers should prevent the plants from flowering so that they can continue leaf harvesting till the end of crop season. The roots can be lifted from mid-autumn to early winter (ShootGardening, 2023).

Root parsley/Tuberous parsley/Hamburg root parsley (*Petroselinum crispum* 'Radicosum group')

Synonym: *Petroselinum crispum* var. *tuberosum*.

Other common names: Large root parsley, Turnip-rooted parsley, Parsley root.

Introduction

Root parsley growing to a height of 30cm with plant spread of 20cm (TerrePromise, 2023) is an upright, clump-forming biennial widely cultivated as an annual vegetable with triangular, pinnate & strongly aromatic edible bright green leaves, flattened clusters of tiny star-shaped yellow-green flowers in summer and enlarged edible roots ready for harvest in mid-autumn (ShootGardening, 2023).

Thickened root, Radicosum Group

By User:MarkusHagenlocher – Own work, CC BY-SA 3.0, https://commons.wikimedia.org/w/index.php?curid=1700723

Uses

Root parsley is very common in CE&S European cuisines where it is used as a snack or a vegetable in many soups, stews and casseroles and as an ingredient for broth (en.wikipedia.org, 2023). The fruits are though aromatic but have found little application though their use in vegetable stews or lentil dishes may however have surprising effects. Since this is an efficient diuretic drug so intake of large amounts may prove hazardous especially for those with kidney weakness. The same holds true but to a lesser extent for the root though not for the leaves (GernotKatzer'sSpicePages, 2023).

References

Delage, Barbara, 2015. Flavonoids. Linus Pauling Institute, Oregon State University, Corvallis, Oregon. Retrieved on 1 June, 2023. https://lpi.oregonstate.edu/mic/dietary-factors/phytochemicals/flavonoids.

en.wikipedia.org, 2023: Parsley. Retrieved on 1 June, 2023. https://en.wikipedia.org/wiki/Parsley.

GernotKatzer'sSpicePages, 2023. Parsley [*Petroselinum crispum* (Mill.) Nyman ex A. W. Hill]. Retrieved on 1 June, 2023. http://gernot-katzers-spice-pages.com/engl/Petr_cri.html.

https://fdc.nal.usda.gov/fdc-app.

https://commons.wikimedia.org/w/index.php?curid=122217.

https://commons.wikimedia.org/w/index.php?curid=1700723.

https://commons.wikimedia.org/w/index.php?curid=18213364.

https://commons.wikimedia.org/w/index.php?curid=29637295.

Huxley, A. (ed.), 1992. New RHS Dictionary of Gardening (vol. 3), 532p. Macmillan, USA.

Kalloo, G. 1993. Parsley and turnip-rooted parsley: *Petroselinum crispum* Mill. Nym., *P. crispum* var. *tuberosum*. Genetic Improvement of Vegetable Crops (eds Kalloo, G. and B.O. Bergh), pp. 573-7. Pergamon, USA. ISBN 9780080408262. https://doi.org/10.1016/B978-0-08-040826-2.50045-X. (https://www.sciencedirect.com/science/article/pii/B978008040826250045X).

Meyer, H., A. Bolarinwa, G. Wolfram and J. Linseisen, 2006. Bioavailability of apigenin from apiin-rich parsley in humans (PDF). *Ann. Nutrit. Metabol.*, **50**(3): 167–72. doi:10.1159/000090736. PMID 16407641. S2CID 8223136.

POWO, 2023: *Petroselinum crispum* (Mill.) Fuss. Retrieved on 1 June, 2023. https://powo.science.kew.org/taxon/urn:lsid:ipni.org:names:60442790-2.

Shankar, E., A. Goel, K. Gupta and S. Gupta, 2017. Plant flavone apigenin: An emerging anticancer agent. *Curr. Pharmacol. Rep..*, **3**(6): 423–46. doi:10.1007/s40495-017-0113-2. PMC 5791748. PMID 29399439.

ShootGardening, 2023: Retrieved on 1 June, 2023. *Petroselinum crispum* var. *tuberosum* (Hamburg parsley).

TerrePromise, 2023: Retrieved on 1 June, 2023. https://terrepromise.ca/en_CA/shop/153-tuberous-parsley-petroselinum-crispum-var-tuberosum-624.

Salsify/Goatsbeard (*Tragopogon* L.)

Family: Asteraceae.

Subfamily: Cichorioideae.

Synonym: *Chromatopogon* F.W.Schmidt (POWO, 2023a).

Introduction and Origin

Generic name *Tragopogon* is derived from Ancient Greek *Tragos* which means 'billy goat' and *pogon* means 'beard', referring to the feathery hairs produced when the plant produces fruit (NCE, 2023a; Neill, 2005). The native range of this genus is Canary Islands, Temperate Eurasia, North Africa to Arabian Peninsula (POWO, 2023a) but several species have been introduced into North America and Australia and have spread widely there. The genus includes the vegetable known as salsify, as well as a number of common wild flowers. Salsifies grow as biennial or perennial plants with a strong taproot having milky sap. These usually have a few upright branches and grass-like leaves. Depending on the species, there is variation in the colour of flower as in some species it may be yellow-flowered and others bronze or purple. Seeds are achenes and are borne in a globe-like dandelion-type but larger and are dispersed by the wind (en. wikipedia.org, 2023a).

Hybrid Speciation

Hybrid speciation has been observed in the genus (Ownbey, 1950; Novak *et al.,* 1991). In the early 1900s, the yellow or western salsify (*T. dubius*), the meadow salsify (*T. pratensis*) and the oyster plant (*T. porrifolius*) were introduced into North America which are now common in urban areas there. Two new species, *T. miscellus* and *T. mirus* were found in 1950s in the regions of Idaho and Washington where the three already known species overlapped. *T. miscellus* is a tetraploid hybrid of *T. dubius* and *T. pratensis* whereas *T. mirus* is an allopolyploid though its ancestors were *T. dubius* and *T. porrifolius*. These new species are usually referred to as 'the Ownbey hybrids' after the botanist who first described these (Soltis *et al.,* 2004).

Uses

The salsify vegetable is usually the root of the purple salsify, *Tragopogon porrifolius*. The young shoots of purple salsify can also be eaten as well as the young leaves (Northern Bushcraft, 2023). Other species are also used in the same way including the **Black** or **Spanish salsify** (*Pseudopodospermum hispanicum*) which is closely related though not a member of the genus *Tragopogon* (en. wikipedia.org, 2023a). Roots and young leaves are edible raw or cooked and can be roasted as coffee substitute. They can be dried and ground. Young stalks and root crowns can be simmered (Northern Bushcraft, 2023).

Common species

Some of the more common species of *Tragopogon* are known in the regions where they are most common by the common names goat's beard, goatsbeard, salsify or common salsify without further qualification. These names are therefore inherently ambiguous and best avoided or reserved for the genus collectively (en. wikipedia.org, 2023a). Species in the Pacific Northwest are Commom or Purple salsify (*Tragopogon porrifolius*), Yellow salsify (*T. dubius*) and Meadow salsify (*T. pratensis*) (Northern Bushcraft, 2023). Yellow salsify is a common weed of cultivated crops, roadsides and waste areas. Meadow salsify and common salsify are known to occur as non-native weeds in North America (Stebbins, 1993).

Common salsify [*Tragopogon porrifolius* L.]

Other names: Salsify, Jack Go To Bed, Jerusalem Star, Oyster Plant, Purple Goat's Beard, Purple Salsify and Vegetable Oyster (NCE, 2023a; en.wikipedia. org, 2023b).

Synonyms: *Tragopogon porrifolius* f. *porrifolius* ; *T.p.* var. *cupani* (Guss.) Fiori; *T.p.* var. *porrifolius* (WFO, 2023).

Subspecies: *Tragopogon porrifolius* subsp. *abbreviatus* (Boiss.) Coşkunç. & Gultepe; *T.p.* subsp. *cupanii* (Guss. ex DC.) I.Richardson; *T.p.* subsp. *eriospermus* (Ten.) Greuter; *T.p.* subsp. *macrocephalus*; *T.p.* subsp. *porrifolius* (POWO, 2023b).

Introduction and Origin

The Latin specific epithet *porrifolius* means 'with leaves like leek' (*Allium porrum*) (Neill, 2005). The common name 'Oyster Plant' refers the flavour of the roots that supposedly tastes like oysters. Purple Goat's Beard references the colour of the flowers and the derivation of the genus name (NCE, 2023a).

The native range of *Tragopogon porrifolius* is Canary Islands, Europe to Mediterranean and Arabian Peninsula (POWO, 2023b) but was introduced elsewhere into the British Isles mainly in C&S England (Stace, 2010), other parts of N. Europe, N. America, southern Africa and in Australia. In the United States it is now found growing wild in almost every state including Hawaii except in the extreme south-east (en.wikipedia.org, 2023b).

Tragopogon porrifolius **at Sibley Volcanic Regional Preserve in Oakland**

By Sarah Stierch - Own work, CC BY 4.0, https://commons.wikimedia.org/w/index.php?curid=19761552

The Purple Goat's Beard through hybridization with the Yellow Goat's Beard (*Tragopogon pratensis*) has produced *Tragopogon* × *mirabilis*. This hybrid has dual-coloured petals ~ the inner petals are yellow while outer edges are purple.

Salsify is an edible plant with a long root similar to parsnip and is often found along roadsides, vacant plots, woodlands and grasslands (NCE, 2023a;

en.wikipedia.org, 2023b). The species grows primarily in the temperate biome and is used as a medicine, has environmental uses and for food (POWO, 2023b).

Botany

The species is reported with chromosome numbers of 2n=2x=12 (Ownbey, 1950; Lim *et al.*, 2008). Salsify is a biennial herbaceous wildflower growing to a height of around 1.2m (Blamey *et al.*, 2003) with large unbranched stem that exudes a milky juice, grass-like uncurled leaves (Heistinger, 2013) that are up to 30cm long, 2cm wide, tapering gradually from the base and not recurved at the tip (WFO, 2023), and a thick and fleshy taproot (NCE, 2023a) with dark brown skin and white or cream coloured flesh (SF, 2023).

Salsify taproots

Source: SF (2023)

The young plant appears in the form of a rosette until the formation of flower stem. The purple or mauve-coloured flowers occur in the second year. In U.K., it flowers from May to September but in warmer areas such as California it can be found in bloom from April. The flowerhead is about 5cm across, each surrounded by green bracts which are longer than the petals (ligules of the ray florets). Dull purple or mauve flowers are ray florets that are 3–5cm across and look like a daisy. Flowers are hermaphrodite with entomophilous pollination (NCE, 2023a; en.wikipedia.org, 2023b).

The fruits looking like a dandelion puffball (NCE, 2023a) are beaked achenes, rod-shaped with light ribs, and have hairs at one end that facilitate wind dispersal (Heistinger and Lerch, 2010). The achenes are 10–17mm long and 1–3mm wide without beak otherwise these are up to 55mm long (Heistinger, 2013). When the fruits are fully formed, the hairs from the fruits give the appearance of a fluffy ball which gives the plant its name 'goatsbeard' (Neiring and Olmstead, 1985). Seeds are in brownish fluffy pappus so are easily dispersed by the wind (NCE, 2023a).

Cultivated Varieties and Seed Production

Cultivated varieties that are generally characterized by larger or better-shaped roots include 'White French', 'Giant French', 'Mammoth Sandwich Island', 'Improved Mammoth Sandwich Island', 'Blauetikett', and 'Luthy' (Lloyd, 1902; en.wikipedia.org, 2023b). To maintain the purity of the cultivar, an isolation distance of 150m has to be maintained from one variety to another. When *T. porrifolius* is grown for seed, it is harvested in the second season from midsummer to early autumn to select for bolt resistance (en.wikipedia.org, 2023b).

Nutritional Value

Salsify, raw	
Nutritional value per 100g	
Energy	343 kJ (82 kcal)
Carbohydrates	18.6 g
Dietary fiber	3.3 g
Fat	0.2 g
Protein	3.3 g
Vitamins	Quantity (%DV†)
Thiamine (B1)	0.08 mg (7%)
Riboflavin (B2)	0.22 mg (18%)
Niacin (B3)	0.5 mg (3%)
Pantothenic acid (B5)	0.371 mg (7%)
Vitamin B6	0.277 mg (21%)
Folate (B9)	26 μg (7%)
Vitamin C	8 mg (10%)
Minerals	Quantity (%DV†)
Calcium	60 mg (6%)
Iron	0.7 mg (5%)
Magnesium	23 mg (6%)
Manganese	0.268 mg (13%)
Phosphorus	75 mg (11%)
Potassium	380 mg (8%)
Sodium	20 mg (1%)
Zinc	0.38 mg (4%)
Other constituents	Quantity
Water	77 g

Salsify, raw
Units: µg = micrograms; mg = milligrams; IU = International units †Percentages are roughly approximated using US recommendations for adults. ***Source:*** USDA FoodData Central, 2023
†Percentages are roughly approximated using US recommendations for adults. ***Source***: USDA FoodData Central

Phytochemicals

Tragopogon porrifolius contains polyphenol phytochemicals which are under preliminary investigation for their potential biological effects (Spina *et al.*, 2008). *Tragopogon* species are being studied at the biochemical level for their novel enzyme forms and genetic polymorphism (Soltis *et al.*, 2014).

Flavour

Fresh salsify has a mild and soft taste that resembles asparagus or oysters from which the plant derives its alternative name of oyster plant, but more insipid with a touch of sweetness. Due to its soft taste, salsify can be combined with many other products (SF, 2023).

Uses

The young root, young shoots and young leaves can be eaten raw or cooked (Lyons, 1956; Heistinger, 2013; Neiring and Olmstead, 1985). The flowering shoots can be used like asparagus, either raw or cooked and the flowers can be added to salad, while the sprouted seeds can be used in salads or sandwiches. Latex derived from the root can be used as a chewing gum (en.wikipedia.org, 2023b).

The roots and leaves are more palatable when collected before formation of floral stalk. Raw young roots can be grated for use in salads but older roots are better cooked and can be added to soups, stews or stir-fries. If the roots are too tough for eating these can be boiled with a pinch of baking soda and a change of water (Elias and Dykeman, 2009).

Salsify root is usually peeled before preparation. While peeling it releases a sticky milky juice, which will discolour the vegetable and hands. This can be prevented by wearing kitchen gloves and peeling the salsify under water with vinegar or by briefly blanching them before peeling. The peeled root can then be boiled with a splash of milk or vinegar to prevent discolouration (SF, 2023; Elias and Dykeman, 2009).

Cultivation

Salsify is grown similar to parsnips and carrots and thus requires similar attention as well (Boswell, 1957). Though the salsify roots are cultivated and eaten most frequently in France, Germany, Italy and Russia, the popularity of this crop as a vegetable has declined and it has tended to be replaced by 'black salsify' or 'Spanish salsify' (*Pseudopodospermum hispanicum*) as a cultivated crop (en.wikipedia.org, 2023b). Salsify seeds germinate in 8-10 days if soil is having ample moisture. In the first year it takes only vegetative growth and during autumn it stores the energy in the form of roots but in the second year (season) the generative purple flowers appear from early to mid summer (Heistinger and Lerch, 2010; Dachler, 2017).

Climate

There is no any specific requirement for cultivation of *T. porrifolius* except that this is successfully cultivated in temperate climatic zones (Heistinger and Lerch, 2010). The seedlings need a temperature of 8–16°C to germinate and the plant will freeze between −1.1 and −1.6 °C. *T. porrifolius* can cope with low temperatures and is resistant to light freezing (Dachler, 2017; Joshi and Shrestha, 2018).

Sowing, sowing season and soil requirement

The seeds of *T. porrifolius* should be sown under full sun in late summer or early winter (NCE, 2023a) to foster an early growth (Lloyd, 1902). Planting can also be done in early spring about 100 days before the first frosts in a well prepared deep and loose soil, preferably a sandy- or silt-loam (Boswell, 1957) which remains moist but not waterlogged and may not dry too much so that taproot development is not hampered. Stony or waterlogged soils have negative effects on yields and hamper harvesting. However, at the time of field preparation, the soil should be mixed with wellrotten cowdung manure @ 10q/ha and then only beds should be prepared for sowing or planting though *T. porrifolius* is a moderate feeder therefore application of fresh manure does not benefit much if soil is normal. Other root crops, legumes and cereals have been mentioned as possible preculture. However, during the main growing period a good water supply prevents potential branching of the taproot (Dachler, 2017).

The seeds are usually sown at the depth of 1.3-2cm. It should be shown at a spacing of around 3cm in rows that are 45-60cm apart. For the cultivar 'Mammoth Sandwich Island', about 12g of seeds per 10m of row is required. A thinning to 5cm between the plants is required when seedlings reach 5cm (Andersen, 2011).

Harvest

The taproots are usually harvested from late autumn onwards (later than October in the Northern Hemisphere) till whole winter. Harvesting after a frost is favoured to improve the taste of the root. After flowering, the taproot becomes stringy and inedible (Dachler, 2017). The root becomes discoloured and spoils quickly if broken which can easily happen since it is difficult to remove from the soil without damage so one should be very careful while lifting the roots. Per hectare root yield is about 15-20t/ha.

Storage

The taproots can be stored in traditional clamps although refrigerated storage has been recommended at 0-1°C and 90–95% relative humidity for 2–4 months (Heistinger and Lerch, 2010; Joshi and Shrestha, 2018). The stored roots neither produce ethylene nor are so sensitive to ethylene (SF, 2023).

Pests and Diseases

Few pests or diseases affect *Tragopogon porrifolius*. **White rust** (*Albugo tragopogonis*) is the most common pathogen of *T. porrifolius* as in case of *T. pratensis*, black salsify (*Scorzonera hispanica*), gerbera and sunflower (*Helianthus annuus*). Its infection at an early stage causes chlorotic spots in the form of small white blisters on leaves and stems. This can be controlled by spraying with 0.2% Dithane M-45 or 0.3% Phytolaxin/Macimin Xoton fortnightly.

Occasional problems have been reported by Thompson (2015) with **rust diseases** (*Puccinia hystericum*, *P. jackyana* and *P. scorzonera*). Mulching with oats and the use of spring vetch (*Vicia sativa*) or blue tansy (*Phacelia tanacetifolia*) as cover crops reduces the infection of the fungi infecting seedlings and roots. To get rid of these pathogens, spraying of 0.2% penconazole, mancozeb or myclobutanil is quite effective. The fungi isolated from seedlings and roots of *T. porrifolius* are *Alternaria alternata*, *Fusarium culmorum*, *F. oxysporum*, *Penicillium* spp., *Rhizoctonia solani* and *Sclerotinia sclerotiorum* (Clements *et al.*, 1999). Close planting may also invite powdery mildew (*Erysiphe cichoriacearum*) (Joshi and Shrestha, 2018). Roots left in the ground over winter invite mice and voles to nibble them (Joshi and Shrestha, 2018).

Western Salsify/Yellow Salsify (*Tragopogon dubius* Scop.)

Other common names: Western goat's-beard, Wild oyster plant, Yellow goat's beard, Goat's beard, Goatsbeard, Common salsify and Salsify (en.wikipedia.org, 2023c).

Synonyms: *Tragopogon major* ssp. *dubius* (Scop.) Nyman and *T. pratensis* ssp. *dubius* (Scop.) Gaut. (POWO, 2023c).

Subspecies: *Tragopogon dubius* ssp. *desertorum* (Lindem.) Tzvelev, *T.d.* ssp. *dubius* and *T.d.* ssp. *major* (Jacq.) Vollm (POWO, 2023c).

Introduction and Origin

Tragopogon dubius grows primarily in the temperate biome and its native range is Europe to Xinjiang and Himalaya (POWO, 2023c). It has been introduced into North America where it has become widespread, being reported from all the continental United States except for a few in the far south-east, and all provinces of Canada except Newfoundland and the northern territories (en.wikipedia.org, 2023c).

Though not closely related to meadow (*Tragopogon pratensis*) or common salsify (*T. porrifolius*), the western salsify hybridizes readily with both, and in North America its hybrids have given rise to the new alloploid hybrid species *T. mirus* and *T. miscellus* (en.wikipedia.org, 2023c).

Similar Species

Yellow salsify seedlings can be mistaken for small grass plants. Meadow salsify (*T. pratensis*) also has yellow flowers and could be confused with yellow salsify. It grows throughout Canada. Unlike yellow salsify, meadow salsify does not have swollen stems below the flower heads, and each of its flowers has only 8 or 9 floral bracts (Royer and Dickinson, 1999). No any other tall and yellow-flowered Asteraceae species in Alaska has milky juice and long, narrow bracts (Klein, 2011).

Goats beard (*Tragopogon dubius*), closeup of flower showing long green sepals and 5-pointed petal ends. At 2040m in the Eastern Sierra Nevada, California, USA (not native).

By Dcrjsr - Own work, CC BY 3.0, https://commons.wikimedia.org/w/index.php?curid=19937748

Botany

Ownbey (1950) reported 2n=2x=12 chromosome number in yellow salsify. Yellow salsify is biennial herb that grows to a height of 20-30cm (occasionally up to 1m) (en.wikipedia.org, 2023c) from a large taproot and its light- to blue-green, alternate, narrow, little fleshy, clasping, hairless and grass-like leaves are up to 30.5cm long. All parts of the plant contain a milky-white juice. Its flowerheads being 2.5-6.25cm across bearing yellow ray flowers that are formed at the end of long but hollow peduncles. Subtending each head are 10 to 14 bracts that are 2.5-5cm long and extend beyond the ray flowers. Previous year leaves are often found at the base of the plant. The fruiting head is globe-shaped, 6.25-10cm in diameter and composed of seeds that are easily and widely dispersed by wind because each seed has a feathery *cum* webbed pappus (Royer and Dickinson, 1999; Whitson *et al.,* 2000).

Yellow flowerhead having 4-6cm diameter is likely to be seen in late spring or early summer. It produces long and tapered blue-green buds. The inflorescence opens early in the morning and closes usually by late afternoon. Later the plant forms a seedhead that resembles to that of dandelions but is distinctly larger. The seeds (achenes) themselves are 2–4cm long and feather weighing about 8mg each. There is some natural variation between the central and peripheral achenes in the seedhead, with the peripheral ones being generally darker and heavier, and having a higher concentration of phenolic compounds and this may enhance their survival potential (en.wikipedia.org, 2023c).

***T. dubius*, large seedhead**

By Dcrjsr - Own work, CC BY 4.0, https://commons.wikimedia.org/w/index.php?curid=41298397

Uses

The basal leaves can be eaten raw or cooked (Neiring and Olmstead, 1985). Native Americans eat well-cooked roots (Reiner, 1969) which are said to taste like oysters (Fagan, 2019). Yellow salsify is used for medicinal purposes (POWO, 2023c).

Cultivation and Propagation

Its cultivation is just similar to common salsify. This reproduces by seeds only. Each plant is capable of producing up to 500 seeds (Royer and Dickinson, 1999). Seeds do not require cold stratification for germination.

Growth requirements

Yellow salsify is adapted to all the soil textures with pH levels ranging from 6.5 to 7.5. It has low nutrient and moisture requirements. This species is shade intolerant, can withstand temperatures as low as -33°C and requires 160 frost-free days for successful growth and reproduction.

Meadow salsify (*Tragopogon pratensis* L.)

Other common names: Jack-go-to-bed-at-noon, showy goat's-beard or meadow goat's-beard (en.wikipedia.org, 2023d).

Subspecies: *Tragopogon pratensis* ssp. *hayekii* (Soo) Ciocarlan, *T. p.* ssp. *leiocarpos* (Trnka) Greuter and *T. p.* ssp. *pratensis* (POWO, 2023d).

Tragopogon pratensis* subsp. *pratensis

CC BY-SA 3.0, https://commons.wikimedia.org/w/index.php?curid=196810

Introduction and Origin

The species name *pratensis* means 'of the meadows' and refers to its typical habitat (NCE, 2023b). Meadow salsify is native from Europe to Central Asia and Turkiye (POWO, 2023d).

Botany

Ownbey reported this species possessing 2n=2x=12 chromosome numbers. This herbaceous biennial plant forms a low-growing rosette during the first year, producing one or more 30-90cm tall, round and erect flowering stems during the second year that branches sparingly above the withered remains of the rosette. Its stems are hairless and often little enlarged at the base of each leaf (NCE, 2023b; Illinoiswildflowers, 2023). The upper leaves are shorter and more erect though lower ones are grey-green, 10-30cm long with white midrib, pointed, lanceolate, keeled lengthwise and hairless (Blamey *et al.*, 2003). Fleshy edible taproot of meadow salsify is conically straight and extend up to 30cm deep into the soil with a diameter of nearly 4cm. The stems, leaves and taproot contain white latex (NCE, 2023b).

Each stem terminates in a long naked stalk bearing a 5cm wide flowerhead that opens only in the morning sunshine hence its name 'Jack go to bed at noon' (en.wikipedia.org, 2023d). It flowers from May to August (NCE, 2023b) and its yellow flowers could be confused with yellow salsify. However, unlike yellow salsify, meadow salsify does not have swollen stems below the flowerheads and each of its flowers has only 8 or 9 floral bracts (Royer and Dickinson, 1999). The achenes are rough and long beaked pappus radiating outwards interwoven like a spider's web of fine white side hairs (referred to as a 'blowball') (Rose, 1981). They are easily carried off by the wind (NCE, 2023b).

Uses

The roots can be boiled and eaten similar to potatoes (Elias and Dykeman, 2009). The young shoots can also be used in diabetic salads. Rural kids in Armenia collect the milky juice released from the torn stems on the walls of a glass and dried to make chewing gum (en.wikipedia.org, 2023d).

The roots of the plant have a sweet flavour due to their inulin content. The older roots are best when cooked. Young leaves and shoots may be served raw in salads or cooked in soups. The leaves taste best in the spring. The flower stems may be cooked and served like asparagus. The plant has been used to treat liver and gallbladder problems because it has a detoxifying effect and stimulates appetite and digestion. Syrups have been made from the root to treat coughs and bronchitis. The root is best harvested in the fall and dried for later use (NCE, 2023b).

Cultivation

Its cultivation is almost similar to common and Western salsify. Meadow salsify performs best in full sun, average to above-average levels of moisture and fertile sandy-loam or clay soils. It is intolerant to shade. The plant reseeds itself from its 'giant dandelion' plume which is easily carried off by the wind (NCE, 2023b; Illinoiswildflowers, 2023).

References

Andersen, Craig R. 2011. Salsify. Home Gardening Series (PDF). Agriculture and Natural Resources – via Elsevier. Retrieved on 28 May, 2023 from https://en.wikipedia.org/wiki/Tragopogon_porrifolius.

Blamey, Marjorie, Richard Fitter and Alistair Fitter, 2003. Wild Flowers of Britain and Ireland, pp. 294-5. A & C Black, London. ISBN 0-7136-5944-0.

Boswell, Victor R. 1957. Vegetables. Yearbook of Agriculture, pp. 692–8. England.

Clements, David R., Mahesh K. Upadhyaya, Shelley J. Bos, 1999. The biology of Canadian weeds. 110. *Tragopogon dubius* Scop., *Tragopogon pratensis* L. and *Tragopogon porrifolius* L. *Canad. J. Pl. Sci.*, **79**(1): 153–63. doi:10.4141/P98-007. Retrieved on 28 May, 2023 from https://en.wikipedia.org/wiki/Tragopogon_porrifolius.

Dachler, Michael, 2017. Bildatlas Samen: Getreide, Pseudocerealien, Gemuse, Olpflanzen, EiweiBpflanzen, Grunlandpflanzen, Grundungung, Faserpflanzen, Farbstoffpflanzen, Arzneipflanzen, Gewurzpflanzen. Frankfurt am Main: DLG-Verlag, Germany. ISBN 978-3-7690-0845-6. Retrieved on 28 May, 2023 from https://en.wikipedia.org/wiki/Tragopogon_porrifolius.

Elias, Thomas S. and Peter A. Dykeman, 2009. Edible Wild Plants: A North American Field Guide to Over 200 Natural Foods, 106p. Sterling Press, New York: ISBN 978-1-4027-6715-9. OCLC 244766414.

en.wikipedia.org, 2023a: en.wikipedia.org/wiki/Tragopogon. Retrieved on 28 May, 2023.

en.wikipedia.org, 2023b: en.wikipedia.org/wiki/Tragopogon_porrifolius. Retrieved on 28 May, 2023.

en.wikipedia.org, 2023c: en.wikipedia.org/wiki/ Tragopogon_dubius. Retrieved on 28 May, 2023.

en.wikipedia.org, 2023d: en.wikipedia.org/wiki/Tragopogon_pratensis. Retrieved on 28 May, 2023.

Fagan, Damian, 2019. Wildflowers of Oregon: A Field Guide to Over 400 Wildflowers, Trees, and Shrubs of the Coast, Cascades, and High Desert, 117p. Guilford, CT: FalconGuides. ISBN 978-1-4930-3633-2. OCLC 1073035766.

Heistinger, Andrea, 2013. The manual of seed saving: harvesting, storing and sowing techniques for vegetables, herbs and fruits. London: Timber Press Portland, USA and London. ISBN 978-1-60469-382-9. Retrieved from "https://en.wikipedia.org/wiki/Tragopogon_porrifolius" on 29 May, 2023.

Heistinger, Andrea and Franziska Lerch, 2010. Handbuch Bio-Gemüse: Sortenvielfalt für den eigenen Garten. Ulmer. ISBN 978-3-8001-6950-4. Retrieved from "https://en.wikipedia.org/wiki/Tragopogon_porrifolius" on 29 May, 2023.

https://commons.wikimedia.org/w/index.php?curid=196810.

https://commons.wikimedia.org/w/index.php?curid=19761552.

https://commons.wikimedia.org/w/index.php?curid=19937748..

https://commons.wikimedia.org/w/index.php?curid=41298397.

https://fdc.nal.usda.gov/fdc-app.html.

Illinoiswildflowers, 2023. Retrieved on 30 May, 2023 via http://www.illinoiswildflowers.info/weeds/plants/yl_goatbeard.htm.

Joshi, Bal Krishna and Renuka Shrestha, 2018. Working Groups of Agricultural Plant Genetic Resources (APGRs) in Nepal (Kathmandu). Proceedings of National Workshop, 21–22 June 2018, Kathmandu. National Agriculture Genetic Resources Center. ISBN 978-9937-0-6000-4. Retrieved 20 November 2019. Retrieved on 28 May, 2023 from https://en.wikipedia.org/wiki/Tragopogon_porrifolius.

Klein, Helen, 2011. Alaska Natural Heritage Program. University of Alaska Anchorage. Last Updated on 8 February, by Helen Klein. https://accs.uaa.alaska.edu/wp-content/uploads/Tragopogon_dubius_BIO_TRDU.pdf.

Lim, K.Y., D.F. Solti, P.S. Soltis, J. Tate, R. Matyasek, H. Srubarova, A. Kovarik, J.P. Chris, Z. Xiong and A.R. Leitch, 2008. Rapid chromosome evolution in recently formed polyploids in *Tragopogon* (Asteraceae). *Plos One*, **3**(10): e3353. https://doi.org/10.1371/journal.pone.0003353.

Lloyd, John W. 1902. Vegetables for a farmer's garden in northern Illinois (PDF). Agricultural Experiment Station, University of Illinois, USA. Circular No. 45: 1–5.

Lyons, C.P. 1956. Trees, Shrubs and Flowers to Know in Washington (1st ed.). J. M. Dent & Sons Publisher, Canada, pp. 153, 196.

NCE, 2023a. *Tragopogon porrifolius*. North Carolina Extension Gardener. Retrieved on 28 May, 2023. https://plants.ces.ncsu.edu/plants/tragopogon-porrifolius/.

NCE, 2023b: *Tragopogon pratensis*. North Carolina Extension Gardener. Retrieved on 28 May, 2023. https://plants.ces.ncsu.edu/plants/tragopogon-pratensis/.

Neill, Amanda, 2005. A Dictionary of Common Wildflowers of Texas & the Southern Great Plains, 155p. TCU Press, Fort Worth, England. Retrieved on 28 May, 2023 from https://en.wikipedia.org/wiki/Tragopogon_porrifolius.

Niering, William A. and Nancy C. Olmstead, 1985. The Audubon Society Field Guide to North American Wildflowers, Eastern Region, 409p. Knopf, USA. ISBN 0-394-50432-1.

Northern Bushcraft, 2023. https://www.northernbushcraft.com/plants/salsify/notes.htm. Retrieved on 30 May, 2023.

Novak, S.J., D.E. Soltis and P.S. Soltis, 1991. Ownbey's Tragopogons Forty Years Later. *Amer. J. Bot.*, **78**:1586–1600.

Ownbey, M. 1950. Natural bybridization and amphiploidy in the genus *Tragopogon. Am. J. Bot.*, **37**(7): 487–99.

POWO, 2023a: *Tragopogon* L. Retrieved on 30 May, 2023. https://powo.science.kew.org/taxon/urn:lsid:ipni.org:names:30102386-2.

POWO, 2023b: *Tragopogon porrifolius* L.. Plants of the World Online. Royal Botanical Gardens, Kew. Retrieved on 30 May, 2023. https://powo.science.kew.org/taxon/urn:lsid:ipni.org:names:256109-1.

POWO, 2023c: *Tragopogon dubius* Scop. Retrieved on 30 May, 2023. https://powo.science.kew.org/taxon/urn:lsid:ipni.org:names:255993-1.

POWO, 2023d: *Tragopogon pratensis* L.. Retrieved on 30 May, 2023. https://powo.science.kew.org/taxon/urn:lsid:ipni.org:names:256112-1.

Reiner, Ralph E. 1969. Introducing the Flowering Beauty of Glacier National Park and the Majestic High Rockies. Glacier Park, Inc., 48p.

Rose, Francis, 1981. The Wild Flower Key. Frederick Warne & Co., U.K., pp. 390–91. ISBN 0-7232-2419-6.

Royer, F. and R. Dickinson, 1999. Weeds of the Northern U.S. and Canada. The University of Alberta press, Canada,. 434p.

SF, 2023. Salsify. Special Fruit NV. Retrieved on 29 May, 2023. https://www.specialfruit.com/en/products/detail/salsify/nzffz.

Soltis, D.E., P.S. Soltis, J.C. Pires, A. Kovarik, J.A. Tate and E. Mavrodiev, 2004. Recent and recurrent polyploidy in Tragopogon (Asteraceae): cytogenetic, genomic and genetic comparisons. *Biol. J. Linn. Soc.*, **82**: 485–501.

Soltis, Pamela S., Xiaoxian Liu, D. Blaine Marchant, Clayton J. Visger and Douglas E. Soltis, 2014. Polyploidy and novelty: Gottlieb's legacy. *Philosophical Trans. Roy. Soc. B: Biol. Sci.*, **369**(1648): 20130351. doi:10.1098/rstb.2013.0351. ISSN 0962-8436. PMC 4071524. PMID 24958924.

Spina, Michele, Massimiliano Cuccioloni, Luca Sparapani, Stefano Acciarri, Anna Maria Eleuteri, Evandro Fioretti and Mauro Angeletti, 2008. Comparative evaluation of flavonoid content in assessing quality of wild and cultivated vegetables for human consumption. *J. Sci. Food Agric.*, **88**(2): 294–304. doi:10.1002/jsfa.3089. ISSN 1097-0010.

Stace, Clive, 2010. New Flora of the British Isles (3rd ed.), 707p. Cambridge University Press., Cambridge, UK. ISBN 978-0-521-70772-5.

Stebbins, L.G. 1993. *Tragopogon:* Goat's Beard. In: The Jepson Manual: Higher Plants of California (ed. Hickman, J.C.), pp. 354-5. U.K.

Thompson, Anthony Keith, 2015. Fruit and Vegetables: Harvesting, Handling and Storage (3rd ed.). Wiley-Blackwell, Chichester, England.. ISBN 978-1-118-65404-0. OCLC 904730845. Retrieved on 28 May, 2023 from https://en.wikipedia.org/wiki/Tragopogon_porrifolius.

Whitson, T.D., L.C. Burrill, S.A. Dewey, D.W. Cudney, B.E. Nelson, R.D. Lee and R. Parker, 2000. Weeds of the West. The Western Society of Weed Science in cooperation with the Western United States Land Grant Universities, Cooperative Extension Services. University of Wyoming. Laramie, Wyoming. Via https://accs.uaa.alaska.edu/wp-content/uploads/Tragopogon_dubius_BIO_TRDU.pdf.

10

Spanish or Black Salsify [*Pseudopodospermum hispanicum* (L.) Zaika, Sukhor. & N.Kilian]

Family: Asteraceae.

Subfamily: Cichorioideae (species.wikimedia.org).

Other names: Black oyster plant, Serpent root, Viper's herb, Viper's grass or Scorzonera.

Synonyms: *Myscolus hispanicus* (L.) Endl., *Scorzonera hispanica* L., *S.h.* var. *genuina* Pau (POWO, 2023).

Subspecies: *Pseudopodospermum hispanicum* ssp. *asphodeloides* (species. wikimedia.org); *P. h.* ssp. *hispanicum* and *P.h.* ssp. *neapolitanum* (Grande) Bartolucci, Galasso & F.Conti (POWO, 2023; species.wikimedia.org).

Introduction and Origin

In Greek the *pseudo* prefix of *Pseudopodospermum* means 'false', *podos* in Greek stands for 'foot or stem' and *sperma* 'the seed' for the stipulated achenes. The specific epithet *hispanicum* comes from *Hispania*, an ancient Latin name for Spain. *Pseudopodospermum hispanicum* is an original and widespread species in an area ranging from southern Europe to southern Siberia, generally found growing in temperate biomes in rocky slopes and steppe-like meadows, in the edges of mature woods, from the sub-Mediterranean to the mountain range (Un Mondo Ecosostenibile, 2023; Granica, 2015). The species has been introduced in Bulgaria, California, Puerto Rico and Switzerland (POWO, 2023).

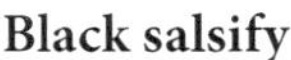

Black salsify

By Rasbak - Own work, CC BY-SA 3.0, https://commons.wikimedia.org/w/index.php?curid=276265

Black salsify is cultivated for its edible taproot mainly in Europe. Belgium, France, the Netherlands and Germany are the world's largest producers (Hernando Bermejo and Leon, 1994). It is also cultivated in Russia and parts of India (www.flowersofindia.net, 2023).

Cultivars

Three common black salsify cultivars are given below (growingtaste.com, 2023):

- **Duplex**: It matures in 80 days. Its roots are 20-30cm. It is said to have good bolting resistance. It is popular among small scale gartdeners in Germany. It is frost-hardy and overwinters well. Its seed are widely available.
- **Enorma**: It is also known as **Lange Jan.** It requires 80 days for maturity. Its roots are 22-27cm long, has good resistance to bolting, is frost-hardy and may overwinter (hardy to -10°C). It has good tolerance against 'soft tops'. Its seeds are widely available.

- **Schwarzer Pfahl (Hoffmann's Schwarzer Pfahl)**: It takes 100-140days for maturity and is widely used by commercial growers in Europe. Its roots are 25-30cm long or even longer. It is frost-hardy even from -15^{0}C to -27^{0}C temperatures. It is stump-rooted with virtually no forking. Its seeds are though uncommon but available.
- **Maxima**: It matures in 100-140 days. Its roots are 30-40cm long. It has improved resistance to bolting, frost-hardy, wintering well and is a high yielder. It is an excellent variety for industrial use. Its seeds are though available but variety is not so common.
- **Noir de Russie**: It matures in 120 days, is frost-hardy and can overwinter. Its seeds are though available but variety is not so common.

Botany

It possesses 2n=14, 28 chromosome numbers (PROTA, 2023). Black salsify is a glabrescent perennial herbaceous plant though is cultivated as an annual for its long, slenderly cylindrical and fleshy taproot that can grow up to 1m long and 2.5cm in diameter with robust black corky inedible skin and edible white flesh (Un Mondo Ecosostenibile, 2023; www.smgrowers.com, 2023). Older replanted plants often have branched roots (backyardlarder.co.uk, 2023).

Black salsify roots

CC BY-SA 3.0, https://commons.wikimedia.org/w/index.php?curid=180802

After a vegetative phase in the first year, the plant flowers in its second year and can reach a height of 60-90cm with a stem diameter of 5cm (en.wikipedia.org, 2023). Broad, lanceolate to ovate and mid-green leaves with entire or wavy margins (Bastgen *et al.*, 2017) are 12-40cm long and 3-6cm broad. The stems are solitary or few in number, usually branched on the upper part (www.flowersofindia.net, 2023).

Rising well above the foliage in late spring is an erect-branching open inflorescence bearing about 5cm wide heads of daisy flowers consisting of bright yellow ray flowers at the tips of long peduncles (www.smgrowers.com, 2023). The flowerheads open only in the evening (Un Mondo Ecosostenibile, 2023). The species is hermaphrodite and the pollination is entomophilous (pfaf.org, 2023).

Nutritional Value

The black salsify contains proteins, fats, asparagine, choline, laevulin as well as minerals such as potassium, calcium, phosphorus, iron, sodium and vitamins A, B1, E and C. The root also contains the polysaccharide inulin instead of starch as storage carbohydrate. When digested it yields fructose and not glucose and is therefore ideally suited for use by diabetics (Hernando Bermejo and Leon, 1994).

Phytochemicals

The compounds detected through aqueous methanol extract method in the aerial parts are flavonoid glycosides (isoorientin, hyperoside, isoquercitrin, miquelianin), luteolin di-*C*-glycoside (*C*-hexoside, *C*-pentoside), quercetin, caffeic acid and caffeic acid derivatives (chlorogenic acid, 4-*O*-caffeoylquinic acid 1,5-*O*-dicaffeoylquinic acid, 3,5-*O*-dicaffeoylquinic acid, 4,5-*O*-dicaffeoylquinic acid, 3-*O*-caffeoylquinic acid, 4-*O*-caffeoylquinic acid, 1,5-*O*-dicaffeoylquinic acid, 3,5-*O*-dicafeoylquinic acid, 4,5-*O*-dicaffeoylquinic acid) (Granica *et al.*, 2015; Granica and Zidorn, 2015).

Through methanol extract of *S. hispanica* subaerial parts the compounds isolated were plugitone, ixerioside D and 3-*O*-angeloyl-11β,13-dihydrodesacylcynaropicrin-8β-D-glucoside (Zidorn *et al.*, 2000). Through ethyl acetate extract of subaerial parts, Granica *et al.* (2015) reported presence of a lignan ((−)-syringaresinol), octadecadienoic acids (linoleic acid, 9-hydroxyocta-(10*E*,12*E*)-decadienoic acid, 8-13-*oxo*-(9*Z*,11*E*)-octadecadienoic acid, 9-*oxo*-(10*E*,12*Z*)-octadecadienoic acid, 13-*oxo*-(9*E*,11*E*)-octadecadienoic acid, and 9-*oxo*-(10*E*,12*E*)-octadecadienoic acid), and sesquiterpenoids (1-*oxo*-bisabola-(2,10*E*)-diene-12-carboxylic acid, 1-*oxo*-bisabola-(2,10*E*)-diene-12-ol, plitostemonol, puliglutone, 1-*oxo*-bisabola-(2,10*E*)-diene-12-carboxylic acid methyl ester, 2,9-epoxycurcumen-12-al, and ixerisoside D), *vis-à-vis* caffeic acid and caffeic acid derivatives (chlorogenic acid, 4-*O*-caffeoylquinic acid, 1,5-*O*-dicaffeoylquinic acid, 3,5-*O*-dicaffeoylquinic acid, and 4,5-*O*-dicaffeoylquinic acid). Petkova (2018) in the roots of *S. hispanica* recorded notable amounts of inulin (over 20% of dry plant material).

Uses

The black salsify flavour is akin to asparagus or oyster so it is used accordingly. Whole plant is edible as its peeled taproots can be consumed after boiling or steaming and is often eaten together with other vegetables such as peas and carrots or similar to those of asparagus in a white bechamel or mustard sauce. Boiled salsify roots may also be deep fried and used (www.flowersofindia.net, 2023). Apart from roots the young leaves can also be eaten usually after boiling and very young, fresh and tender leaves can be eaten raw. Young shoots can be used much in the same way as asparagus. The flowers are edible and can be added to salads as it is said that these have an aroma reminiscent of cocoa (growingtaste. com, 2023).

The inulin present in roots is used as a dietary fibre and may have a positive effect on the digestive system, blood circulation, kidneys and inhibits cancerous processes (Hyun-Na, 2003). Traditionally the subaerial parts of the plant have been used to enhance digestion and perspiration, as a diuretic agent and as a remedy for snakebites ~ and due to this historic usage its vernacular name is serpent's root (Granica *et al.,* 2015).

Preparation Method

The thick black skin of the root is usually considered inedible and can be removed before or after steaming or boiling. If the skin is removed prior to boiling, the peeled root should be immediately immersed in water mixed with vinegar or lemon juice in order to prevent discolouring. Since the root sap is an extremely sticky latex, it is often more convenient to peel it only after boiling for 20-25 minutes (www.flowersofindia.net, 2023; www.smgrowers.com, 2023).

Cultivation

Propagation

Black salsify is usually grown from seeds through direct sowing in the prepared field or raised in pots and planted out in spring, preferably at the end of April (Vulsteke, 1979). Seedlings can take up to three weeks to appear (www. growveg.com, 2023). It can also be propagated by division of older plants in autumn or at the vegetative restart in spring. Larger divisions can be planted directly at their permanent locations (Un Mondo Ecosostenibile, 2023). Growers should use the freshest seed possible. Storage of seeds is also a critical point, as germination is usually only guaranteed for 1-2 years. Seeds should be stored at dry and cool area (en.wikipedia.org, 2023).

Climate

Black salsify can grow in a variety of conditions. It is very hardy and frost-resistant (Lieberei and Reisdorff, 2012) and will grow well in most cool-temperate climates since it is cold hardy to about -20°C temperature (backyardlarder.co.uk, 2023) and plantable down to USDA zone 5 (www.smgrowers.com, 2023). It has potential to be grown as a cash crop in a less favourable marginal environment in temperate zones (Pereira *et al.*, 2014; Hernando Bermejo and Leon, 1994). The plant needs plenty of sunshine to produce its long slender taproots, however, it can also be grown in partial shade.

Soil

Ideal conditions for long root production are an open situation with plenty of sunshine having moist but well-drained sandy or loamy soil. However, it will still produce usable shorter roots in a clay soil and will tolerate a partly shady location too (backyardlarder.co.uk, 2023) provided fresh morning sun is available in plenty. It is also tolerant of poor soils having dry nutrients (www.smgrowers.com, 2023). The roots develop best in very light textured loose sandy-loam soils with pH values being between 5.8 and 6.5. High humus content is beneficial. Any stones or gravel in the ground can cause root deformation. More acidic soils can be improved through addition of lime (en.wikipedia.org, 2023).

Field Preparation and Sowing

The sowing area should be well worked to remove residues so that the roots grow freely and do not run the risk of forking. For field preparation some three deep ploughings are required and at third ploughing some 10q/ha of FYM or well-rotten cow dung manure should thoroughly be mixed in the soil. The seeds are sown directly into shallow furrows at a depth of 1.5–2cm at a distance of 8-10cm in the row spaced at 30-50cm. For optimum yield, the plant densitry can be maintained at 40-50/m^2. Machine showing is not feasible due to unusual shape of the seeds (en.wikipedia.org, 2023). Sowing takes place in April-May and the harvest begins in October (Un Mondo Ecosostenibile, 2023). Too early sowing can result in early bolting and loss of yield.

Irrigation and Cultural Practices

If the weather is dry after sowing, *vis-à-vis* moisture level in the soil at sowing is poor, it is advisable to water the seedbed immediately after seed sowing otherwise germination may be hampered or delayed. In summer it is necessary to water if the soil is very dry (Un Mondo Ecosostenibile, 2023). The crop should be watered lightly but frequently. If weeds start coming up by about 20 days of sowing, these are removed thoroughly and the soil is gently hoed for aeration. Since it a summer crop in temperate areas so watering may be required at 8-10 days though this is very dependent on prevailing weather conditions. A 3-4 year crop rotation with non-compositae crops should be followed.

Harvest and Yield

As with most other root crops, the taste of black salsify too gets improved notably after at least one and preferably a few frosts (it has to do with temperature-triggered conversion of starches to sugars). They can readily stand in the garden right through the winter and beyond and can be dug as per requirement after the first frost or 2nd frost. They should not be dug before they have been through at least one good frost (growingtaste.com, 2023). In British gardens it is common to profit from its perennial character by leaving it in the ground until its roots have grown to sufficient size for harvesting and this requires two years (en.wikipedia.org, 2023).

Roots may be harvested at the end of the first year from October to March, the optimum time for high quality being end of December (Vulsteke, 1979). Roots should be lifted carefully after loosening the soil using a fork. They are quite fragile and a broken root loses it freshness very quickly (en.wikipedia.org, 2023). The root yield per hectare is some 15-20 quintals.

At harvest, when plants are being dug up, the growers can remove most of the roots though a patch can be replanted with the root crowns leaving a 5 cm stub to grow on for future harvests in the following year (backyardlarder.co.uk, 2023).

Storage

Entire roots due to their robust black corky skin will keep fresh all the winter if stored in a cool dark place. In root cellars they may keep fresh well into springtime. The roots can also be washed, peeled, cut and then frozen (Hernando Bermejo and Leon, 1994).

Pests and Diseases [as per en.wikipedia.org (2023) and Rana (2017)]

Pests

- **Carrot rust fly** (*Psila rosae*): The maggots of the carrot rust fly are the nuisance to the root crops as these feed on the plant roots and entrench themselves in the roots. *Psila* thereby inhibits plant growth and facilitates infection by soft-rot bacteria. To control this, insecticides should be used when absolutely necessary. In the areas of its cultivation if it is a perennial problem, in the soil at the time of sowing 0.5kg/ha of Carbofuran or

Phorate should be mixed. However, organically neem cake @ 5q/ha will be quite effective if mixed in the soil before sowing.

- **Wireworms** (*Selatosomus destructor* and *Ctenicera pruinina*): The wireworms are the larvae of click beetles which remain in the soil as larvae for 2-5 years. They live in the soil and feed on roots. The wireworms cause a lot of damage when they bore directly into the roots, facilitating entry of pathogens through their inflicted wounds, and so this causes root rotting. Their infestation can be minimized by dusting the seedlings with lindane or pirimiphos methyl.
- **Root-knot nematodes** (*Meloidogyne incognita, M. javanica, M. hapla* and *M. arenaria*): Their infestation causes great damage to salsify roots. Nematode activity damages growing root tips and causes forking, distortion, deformation and stunting of the taproot. Its root-infestation results into general stunting, leaf chlorosis (yellowing) and wilting of plants aboveground. Though there is no effective chemical control for soil-dwelling nematodes but at the time of field preparation neem cake @ 10q/ha may prove effective if it has thoroughly been mixed in the soil. Carbofuran granules @ 1kg/ha mixing in the soil at the time of sowing may prove quite effective.

Diseases

- **Bacterial soft rot** (*Bacillus carotovorus*): The bacterium invades the root core which results into a soft, watery and slimy root rot. In turn, its invasion also facilitates infection of vertical fungi or bacteria. The ideal temperature for bacterial soft rot is 21-27°C but it can survive and cause damage at temperatures between 0°C - 32°C. For this reason, the cellars should be properly ventilated when storing the roots, and while transporting these crops in tropical and warm environments there should be ample aeration. Since its infection is quite rapid so its control is very difficult. However, before storage or transportation only healthy looking and undamaged roots with surface-dry skins should be selected.
- **White blister** (*Albugo* spp.): It is a foliar disease that favours wet weather for infection and further spread. It is related to downy mildew and even both can also occur together. In its infection the fungus forms blister-like white pustules on the leaf underside, forming a purple pigmentation around the affected areas. This reduces plant growth and heavily infected plant parts may shrivel and die. To check its further spread, infected plant parts are removed and destroyed. If available, resistant or tolerant varieties should be grown.

References

Backyardlarder.co.uk/plants/scorzonera/. Retrieved on 28 May, 2023.

Bastgen, Christa, Berko Schröder and Stefanie Zurlutter, 2017. Welche essbare Wildpflanze ist das? (1st ed.), 108p. Franckh-Kosmos Verlags-GmbH & Co.KG, Stuttgart. ISBN 978-3-440-15193-8.

en.wikipedia.org/wiki/Scorzonera_hispanica. Retrieved on 28 May, 2023.

Granica, Sebastian, 2015. Qualitative and quantitative analyses of secondary metabolites in aerial and subaerial of *Scorzonera hispanica* L. (black salsify). *Food Chem.*, **173**: 321–31. doi:10.1016/j.foodchem.2014.10.006. PMID 25466029.

Granica, S. and C. Zidorn, 2015. Phenolic compounds from aerial parts as chemosystematic markers in the *Scorzonerinae* (*Asteraceae*). *Biochem. Syst. Eccl.*, **58**: 102–13.

Granica, S., U. Lohwasser, K. Johrer and C. Zidorn, 2015. Qualitative and quantitative analyses of secondary metabolites in aerial and subaerial of *Scorzonera hispanica* L. (black salsify). *Food Chem.*, **173**: 321–31.

growingtaste.com/vegetables/scorzonera.shtml. Retrieved on 28 May, 2023.

Hernando Bermejo, J.E. and J. Leon, 1994. Neglected Crops: 1492 from a Different Perspective. *Pl. Prod. Prot. Series*, **206**: 303–32.

Hyun-Na, Koo, 2003. Inulin stimulates NO synthesis *via* activation of PKC-₹and protein tyrosine kinase, resulting in the activation of NF-₹B by IFN-₹-primed RAW 264.7 cells. *J. Nutrit. Biochem.*, **14**: 598–605. doi:10.1016/j.jnutbio.2003.07.002. PMID 14559111 – via Elsevier Science Direct.

Lieberei, Reinhard and Christoph Reisdorff, 2012. Nutzpflanzen (8, überarb. Aufl ed.). Thieme, Stuttgart., pp. 271–2. ISBN 978-3-13-530408-3.

Pereira, I..P., A.S. Dias and L.S. Dias, 2014. Responses of germination and early growth of scorzonera (*Scorzonera hispanica* L.) to pH, mineral deficiencies and growth substrates. *Bulgarian J. Agric. Sci.*, **20** (5): 1195–1201.

Petkova, N. 2018. Characterization of inulin from black salsify (*Scorzonera hispanica* L.) for food and pharmaceutical purposes. *Asian J. Pharm. Clin. Res.*, **11**: 221–5.

pfaf.org/user/Plant.aspx?LatinName=Scorzonera+hispanica#:~:text=Seed%20%2D%20sow%20spring%20in%20situ,as%20growth%20commences%20in%20spring. Retrieved on 28 May, 2023.

PROTA, 2023. *Scorzonera hispanica*, Plant Resources of Tropical Africa. https://uses.plantnet-project.org/en/Scorzonera_hispanica_(PROTA) Retrieved on 29 May, 2023.

POWO, 2023. Plants of the World Online. Zaika, Sukhor and N. Kilian, 2023. *Pseudopodospermum hispanicum* (L.). Royal Botanical Gardens, Kew. Retrieved on 20 May, 2023. https://powo.science.kew.org/taxon/urn:lsid:ipni.org:names:77204090-1.

Rana, M.K. 2017. *Vegetable Crop Science* (1st ed.). CRC Press, USA.

species.wikimedia.org/wiki/Pseudopodospermum_hispanicum.

Un Mondo Ecosostenibile. Retrieved on 28 May, 2023. https://antropocene.it/en/2023/03/07/pseudopodospermum-hispanicum-2/.

Vulsteke, G. 1979. Harvest-time in relation to sowing time of Scorzonera (*Scorzonera hispanica* L.). *Acta Hort.*, **93**: 101–12. doi:10.17660/ActaHortic.1979.93.12.

www.flowersofindia.net/catalog/slides/Black%20Salsify.html. Retrieved on 28 May, 2023.

www.growveg.com/guides/unusual-roots-how-to-grow-salsify-and-scorzonera/. Retrieved on 28 May, 2023.

www.smgrowers.com/products/plants/plantdisplay.asp?plant_id=4727. Retrieved on 28 May, 2023.

Zidorn C, E.P. Ellmerer-Müller and H. Stuppner, 2000. *Sesquiterpenoids* from *Scorzonera hispanica* L. *Pharmazie*, **55**: 550–51.

Kudzu [*Pueraria montana* var. *lobata* (Willd.) Maesen & S. M. Almeida ex Sanjappa & Pradeep]

Family: Fabaceae.

Subfamily: Faboideae.

Synonyms: *Dolichos hirsutus* Thunb., *Dolichos lobatus* Willd., *D. stipulaceus* Lam., *Neustanthus chinensis* Benth., *Pachyrhizus thunbergianus* Siebold & Zucc., *Phaseolus aconitifolius* Roxb., *P. cornutus* Blume ex Miq., *P. ficifolius* Schrank, *P. lobatus* (Willd.) Roxb. ex Wight & Arn., *Pueraria argyi* H.Lev. & Vaniot, *P. bodinieri* H.Lev. & Vaniot, *P. caerulea* H.Lev. & Vaniot, *P. chinensis* (Benth.) Ohwi, *P. harmsii* Rech., *P. hirsuta* (Thunb.) C.K.Schneid., *P. koten* H.Lev. & Vaniot, *P. lobata* (Willd.) Ohwi, *P. montana* var. *chinensis* (Ohwi) Maesen & S.M.Almeida ex Sanjappa & Predeep, *P. novoguineensis* Warb., *P. thunbergiana* (Siebold & Zucc.) Benth., *P. triloba* (Aiton) Makino, *P. triloba* var. *leucostachya* Honda, *P. trilobus* Aiton, *P. volkensii* Hosok., *Vigna lobata* (Willd.) Endl. ex Miq., *Vigna stipulacea* (Lam.) Kuntze (powo.science.kew.org)

Other common names: East Asian arrowroot, Japanese arrowroot.

Regional name: In Japanese: Kuzu; in Korean: Chik or Gal; in Chinese: ge (en.wikipedia.org); in **Hindi**: Sisali, Mugani, Mugawana, Mungani; in **Sanskrit**:

काकमुग्द् kakamudga, मार्जारगन्धिका marjaragandhika, मुग्द्पर्णि mudgaparni; in Kannada: Kohasaru; in Khasi: Suting, Suting-rit; in Marathi: Arkamath, Arkmut, Jangli mung; in Tamil: Cirupayaru, Narip payaru, Naripayar; in Telugu: Pillippersara; in Mizo: Hruihmul (flowersofindia.net).

Introduction and Origin

The genus name *Pueraria* honors Swiss botanist Marc Nicolas Puerari (176-1845) and the specific epithet *montana* means pertaining to mountains.

***Pueraria montana* var. *lobata* (leaves and flowers)**

By Forest & Kim Starr, CC BY 3.0, https://commons.wikimedia.org/w/index.php?curid=6120413

Pueraria montana is native to East Asia, primarily subtropical and temperate regions of China, Japan and Korea (Harrington *et al.,* 2003; Hickman *et al.,* 2010). It is closely related to other species in the genus *Pueraria* (*P. edulis* and *P. phaseoloides*) and the common name kudzu is used for all of these species and hybrids between them. Although there are subtle morphological differences between them, they can breed with each other, and it appears that introduced kudzu populations in the United States have ancestry from more than one of the species (Jewett *et al.,* 2003; Sun *et al.,* 2005). *Pueraria montana* var. *lobata* is one among at least three known subspecies or botanical varieties of the species *P. montana*, the others are *P.m.* var. *chinensis* (Ohwi) Sanjappa & Pradeep and *P.m.* var. *thomsonii* (Benth.) Wiersema ex D.B. Ward (en.wikipedia.org/wiki/Pueraria_montana).

Pueraria montana var. *lobata* is a perennial vine native to India (Assam), Bangladesh, Cambodia, Caroline Is., NC,SC&SE China, East Himalaya, Hainan, Inner Mongolia, Japan, Jawa, Korea, Laos, Malaya, Maluku, Manchuria, Myanmar, Nansei-shoto, Nepal, New Guinea, Northern Territory, Ogasawara-shoto,

Philippines, Primorye, Solomon Is. and Taiwan (powo.science.kew.org). Its preferred habitats are open sunny areas such as forest edges, abandoned fields, roadsides and disturbed areas. The plant was introduced to the United States from Japan in 1876 at the Philadelphia Centennial Exposition and in early 1900s it was promoted as a forage crop and planted throughout the SE U.S. It was recognized as a pest weed in the 1950s and in 1998 it was listed as a federal noxious weed by the U.S. Congress (invasive.org). When grown in warmer climates than Britain the root can be invasive and plants usually become weeds (https://pfaf.org).

Botany

Kudzu possesses 2n=2x=22 chromosome numbers. It is a large, tuberous and deciduous twining vine that establishes huge taproots and forms a crown at the soil level from which up to 30 vines may grow, and under optimum conditions, each vine can grow up to 30cm per day and to 30m in a single growing season. The hairy, herbaceous, dark brown vines grow up to 10cm in diameter and are clad with compound and trifoliate (sometimes lobed) green leaves. The vines spread through rhizomes and by stems that root at the nodes. When left unchecked, particularly in warm winter climates, vines can cover shrubs and tree canopies, choking out and killing entire plant communities. In late summer to early fall, 20cm long elongated and pendant racemes bearing purple pea-like fragrant flowers 2–2.5cm wide are produced at the leaf axis on plants growing under sun. However, flowers are uncommon in the northern parts of its growing range. Bean-like fruit is a flat and hairy pod about 5cm long with three seeds that mature in Autumn (https://www.missouribotanicalgarden.org).

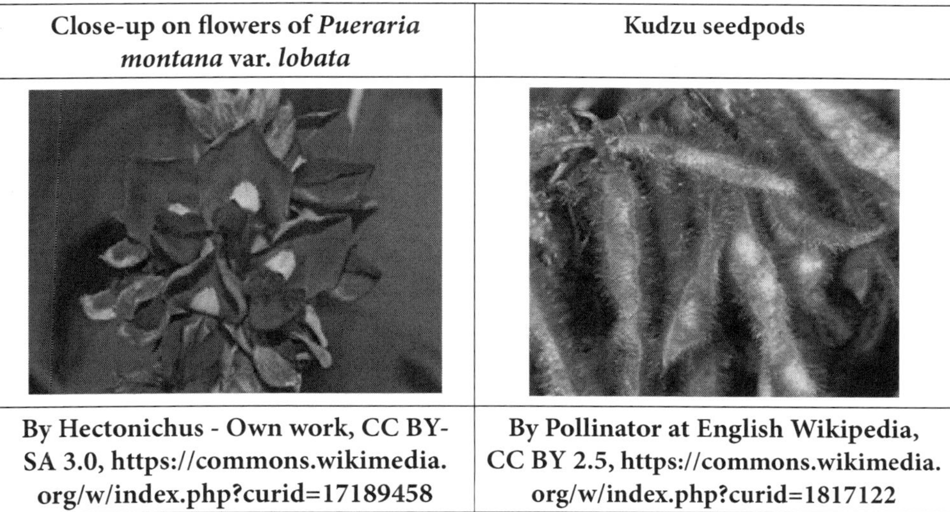

Close-up on flowers of *Pueraria montana* var. *lobata*	**Kudzu seedpods**
By Hectonichus - Own work, CC BY-SA 3.0, https://commons.wikimedia.org/w/index.php?curid=17189458	**By Pollinator at English Wikipedia, CC BY 2.5, https://commons.wikimedia.org/w/index.php?curid=1817122**

Composition

The kudzu root contains about 10% starch. The peeled root contains about 2.1% protein, 0.1% fat, 27.1% carbohydrate and 1.4% ash (https://pfaf.org).

Per 100g fresh kudzu root cantains approximately	
Water	68.6%
Calories	113 per 100g
Protein	2.1 g
Fat	0.1 g
Carbohydrate	27.1 g
Fibre	0.7 g
Ash	1.4 g
Minerals	
Calcium	15 mg
Phosphorus	18 mg
Iron	0.6 mg
Magnesium, Sodium, Potassium, and Zinc	0 mg
Vitamins - A: 0 mg; Thiamine (B1): 0 mg; Riboflavin (B2): 0 mg; Niacin: 0 mg; B6: 0 mg; C: 0 mg	

Source: (https://pfaf.org)

Per 100g starch of the roots contains 340 calories energy, 16.5% moisture, 0.2g protein, 0.1g fat, 83.1g total carbohydrate, 0.1g ash, 35mg Ca, 18 mg P, 2mg Fe and 2mg Na (https://pfaf.org).

Per 100g of cooked leaves contain 36 calories energy, 89% moisture, 0.4g protein, 0.1g fat, 9.7g total carbohydrate, 7.7g fibre, 0.8mg fat, 34mg Ca, 20mg P, 4.9mg Fe, 0.03mg thiamin, 0.91mg riboflavin and 0.8 mg niacin (https://pfaf.org).

Phytochemicals

Kudzu root comprises more than 70 phytochemicals (Wong *et al.,* 2011) and the major groups of phytochemical compounds found are polyphenols, particularly isoflavonoids and triterpenoids with their glycosides (Arao *et al.,* 1995, 1997; Choi *et al.,* 2010). Aromatic glycosides and minor compounds such as choline chloride, acetylcholine chloride, allantoin and D-(+)-pinitol are also reported (Niiho *et al.,* 2010; Miyazawa and Kameoka, 1988). The methanolic extract of the roots contained isoflavone derivatives, *viz.* daidzein (4′,7-dihroxyisoflavone), daidzin (daidzein 7-glucoside), daidzein diacetate, daidzein dimethyl ether, genistein triacetate, and apigenin triacetate (Shibata *et al.*, 1959). Puerarin (8-d-glucopyranosyl-4′,7-dihydroxyisoflavone or 8-d-glucopyranosyldaidzein) isolated from the roots are used as a crude drug in Japan and China (Murakami *et al.*, 1960). Formononetin, pueraria glycoside 1-6 (PG 1-6), puerarol and several other isoflavonoid compounds were isolated from the root of *P. lobata* (Ohshima *et al.*, 1988). Besides these, the isoflavonoids and their glycosides that were isolated from the *P. lobata* root purchased from Shanghai (China) were puerarin-4′-O-d-glucoside, 3'-hydroxypuerarin, 3'-hydroxypuerarin-4'-O-deoxyhexoside, 3′-methoxypuerrin, 6′-O-d-xylosylpuerarin, 3'-methoxy-6′-O-d-xylosylpuerarin, 3'-methoxydaidzin, 3′-methoxydaidzein, daidzein-7-O-methyl ether, 3'-methoxydaidzein-7-O-methyl ether or 3′-methoxyformononetin and biochanin A (Rong *et al.*, 1998a, b). Two new isoflavone diglucosides—formononetin 8-C-(β-d-apiofuranosyl-(1→6)-β-d-glucopyranoside) and formononetin 8-C-(β-d-xylopyranosyl-(1→6)-β-d-glucopyranoside)—as well as the known compounds were separated from the dried roots of *P. lobata* from China (Sun *et al.*, 2008). Twelve phenolic compounds from the crude drug of root determined by a microwave-assisted extraction (MAE) and ultra-high-performance liquid chromatography, together with diode array detection and time-of-flight mass spectrometry were puerarin, puerarin-3′-methoxy-4′-O-glucoside, puerarin-4′-O-glucoside, 3′-methoxypuerarin, daidzein, daidzein-4′,7-O-glucoside, daidzin, genistein, genistin, mirificin, sophoraside A, and ononin (Du *et al.*, 2010). Using high-performance liquid chromatography (HPLC) analysis coupled with electrospray ionization (ESI)-ion trap mass spectrometry analysis, Zhang *et al.* (2010) reported seven isoflavones from the kudzu roots, of which two were the new isoflavonoids: neopuerarin A (8-C-α-glucofuranosyl-7,4′-dihydroxyisoflavone) and neopuerarin B (8-C-β-glucofuranosyl-7,4′-dihydroxyisoflavone) (Zhang *et al.*, 2010).

Using ultra-high-performance liquid chromatography (UPLC) analysis coupled with quadrupole time-of-flight mass spectrometry, 25 isoflavones, 13 saponins and 3 flavones were detected in *P. lobata* flower (Lu *et al.*, 2013). The flower contains a large amount of isoflavones (1.84%–2.86%), including kakkalide, irisolidone and triterpenoidal saponins (0.43%–2.5%), which comprise kaikasaponin and soyasaponin (Niiho *et al.*, 2010). Some isoflavones such as puerarin, daidzin, daidzein, genistin, genistein and biochanin A that were found in the root were also detected in the flower in small quantities (Lu *et al.*, 2013).

Some of the several chemical components in kudzu leaves are triterpenoidal saponins (kaikasaponin III) and isoflavones, *viz.* daidzin, genistin, rutin, kakkalide, robinin (kaempferol 3-O-rhamnosyl (1→6) galactosyl-7-O-rhamnoside) and nicotiflorin (kaempferol-3-O-rutinoside) (Lau *et al.*, 2005; Kinjo *et al.*, 1988).

Uses

In Asia, kudzu has long been cultivated for its starchy roots (called Gegen or Radix *Puerariae lobatae*) and hemplike fibres. It has long been used as food, herbal medicine and as cosmetic ingredient in Asia since ancient times. This plant is a staple food in Japan.

Culinary Use

The tuberous roots have been utilized as a source of starch which is an essential material for cooking, and also used as food in Japan (Kayano *et al.*, 2012). The starch extracted from the kudzu root can be used as a crispy coating in deep fried foods, for thickening soups, can be made into noodles or used like agar or gelatine as a gelling agent for salads (https://pfaf.org). Kudzu flowers also called Gehua or Flos Puerariae, have a sweet vanilla scent and are cooked or made into pickles (https://pfaf.org). Leaves, buds and sprouts of this plant have been consumed as a vegetable in Korea (Son *et al.*, 2019). Stems and young leaves are consumed raw or cooked. The fresh young shoots taste like a cross between a bean and a pea (https://pfaf.org).

Medicinal Uses

Starch powder made from kudzu root is used to make arrowroot tea in traditional medicines of China, Japan and Korea (Chun, 2004). The roots and flowers have been utilized as traditional medicine for treatments of dysentery, alcohol intoxication and alcohol abuse (Wong *et al.*, 2011).

Kudzu vine, known as 'Ge Gen' in China, is commonly used in Chinese herbalism where it is considered as one of the 50 fundamental herbs. The compounds daidzin and daidzein present in kudzu roots and flowers have been proved to be a safe and effective method for treating alcohol abuse. The flowers and the roots are antidote, antiemetic, antipyretic, antispasmodic, demulcent, diaphoretic, digestive, febrifuge, hypoglycaemic and hypotensive. A concoction of the flowers and tubers is used to treat alcoholism, fever, colds, diarrhoea, dysentery, acute intestinal obstruction, *etc.* It is useful in the treatment of angina pectoris and migraine. The root is frequently used as a remedy for measles, often in combination with *Cimicifuga foetida*. The root contains puerarin which increases the blood flow to the coronary artery and protects against acute myocardial ischaemia caused by injection of pituitrin. The stems are galactogogue and are also applied as a poultice to incipient boils, swellings, sore mouths, *etc.* Seed is used in the treatment of hangover and dysentery and the leaves are styptic (https://pfaf.org).

Other Uses

Legumes, such as kudzu, are beneficial in nitrogen fixation and improvement of animal diets. Kudzu is known to produce forage of high quality that contains a crude protein concentration of 15–18% and a total digestible nutrient value of over 60 (https://jewlscholar.mtsu.edu).

A tough and strong fibre from the stems is used to make ropes, cables, coarse cordage and textiles. Kudzu is considered to be a beneficial species within its native area, fixes atmospheric nitrogen and its deep taproots bring mineral nutrients from the subsoil to the topsoil thus enhancing soil fertility. It is used as green manure as well as anti-erosion cover on hillslopes. In the Central Amazon Basin, it has been reported to improve soil pore space in heavy clays (Chauvel *et al.*, 1991).

Kudzu is also popular as an ornamental climber with fragrant flowers (https://apps.worldagroforestry.org). It can survive in full sun or partial shade; however, it prefers forest regrowth and edge habitats with high sun exposure (Forseth and Innis, 2004). These attributes of kudzu make it attractive as an ornamental plant for shading porches in the SE United States but they facilitated the growth of kudzu as it became a 'structural parasite' of the southern states enveloping entire structures when untreated (Mitich, 2000).

Anti-quality and Anti-nutritional Factors of Kudzu

Kudzu contains a variety of toxic and nontoxic secondary metabolites that vary in concentration and can interfere with nutrient intake, absorption and utilization. Alkaloids, cyanogenic glycosides, toxic amino acids, saponins and isoflavones are toxic compounds present in low concentrations and have negative effects when absorbed by an animal including neurological problems, reproductive failure, gangrene and potential fatalities. Lignin, tannin, cutin, biogenic silica and volatile terpenoids make up the non-toxic compounds present in high concentrations. Decreased digestibility and palatability can result from these compounds (Reed *et al.*, 2000; Nepomuceno *et al.*, 2013; Allen and Segarra, 2001).

Organic acids and reducing sugars are the remaining secondary metabolites that have large concentrations in kudzu. Organic acids can bring about precipitation of calcium ions in the blood leading to muscle weakness, nephritis, kidney stones, gastrointestinal irritation and hypocalcemia syndrome in grazing ruminants and horses (Nepomuceno *et al.*, 2013). In large concentrations, reducing sugars can be problematic for equines if they are fed with a high concentrate diet. They will produce excess lactic acid which will result in water retention and decreased pH values in the lumen of the digestive tract and this risks the possibility of digestive disorders including osmotic diarrhoea and colic (Cohen *et al.*, 1999; Nepomuceno *et al.*, 2013). Additional secondary metabolites such as coumarin by-products, depsides & depsidones, alkaloids, steroids, triterpenoids, flavonoids, and cardiac glycosides can also cause negative effects when used as a feedstock (Nepomuceno *et al.*, 2013).

Invasive Species

Kudzu vigorously competes with native flora for light. It grows over them and shades them with its leaves, and tends to block their access to sunlight so native plants may then die (Cain, 2011).

Pueraria montana **(Kudzu) growing over trees in a forest**

By Scott Ehardt - Own work, Public Domain, https://commons.wikimedia.org/w/index.php?curid=702909

When kudzu invades an ecosystem, it makes the leaf litter more labile, thereby lessening the carbon sequestration ability of the soil. This feeds climate change (Tamura and Tharayil, 2014).

Control

An interesting way to control the spread of kudzu is heavy grazing. It can be mechanically controlled by cutting off crowns from roots usually just below the ground level which kills this plant immediately. All removed crown material must be destroyed but should not be buried as the buried crowns can regenerate into healthy kudzu. Transporting crowns in soil removed from a kudzu infestation is one common way that kudzu unexpectedly spreads in new locations (en.wikipedia.org/wiki/Kudzu).

A systemic herbicide such as glyphosate, triclopyr or picloram can be applied directly on cut stems which is an effective means of transporting the herbicide into the kudzu's extensive root system. Herbicides can also be used after mowing, grazing or burning which can allow for an easier application of the chemical to the weakened plants. In large-scale forestry infestations, soil-active herbicides have been shown to be highly effective (en.wikipedia.org/wiki/Kudzu).

Cultivation

Propagation

Kudzu's primary method of reproduction is asexual vegetative spread (cloning) which is aided by the ability to root wherever a stem is exposed to soil. Kudzu spreads by vegetative reproduction *via* stolons (runners) that root at the nodes to form new plants, and through rhizomes (en.wikipedia.org/wiki/Kudzu). For sexual reproduction, kudzu is entirely dependent on pollinators (Harrington *et al.*, 2003).

Young shoots are removed in the spring with some of the underground part of the stem, preferably with some roots already formed. They are potted up and will usually develop new roots from the nodes. They are planted out in the summer if growth is sufficient otherwise they are grown in pots for a year and planted out late in the following spring (https://pfaf.org).

Hard-coated seeds can remain viable for many years and can germinate only when temperature is above 20^{0}C and soil is persistently soggy for 5–7 days (en.wikipedia.org/wiki/Kudzu). Seeds can directly be sown outside under favourable conditions or can be started off in a greenhouse and planted out after the last frosts. It is recommended to pre-soak the seed in warm water for 12h before sowing. They are sown in a warm greenhouse in early spring. Germination usually takes place within 2 weeks and when the seedlings become large enough to handle, they are pricked out into individual pots and planted out after the last expected frosts. The young plants should be covered with a frame or cloche until they are growing away well (https://pfaf.org).

Once germination has occurred the saplings are transferred in a well-drained medium that retains high moisture. During this stage of growth, kudzu must receive as much sunlight as possible. Kudzu saplings are sensitive to mechanical disturbance and are damaged by chemical fertilizers. They do not tolerate long periods of shade or high water tables (en.wikipedia.org/wiki/Kudzu).

Factors Effecting Growth

Kudzu grows well in regions where winters are mild, summer temperatures are above 26^{0}C and annual rainfall is 1000mm or more (www.invasive.org), however, these plants can tolerate an annual precipitation in the range of 970 to 2140mm, an annual mean temperature in the range of 12.2-26.7°C and a pH of 5-7.1. Though kudzu grows well in many soil types and wide range of climatic conditions, it grows well under sunny position on fertile but well-drained loamy soil (https://pfaf.org). Poor sandy and poorly drained heavy clays are unfit for its growing (https://apps.worldagroforestry.org).

An established plant is quite drought resistant but does not stand waterlogging. Kudzu plants are hardy to about -15 ^{0}C and resprout from the base if their aerial parts die down by frosts (https://pfaf.org). In the first year of planting, the field is kept free of weeds (https://apps.worldagroforestry.org).

Harvest

Kudzu roots can be harvested from autumn to spring after one year of planting the cuttings and can be used fresh or dried. If left longer in the soil the roots become very large. For fodder production, the first harvest is possible in the second year though full production is reached from the third year onwards (https://apps.worldagroforestry.org). Fully open flowers are harvested as soon as these are fully open and dried for later use (https://pfaf.org).

Pests and Diseases

Root-knot nematodes (mainly *Meloidogyne* spp.) have been reported to attack the roots. To get rid of these, the soil is required to be solarized at least for one month during summer. Integration of neem cake + *Trichoderma harzianum* or *Paecilomyces lilacinus* or *Glomus mosseae* will drastically reduce their population in the soil. **Velvetbean caterpillars** (*Anticarsia gemmatilis*) have been found feeding on the leaves (https://apps.worldagroforestry.org) which can be controlled through spraying with 4% NSKE or 0.04% Indoxacarb 14.5SC whenever these are seen feeding on the leaves.

Kudzu is infected with various pathogens as enlisted through https://apps.worldagroforestry.org.

Leaf spot disease (*Alternaria* spp.) on kudzu plants occurs as grey or brown and usually circular necrotic spots on leaves which after coalescing cause larger dead areas on leaves. Only in exceptional cases leaves are killed.`To control this problem, affected leaves should be removed and destroyed, field sanitation should strictly be followed. Mancozeb or Carbendazim at 0.2% can be sprayed on the plants fortnightly so long the disease is persisting.

Anthracnose (*Colletotrichum lindemuthianum*) causes sunken and brown longitudinal marks on stems. Leaf veins also show red-colouration which may soon turn brown, withering and dieing off. Pods may also exhibit reddish-brown spotting. During wet weather, a pink-coloured slimy fungal growth may occur. Its fungus is usually seed-borne and infection to other plants occur through rain or overhead watering. The seeds of such plants should not be collected and infected plant or plant parts should immediately be removed and destroyed. Chemical control is not very effective, however, if available resistant cultivars should be used.

Stem rot (*Fusarium* spp.) causes blackening of stems followed by leaf yellowing and rotting of the stems. Dry soil may aggravate the situation. There is no effective chemical control measures against this disease, however, infected stems should be sheared off and destroyed to check its further spread. Where this crop has been grown, again up to further 5 years the same crop or closely related ones should not be grown.

In **damping off** (*Pellicularia solani*) disease the seedlings collapse from the base as base shows water-soaking. This may be due to infection of various fungi such as *Pellicularia, Phytophthora, Pythium, Rhizoctonia* and many more. Occurrence of the disease is aggravated due to prolonged high temperatures coupled with over-wet compost and also when these are very closely sown. Generally this disease occurs due to unhygienic conditions have been used such as quite stagnant water, unsterilized compost, *vis-à-vis* uncleaned pots, trays and implements. To get rid of this problem the soil should be drenched with copper-based fungicide at the sowing time and frequently after until the transplanting of the seedlings.

Bacterial blight (*Pseudomonas* spp.) is caused on leaves and other aerial parts of the plant in round, oval or irregular form of reddish-brown colouration surrounded by darker brown margin which after coalescing produce blighted appearance. It can be controlled by spraying the plants with 0.2% Mancozeb and 0.06% Bactrinashak fortnightly.

References

Cain, M. L., William D. Bowman, Sally D. Hacker, 2011. *Ecology*. Sinauer Associates, Inc., 246p.

Allen, V.G., and E. Segarra, 2001. Anti-quality components in forage: overview, significance, and economic impact. In: Anti-quality factors in rangeland and pastureland forages (ed. Launchbaugh, K.). Station Bulletin No. 73 of the

Idaho Forest, Wildlife, and Range Experiment Station, University of Idaho, USA, pp.1-4.

Arao, T., J. Kinjo, T. Nohara and R. Isobe, 1995. Oleanene-type triterpene glycosides from Puerariae radix. II. Isolation of saponins and the application of Tandem Mass Spectrometry to their structure determination. *Chem. Pharm. Bull.*, No. 43, pp. 1176–9.

Arao, T., J. Kinjo, T. Nohara and R. Isobe, 1997. Oleanene-type triterpene glycosides from Puerariae radix. IV. Six new saponins from *Pueraria lobata. Chem. Pharm. Bull.,* No. 45, pp. 362–6.

Chauvel, A., M. Grimaldi and D. Tessier, 1991. Changes in soil pore-space distribution following deforestation and revegetation: An example from the Central Amazon Basin, Brazil. *For. Ecol. Manage.*, **38**(3-4): 259-67. https://www.feedipedia.org/node/258.

Choi, Y.-H., S.S. Hong, Y.S. Shin, B.Y. Hwang, S-Y. Park and D. Lee, 2010. Phenolic compounds from *Pueraria lobata* protect PC12 cells against Aβ-induced toxicity. *Arch. Pharmacal. Res.*, **33**: 1651–4.

Chun, Hui-jung, 2004. Korean Food Guide 800 (ed.Yoon, Ho-mi). The Korea Foundation, Seoul, 208p. ISBN 978-89-89782-10-0. Retrieved 23 February 2018 – via issuu.

Cohen, N.D., P.G. Gibbs and A.M. Woods, 1999. Dietary and other management factors associated with colic in horses. *J. Amer. Vet. Med. Assn.*, **215**: 53-60.

Du, G., H. Zhao, Q. Zhang, G. Li, F.-Q. Yang, Y. Wang and Y.A. Li, 2010. A rapid method for simultaneous determination of 14 phenolic compounds in Radix Puerariae using microwave-assisted extraction and ultra high performance liquid chromatography coupled with diode array detection and time-of-flight mass spectrometry. *J. Chromatogr.,* **12**(17): 705–14.

en.wikipedia.org/wiki/Kudzu.

en.wikipedia.org/wiki/Pueraria_montana..

flowersofindia.net/catalog/slides/Kudzu%20Vine.html.

Harrington, Timothy B.; Laura T. Rader-Dixon and John W. Taylor, 2003. Kudzu (*Pueraria montana*) community responses to herbicides, burning, and high-*density loblolly pine. Weed Sci.,* **51**(6): 965–74. doi:10.1614/02-142. S2CID 2356816.

Hickman, Jonathan E.; Shiliang Wu, Loretta J. Mickley and Manuel T. Lerdau, 2010. Kudzu (*Pueraria montana*) invasion doubles emissions of nitric oxide and increases ozone pollution. *Proc. Nat. Acad. Sci.,* **107** (22): 10115–9. Bibcode:2010PNAS..10710115H. doi:10.1073/pnas.0912279107. PMC 2890437. PMID 20479252.

https://apps.worldagroforestry.org/treedb/AFTPDFS/Pueraria_montana.PDF.

https://jewlscholar.mtsu.edu/server/api/core/bitstreams/7fb115d9-5fb3-4eeb-8804-5d4fc4c9b097/content.

https://pfaf.org/user/cmspage.aspx?pageid=174.

https://www.missouribotanicalgarden.org/PlantFinder/PlantFinderDetails.aspx?taxonid=263642.

invasive.org/alien/pubs/midatlantic/pumol.htm.

Jewett, D.K.; C.J. Jiang, K.O. Britton, J.H. Sun and J. Tang, 2003. Characterizing specimens of Kudzu and related taxa with RAPD's. *Castanea*, **68** (3): 254–60. JSTOR 4034173.

Kayano, S.-I., Y. Matsumura, Y. Kitagawa, M. Kobayashi, A. Nagayama, N. Kawabata, H. Kikuzaki, Y. Kitada, 2012. Isoflavone C-glycosides isolated from the root of kudzu (*Pueraria lobata*) and their estrogenic activities. *Food Chem.*, **134**: 282–7.

Kinjo, J., T. Takeshita, Y. Abe, N. Terada, H. Yamashita, M. Yamasaki, K. Takeuchi, K. Murakami, T. Tomimatsu and T. Nohara, 1988. Studies on the constituents of *Pueraria lobata*. IV. Chemical constituents in the flowers and the leaves. *Chem. Pharm. Bull.*, **36**: 1174–9.

Lau, C.S., D.J. Carrier, R.R. Beitle, L.R. Howard, Jr., J.O. Lay, R. Liyanage and E.C. Clausen, 2005. A glycoside flavonoid in kudzu (*Pueraria lobata*): identification, quantification, and determination of antioxidant activity. *Appl. Biochem. Biotechnol.*, **121**: 783–94.

Lu, J., Y. Xie, Y. Tan, J. Qu, H. Matsuda, M. Yoshikawa and D. Yuan, 2013. Simultaneous determination of isoflavones, saponins and flavones in Flos Puerariae by ultra performance liquid chromatography coupled with quadrupole time-of-flight mass spectrometry. *Chem. Pharm. Bull.*, **61**: 941–51.

Mitich, Larry W. 2000. Kudzu [*Pueraria lobata* (Willd.) Ohwi]. Weed Techn., **14**(1): 231–5. doi:10.1614/0890-037X(2000)014[0231:KPLWO]2.0.CO;2. JSTOR 3988532. S2CID 86120103.

Miyazawa, M. and H. Kameoka, 1988. Volatile flavor components of Puerariae radix (*Pueraria lobata* ohwi). *Agric. Biol. Chem.*, **52** (4): 1053–5.

Murakami, T., Y. Nishikawa and T. Ando, 1960. Studies on the constituents of Japanese and Chinese crude drugs. IV. On the Constituents of *Pueraria* root. II. *Chem. Pharm. Bull.*, **8**: 688–91.

Nepomuceno, D.D., J.C.C. Almeida, M.G. Carvalho, R.D. Fernandes and F.E.A. Catunda, 2013. Classes of secondary metabolites identified in three legume species. *Revista Brasileira de Zootecnia*, **42**(10): 700-705. doi:10.1590/S1516-35982013001000002.

Niiho, Y., Y. Nakajima, T. Yamazaki, M. Okamoto, R. Tsuchihashi, M. Kodera, J. Kinjo and T. Nohara, 2010. Simultaneous analysis of isoflavones and saponins in *Pueraria* flowers using HPLC coupled to an evaporative light scattering detector and isolation of a new isoflavone diglucoside. *J. Nat. Med.*, **64**: 313–20.

Ohshima, Y., T. Okuyama, K. Takahashi, T. Takizawa and S. Shibata, 1988. Isolation and high performance liquid chromatography (HPLC) of isoflavonoids from the *Pueraria* root. *Planta Med.*, **54**: 250–54.

powo.science.kew.org/taxon/urn:lsid:ipni.org:names:967441-1.

Reed, J.D., C. Krueger, G. Rodriguez and J. Hanson, 2000. Secondary plant compounds and forage evaluation. In: Forage Evaluation in Ruminant Nutrition (eds Givens, D.I., E. Owen, R.F.E. Axford and H.M. Omed), pp. 433-45.

Rong, H., D. De Keukeleire, L. De Cooman and W.R.G. Baeyens, 1998a. Narrow-bore hplc analysis of isoflavonoid aglycones and their o- and c-glycosides from *Pueraria lobata. Biomed. Chromatogr.*, **12**: 170–1.

Rong, H., J.F. Stevens, M.L. Deinzer, L. De Cooman and D. De Keukeleirel, 1998b. Identification of isoflavones in the roots of *Pueraria lobata. Planta Med.*, **64**: 620–27.

Shibata, S., T. Murakami, Y. Nishikawa and M. Harada, 1959. The Constituents of *Pueraria Root. Chem. Pharm. Bull.*, **7**: 134–6.

Son, E., J. Yoon, B. An, Y.M. Lee, J. Cha, G. Chi and D. Kim, 2019. Comparison among activities and isoflavonoids from *Pueraria thunbergiana* aerial parts and root. *Molecules*, 24: 912.

Sun, J.H., Z.-C. Li, D.K. Jewett, K.O. Britton, W.H. Ye and X.-J. Ge, 2005, Genetic diversity of *Pueraria lobata* (kudzu) and closely related taxa as revealed by inter-simple sequence repeat analysis, *Weed Res.*, **45**(4): 255. doi:10.1111/j.1365-3180.2005.00462.x.

Sun, Y.-G., S.-S. Wang, J.T. Feng, X.-Y. Xue and X.-M. Liang, 2008. Two new isoflavone glycosides from *Pueraria lobata. J. Asian Nat. Prod. Res.*, **10**: 719–23.

Wong, K.H., G.Q. Li, K.M. Li, V. Razmovski-Naumovski, K. Chan, 2011. Kudzu root: Traditional uses and potential medicinal benefits in diabetes and cardiovascular diseases. *J. Ethnopharmacol.*, **134**: 584–607.

Zhang, H.-J., X.-P. Yang and K.-W. Wang, 2010. Isolation of two new C-glucofuranosyl isoflavones from *Pueraria lobata* (Wild.) Ohwi with HPLC–MS guiding analysis. *J. Asian Nat. Prod. Res.*, **12**: 293–9.

Forseth, Irwin, N. and Anne F. Innis, 2004. Kudzu (*Pueraria montana*): History, Physiology and Ecology combine to make a major ecosystem threat. *Critical Reviews in Plant Sciences*, **23**(5): 401–13. doi:10.1080/07352680490505150. ISSN 0735-2689. S2CID 84404633.

Tamura, Mioko and Nishanth Tharayil, 2014. Plant litter chemistry and microbial priming regulate the accrual composition and stability of soil carbon in invaded ecosystems. *New Phytologist*, **203**(1): 110–24. doi:10.1111/nph.12795. PMID 24720813.

www.invasive.org

Yam bean (*Pachyrhizus* Rich. ex DC.)

Family: Faboideae.

Subfamily: Phaseoleae.

Synonyms: *Cacara* Rumph. ex Thouars, *Pachyrrhizos* Spreng., *Robynsia* M.Martens & Galeotti, *Taeniocarpum* Desv.

Introduction

The name *Pachyrhizus* comes from the Greek *Pachys* meaning 'thick' and *rhiza* meaning 'root'. Among the Neotropical legume genera with thickened edible roots such as *Apios*, *Pachyrhizus*, *Pediomelum* and *Vigna* the yam bean (*Pachyrhizus*) is the only genus that is extensively cultivated as a garden crop where *P. erosus* is exclusively cultivated on a large scale for export. *P. ahipa* and *P. tuberosus* are also cultivated (Sorensen *et al.*, 1996) but only for home consumption. *Pachyrhizus* is a small genus of five species of tropical and subtropical plants growing from large taproots (en.wikipedia.org1). The native range of the genus is Mexico to Venezuela and NW Argentina and its accepted species are *P. ahipa, P. erosus, P. ferrugineus, P. panamensis* and *P. tuberosus* (POWO1, 2023).

Botany

Ahipa, jicama and other members of the genus *Pachyrhizus* are collectively known as 'yam beans' since these beans form edible storage roots though these have no close botanical relationship to yams. As a legume, yam beans do not have convenient feature of fixing nitrogen (Cultivariable, 2023).

Self-pollinating yam beans are diploids with 2n=22 chromosome numbers though up to 8% cross-pollination opccurs (Zanklan, *et al.*, 2018). *Pachyrhizus* species are vines or semi-erect herbaceous to somewhat lignified perennial plants with one or more tuberous roots, trifoliolate leaves with stipules and pinnately arranged leaflets having caducous stipels. Inflorescence is a complex to simple raceme and the flowers have a tubular calyx and a papilionaceous corolla. The ovary has a basal crenate disc-formed nectarium, the recurving style is ventrally ciliated and the vertical surface of the stigma is subglobose. The straight legume is septate between the seeds and the seeds may be more or less flattened, square or a rounded kidney shape with olive-green and deep maroon to black or from black and white to mottled cream (Sorensen *et al.*, 1996).

Though all the species bear self-pollinating bisexual flowers but some crossbreeding may occur depending on the availability of pollinators, *viz.* mainly various bumblebee species. Cross-breeding is limited when cultivation takes place under irrigation during the dry season owing to the lack of pollinators. With possible exception of *P. ferrugineus*, all the species have been found to be compatible which results into fertile interspecific hybrids. However, no naturally occurring hybrids have been recorded in areas shared by two species as the flowering period of different species does not overlap under natural conditions. The cultivated species are propagated by seed, with the exception of the ahipa cultivar group of *P. tuberosus* (Sorensen *et al.*, 1996).

Reproductive Pruning

All the cultivated species are usually grown as annual crops though the plants have a perennial habit. In *P. erosus* and *P. ahipa* both, the reproductive pruning (manual removal of fertile shoots) is usually practiced to increase storage root growth. In *P. tuberosus* only the cultivar group from the province of Manabi and Ecuador is reproductively pruned. In this cultivar group, however, not only are the flowering shoots pruned but up to 1/3rd of the vegetative shoots are also removed. In areas with high precipitation rates during growth period or in flood-irrigated fields, the crops are usually cultivated on ridges in order to facilitate drainage (Sorensen *et al.*, 1996).

Diseases and Pests

Though diseases and pests for the three cultivated species have been mentioned separately under cultivation section of each species, majority of both the diseases and insect-pests may be common to all the three species and the differences recorded between the species are thus due to geographic, climatic, ecological and edaphic conditions. In field trials in Tonga (Costa Rica), Ecuador and Thailand when all the five species were cultivated at one location, bean common mosaic virus (BCMV) infected all the three cultivated species as well as the wild species *P. panamensis* though with some delays due to the prominent hairiness of all vegetative parts which have a repelling effect on the aphid vectors. Only the 5th remaining wild species, *P. ferrugineus*, appears to possess any resistance to this virus (Sorensen *et al.,* 1996). In all the species, the most serious pests endangering cultivation are various bruchids, which may destroy as much as 80% of the locally stored seeds (Sorensen *et al.,* 1996).

Yield

The potential yield of yam bean varies between species. *P. ahipa* was reported to have fresh storage root yield of 29-50t/ha. The weight of fresh storage roots of *P. tuberosus, i.e.* 7 to 10t/ha is lower than other two cultivated species though in fertile soils 30t/ha yield was recorded. Yield performance of *P. erosus* has been repeatedly obtained in the range of 100-145t/ha fresh storage roots (Heredia and Garcia, 1994; Nielsen, 1995; Morera, 1994). However, in India, Ramaswamy *et al.* (1980) while working with two Mexican lines recorded, fresh storage root yield of 140 to 186.6t/ha.

Jicama or Yam bean [*P. erosus* (L.) Urb.]

Other names: Mexican potato, Mexican turnip, Mexican yam and Potato bean. In India: Mishrikand, Ram Kaseru, Sankalu/Shankalu or Sankeh alu (Root Crops, 2023; Global Bhasin, 2012; en.wikipedia.org3).

Synonyms: *Cacara erosa* (L.) Kuntze; *C. palmatiloba* (DC.) Kuntze; *Dolichos articulatus* Lam.; *D. bulbosus* L.; *D. erosus* L.; *D. mammosus* Noronha; *D. palmatilobus* DC.; *D. trilobus* Lour. ; *Pachyrhizus angulatus* f. *integrifolius* Wawra; *P. angulatus* Rich. ex DC.; *P. articulatus* Duchass. ex Walp.; *P. bulbosus* Kurz; *P. erosus* var. *palmatilobus* (DC.) R.T.Clausen; *P. jicamas* Blanco; *P. palmatilobus* (DC.) Benth. & Hook.f. ex Hemsl.; *P. strigosus* R.T.Clausen; *P. trilobus* DC.; *Phaseolus tuberosus* Lour.; *Robynsia lobata* M.Martens & Galeotti; *R. macrophylla* M.Martens & Galeotti; *Taeniocarpum articulatum* Desv. (POWO3, 2023)

Introduction and Origin

Jicama is a fleshy root vegetable that looks similar to a turnip or a globular radish with a taste and texture similar to the water chestnut (Global Bhasin, 2012). The native range of this species is Mexico to Central America. It grows primarily in the seasonally dry tropical biome (POWO3, 2023). It has been found at archaeological sites in Peru dating to 3000 BC which was introduced to Asia by the Spanish. *P. erosus* is cultivated in Mexico, Guatemala, El Salvador and to a limited extent in Honduras. It has been introduced to different pantropical regions with notable success in SE Asia (Sorensen *et al.*, 1996). It is now cultivated and naturalized in the Philippines, Cambodia, China, Indonesia, northern India as well as in western and northern South America and the Caribbean (Sanderson, 2005; Root Crops, 2023).

By Alex Pronove - Me. I took this photo last week in Quezon, Palawan, Philippines., CC BY-SA 4.0, https://commons.wikimedia.org/w/index.php?curid=14577948

Jicama Types/Varieties

P. erosus is widely grown for its large (10–15cm diameter and up to 20kg weight), spherical or elongated taproot. It has two cultivated forms that are named for the consistency of their juice in tuberous roots (en.wikipedia.org3; www.britannica.com).

- *Jicama de agua*: It includes varieties having top-shaped to oblate roots that are thin skinned with a light brown to almost whitish colour and watery translucent sap. It is the preferred form for the market.

- *Jicama de leche*: It includes varieties having elongated root with thicker *cum* dark brown skin and a rather milky sap.

Botany

Jicama is a hairy perennial herbaceous vine that is woody at the base. It is trailing or climbing plant with great variation in the outline of the leaflets from dentate to palmate. It is typically grown as an annual since the plant is killed during the harvest. However, if left to grow the plant can reach 5m in length. Its petioles are 3-18cm long, compound deep green leaves are alternate and composed of 3 ovate or rhomboidal leaflets which are toothed or lobed and about as broad as long and usually large from 4 to 20cm. The blue/violet or white flowers are produced in long axillary racemes, 1-5 in each of several clusters along the peduncle, the petals are 1.5-2cm long and broad, the pods are 7.5-15cm long and about 1.5cm broad, flattened, almost smooth at maturity which contain 4-12 seeds which may be yellow, brown or red, almost square and flattened and 5-10mm in diameter. Flowers and young pods are usually removed to promote root growth (www.britannica.com; Root Crops, 2023).

Its solitary or clustered root is usually turnip-shaped but irregularly globular with 10-15cm in diameter. Its skin is brown papery and white with crisp and juicy flesh. If not harvested, the roots go on increasing in size, reaching up to 30cm in diameter and 5-18kg in weight but then these become tough and unappetizing (Root Crops, 2023). The heaviest jicama root ever recorded weighed 23kg in the Philippines in 2010 (news.abs-cbn.com).

Fresh jicama

By Eric in SF - Own work, CC BY-SA 3.0, https://commons.wikimedia.org/w/index.php?curid=4324665

Nutritional Value

The tuberous roots contain both starch and sugar and are a moderately good source of ascorbic acid. Nutritional value per 100g of raw jicama root as per fdc.nal.usda.gov is given below:

Jicama, Raw	
Nutritional value per 100g	
Energy	159 kJ (38 kcal)
Carbohydrates	8.82 g
Sugars	1.8 g
Dietary fiber	4.9 g
Fat	0.09 g
Protein	0.72 g
Vitamins	**%DV† (Quantity)**
Thiamine (B1)	2% (0.02 mg)
Riboflavin (B2)	2% (0.029 mg)
Niacin (B3)	1% (0.2 mg)
Pantothenic acid (B5)	3% (0.135 mg)
Vitamin B6	3% (0.042 mg)
Folate (B9)	3% (12 μg)
Choline	3% (13.6 mg)
Vitamin C	24% (20.2 mg)
Minerals	**%DV† (Quantity)**
Calcium	1% (12 mg)
Iron	5% (0.6 mg)
Magnesium	3% (12 mg)
Manganese	3% (0.06 mg)
Phosphorus	3% (18 mg)
Potassium	3% (150 mg)
Sodium	0% (4 mg)
Zinc	2% (0.16 mg)
Other constituents	**Quantity**
Water	90 g

Units: μg = micrograms, mg = milligrams, IU = International units
†Percentages are roughly approximated using US recommendations for adults.
Source: USDA FoodData Central; Table source: fdc.nal.usda.gov via en.wikipedia.org3

Approximately 65 per cent of the carbohydrate is starch, 20 per cent non-reducing sugars and 15 per cent reducing sugars. Mature tubers yield a greyish-white starch, consisting of polyhedral or semi-polyhedral grains of 8-35 microns diameter. Non-protein nitrogen may be as high as 80 per cent of the total nitrogen as indicated by the given crude protein figure (Root Crops, 2023).

Phytochemicals

Following is the list of phytochemicals identified in *P. erosus* and their potential antioxidant properties (Jaiswal *et al.*, 2021):

Sr.	Name	Type	Amount/Yield	Sample	Activity	References
1	12a-Hydroxydolineone	Rotenoids	31 µg/g	Seeds	Against HSV types 1 and 2	Phrutivorapongkul *et al.*, 2002; Kardono *et al.*, 1994
2	12a-Hydroxyerosone	Rotenoids	6 µg/g	Seeds	NA	Kardono *et al.*, 1994
3	12a-hydroxypachyrhizone	Rotenoids	36 µg/g	Seeds	Against HSV types 1 and 2	Phrutivorapongkul *et al.*, 2002; Kardono *et al.*, 1994; Necha*et al.*, 2004
4	12a-Hydroxyrotenone	Rotenoids	9 µg/g	Seeds	Anticancer and against HSV types 1 and 2	Phrutivorapongkul *et al.*, 2002; Kardono *et al.*, 1994
5	Dehydroneotenone	Isoflavone	4 µg/g	Seeds	Antifungal activity	Phrutivorapongkul *et al.*, 2002; Kardono *et al.*, 1994; Necha *et al.*, 2004
6	Dolineone	Rotenoids	60 µg/g	Seeds	Antifungal activity	Phrutivorapongkul *et al.*, 2002; Necha *et al.*, 2004; Estrella-Parra *et al.*, 2014
7	Erosone	Isoflavonoids		Seeds	Antifungal activity	Necha *et al.*, 2004
8	Munduserone	Rotenoids		Seeds	NA	Kardono *et al.*, 1994
9	Neodulin	3-arylcoumarin (pterocarpan)	3.4% µg/g	Seeds	NA	Kardono *et al.*, 1994
10	Neotenone	Isoflavonoids	41 µg/g	Seeds	NA	Phrutivorapongkul *et al.*, 2002
11	Pachyrrhizine	Isoflavonoids	35 µg/g	Seeds Seeds Seeds	Antifungal activity	Phrutivorapongkul *et al.*, 2002; Kardono *et al.*, 1994; Necha *et al.*, 2004; Estrella-Parra *et al.*, 2014
12	Pachyrrhizone	Rotenoids	117 µg/g	Seeds	Antifungal activity	Phrutivorapongkul *et al.*, 2002; Necha *et al.*, 2004
13	Pachyrrhisomene	Pterocarpan	04.7 µg/g	Seeds	NA	Kardono *et al.*, 1994
14	Rotenone	Rotenoids	153 µg/g	Seeds	Anticancer and antifungal activity	Kardono *et al.*, 1994; Necha *et al.*, 2004; Estrella-Parra *et al.*, 2014
15	(8,9)-Furanyl-pterocarpan-3-ol	Pterocarpan	141.05 µg/g	Tuber	Antioxidant and skin whitening	Lukitaningsih *et al.*, 2013, Lukitaningsih, 2014

Sr.	Name	Type	Amount/Yield	Sample	Activity	References
16	2-Butoxy-2,5-bis(hydroxymethyl)-tetrahydrofurane-3,4-diol	Other		Tuber	Antioxidant and skin whitening	Lukitaningsih *et al.*, 2013
17	4-(2-(Furane-2-yl)ethyl)-2-methyl-2,5- dihydro-furane-3-carbaldehyde	Other		Tuber	Antioxidant and skin whitening	Lukitaningsih *et al.*, 2013
18	5-Hydroxy-daidzein-7-O-*β*-glucopyranose	Isoflavonoids		Tuber	Skin whitening	Lukitaningsih, 2014
19	Daidzein-7-*O*-β- glucopyranose / Daidzin	Isoflavonoids	329.93 ± 10.00 μg/g	Tuber	Antioxidant and skin whitening -	Chung *et al.*, 2019; Lukitaningsih, 2014; Yahara *et al.*, 1994; Lukitaningsih *et al.*, 2013
20	Genistin	Isoflavonoids		Tuber	Antioxidant and skin whitening	Lukitaningsih *et al.*, 2013
21	Kaikasaponin III	Triterpenoid glycosides (Saponin)	69 μg/g	Tuber	NA	Yahara *et al.*, 1994
22	Daidzein	Isoflavonoids	2356.10 ± 38.47 μg/g	Tuber juice	Antioxidant and skin whitening	Chung *et al.*, 2019; Lukitaningsih, 2014; Lukitaningsih *et al.*, 2013; Primiani *et al.*, 2013
23	Genistein	Isoflavonoids	1655.30 μg/g	Tuber juice	NA	Primiani *et al.*, 2013
24	Formononetin	7-hydroisoflavones	65.13 ± 3.01 μg/g	Leaves	NA	Chung *et al.*, 2019
25	Gentisic acid	Dihydroxybenzoic acid	69.03 ± 1.00 μg/g	Leaves	NA	Chung *et al.*, 2019
26	l-Phenylalanine	Amino acid	16,654.43 ± 43.52 μg/g	Leaves	NA	Chung *et al.*, 2019
27	Malonyl genistin	Glycosyloxyisoflavone	683.77 ± 12.69 μg/g	Leaves	NA	Chung *et al.*, 2019
28	Malonyldaidzin	Glycosyloxyisoflavone	9853.90 ± 44.28 μg/g	Leaves	NA	Chung *et al.*, 2019
29	*p*-Coumaric acid	Aromatic acid	612.83 ± 2.02 μg/g	Leaves	NA	Chung *et al.*, 2019

Sr.	Name	Type	Amount/Yield	Sample	Activity	References
30	*p*-Hydroxybenzoic acid	Aromatic acid	161.63 ± 1.52 µg/g	Leaves	NA	Chung *et al.*, 2019
31	Salicylic acid	Aromatic acid	58.83 ± 2.57 µg/g	Leaves	NA	Chung *et al.*, 2019
32	Vitexin	Flavones	1970.12 ± 9.22 µg/g	Leaves	NA	Chung *et al.*, 2019
33	Isobavachalcone	Flavonoids	34.3 µg/g	Leaves	Anticancer	Minh *et al.*, 2021
34	Wighteone	Flavonoids	2.9 µg/g	Leaves	Anticancer	Minh *et al.*, 2021
35	Prunin	Flavonoids	2.7 µg/g	Leaves	NA	Minh *et al.*, 2021
36	Orientin	Flavonoids	50 µg/g	Leaves	NA	Minh *et al.*, 2021
37	Erosusone	Flavonoids	2 µg/g	Leaves	Anticancer	Minh *et al.*, 2021
38	3-episedumoside F_1	Megastigmane glycoside epimer	20 µg/g	Leaves	NA	Minh *et al.*, 2021

Toxicity

Except the enlarged roots other plant parts are very poisonous. Fungicidal compounds have been isolated from the stem. Neodunal is the major component of the four pterocarpan derivates isolated from stem. Though young pods are edible but the mature pods, mature seeds, leaves and stems contain rotenone ~ a highly toxic chemical used as an insecticide (Duke, 1992; www.britannica.com; Root Crops, 2023).

Uses

The flavour of young tuberous roots is sweet and starchy, reminiscent of some apples or raw green beans. It is usually eaten raw in salads, sometimes with salt, lemon or lime juice and chili powder. It can be cooked as a vegetable, in soups and stir-fried dishes or in chutney and pickles. Flour can be made from the tuberous roots by slicing, drying and then grinding. As the roots mature, their starch content increases and older roots are sometimes used as a source of starch or for animal feeding. In China, the dried roots are used as a cooling food for people with fever (Root Crops, 2023; United States, 2009; en.wikipedia.org3).

Propagation

Although jicama is grown principally from seed, it can be propagated even from sprouted roots saved from the previous crop. Vegetative propagation is mainly to maintain desirable characteristics in the plants. In the Philippines, 20-25 kg/ha of seed is used for planting, assuming a germination rate of 90-95 per cent. In India, a higher seed rate, *i.e.* 50-70kg/ha is preferred (Root Crops, 2023).

Climate

Though jicama is tolerant of various climatic conditions but grows best under full sun in warm to hot climates with a long frost-free period as temperatures below 10^{0}C may damage the crop (digitalcommons.usu.edu), and a moderate to high rainfall. Usually it is grown at altitudes below 1000m. It in Mexico is grown under irrigation where it has longer vegetative cycle due to areas being cooler (Root Crops, 2023). Jicama has photoperiod dependency for formation of tuberous roots. It takes good vegetative growth under 14-16h photoperiod but on the cost of storage roots, though short day-lengths provide smaller *cum* bushier plants but with good storage root development (Root Crops, 2023).

Soil

It grows well in a well-cultivated sandy-loam soil having a soil pH of 6.5-8 provided there is adequate drainage. It can tolerate even more acidic soils (digitalcommons.usu.edu). It tolerates even clay soils if it is well-drained though not the heavy soils that are likely to become waterlogged (Root Crops, 2023).

However, if it is not possible to have proper drainage facilities, better the crop is planted on 15-20cm raised beds.

Field Preparation and Sowing or Planting

The planting area should be a site that receives full sun. Before sowing, the soil must be properly loosened to a depth of 15–25cm and thoroughly cleaned of weeds and stones. Before planting, growers must determine fertilizer needs with a soil test and then follow the recommendations given with the test report. If fertilizer applications are needed, the fertilizer should be mixed into the top 15cm of soil. If compost is being used, no more than 2.5cm of well-composted organic matter per 9 square metre of garden area should be applied. Growers can make 15-20cm high raised beds to improve drainage and to create a place for root development.

If jicama is being grown directly in the field, usually 2-3 seeds are placed in each hole and the plants are thinned out as necessary or the seeds may be sown singly by drill (Root Crops, 2023). In case transplants are to be used, they should be allowed 4-10 weeks to grow so that they become at least 7-8cm tall with a well-developed root system before planting. Seeds should be germinated at 26-27°C and should be allowed for emergence of roots there and then these are transferred gently to sterile seeding mix to grow at 20-30°C. Sowing should usually be done at the beginning of rainy season (Sorensen *et al.*, 1996). Adequate light is essential to produce quality transplants. Cool white fluorescent tubes placed 5-7.5cm above the plants and allowed to lit for 14-16h per day will ensure plants growing large and healthy. Regular watering and feeding weekly with a half strength complete soluble fertilizer is required before planting into the garden (digitalcommons.usu.edu).

Jicama should be transplanted when soils are warmer than 10°C and all frost danger is past. Transplants should be planted 20-30cm apart within the row with rows spaced 60-90cm apart. Regular watering is required after planting to help the plants establish properly (digitalcommons.usu.edu). However, the spacing given in India is 15cm along rows spaced 50cm apart and in the Philippines it is 10cm in rows spaced 15-20cm apart though 15×15cm spacing doubles the root production (Root Crops, 2023).

Although not essential but some 2.5m high trellises (Root Crops, 2023) can be used to support jicama vines and to control them from spreading into other garden areas (digitalcommons.usu.edu).

Irrigation

Jicama grows well if this is watered deeply, *i.e.* 2.5cm deep per week. If possible, drip irrigation is the best option. Jicama is quite drought-tolerant so over-watering should be avoided. Irrigation should be stopped 2 weeks before harvest since late watering causes root cracking.

Weeding and Mulching

The crop should be kept free from weeds throughout, however at the stage of initial growth in no case there should be any weed in its field, however, at later stage when plants are well established they are able to out-compete these. Black plastic mulch warms the soil, conserves the moisture and helps controlling the weeds that apart from robbing the plants of their nutrients and water also harbour the insect-pests and diseases. Moreover, plastic mulches also allow earlier planting. After laying out the mulch, the edges should be secured with soil and cut holes should be made for transplants. To avoid heat injury to the transplant, the stem should not touch the plastic mulch. Organic mulch such as grass clippings, straw, newspapers, *etc.* should be avoided until soils become warmer than 24°C (digitalcommons.usu.edu).

Row Covers

Row covers negate nature's adversities to some extent and enhance early growth. Hot caps, plastic tunnels, fabric covers and other devices help protect transplants from cool air temperatures. Plants grown under row covers require ventilation when air temperatures exceed 29°C. A thermometer can be used to help determine the temperature under the row covers. Covers can be removed when weather has stabilized (digitalcommons.usu.edu).

Pruning

In order to encourage vegetative growth and to increase tuber yield *cum* flavour, both the vegetative part of the plant and the reproductive shoots are pruned once or twice, with the first pruning taking place after about 1.5-2 months, and if again required second pruning can be done about one month after first pruning leaving only a few flowers for seed production for next crop. If grown for starch production the crop should be left unpruned (Sorensen *et al.*, 1996).

Intercropping and Crop Rotation

Almost all the traditional cropping systems involving *P. erosus* are of the multiple or intercropping type. The crop is traditionally intercropped with maize (*Zea mays*) and common bean (*Phaseolus vulgaris*) or monocropped when grown on a larger commercial scale (Sorensen *et al.*, 1996).

Harvest

In order to obtain optimal root quality the crop is harvested before ripening of seeds (Sorensen *et al.*, 1996). Jicama requires 3-4 months to grow an edible storage root but usually for root digging the crop matures in 5-8 months, however, in the warmer parts of Mexico a commercial crop is obtained only in about 3 months. If a seed crop is required, the growing period is approximately 10 months. For the best yield, harvesting of roots should be done when storage roots are 10-15cm in diameter and 2-2.5kg in weight. Root harvesting should certainly be done when foliage starts to turn yellow or after the first frost when leaves have been damaged. At this stage they are crisp and succulent with a pleasant sweet flavour (Root Crops, 2023; digitalcommons.usu.edu).

The swollen roots are usually dug manually using a spading fork or shovel though with large-scale production they are sometimes ploughed out. Growers should be careful not to bruise, cut or damage the roots. The tops are trimmed or removed entirely and the roots washed, surface-dried and packed in baskets for market.

Postharvest

Within 24 hours the creamy colour of the skin changes to a purplish-brown but this can be arrested if the roots are stored in the dark at 9-10°C. Roots store best when cured for 1-2 weeks at 27°C and then stored in a dark, cool and dry location at 9-10°C. When properly cured, jicama can be stored for 3-4 months. Jicama roots should not be stored at temperatures less than 9°C to avoid chilling injury (Root Crops, 2023; digitalcommons.usu.edu).

Jicama roots can also be 'field-stored'. In Mexico, the normal practice is to withhold irrigation water thus stopping growth and the roots remain in good condition in the soil for 2-3 months. Just prior to lifting they are irrigated to absorb water and then marketed in the usual way (Root Crops, 2023).

Pests and Diseases

There are very few pest problems with jicama as mentioned below:

Insect	*Identification*	*Control*
Aphids	Green or black soft-bodied insects that feed on underside of leaves. Leaves become crinkled and curled. May transmit 'bean common mosaic virus'. Secreted honeydew makes plants appear shiny, wet and sticky	Use insecticidal soaps or strong water stream to dislodge insects.
Bean weevil	It is a brown-flecked beetle with a short, broad snout. Adult females lay eggs on young bean pods and the larvae burrow into the pod and feed on the seed.	Apply appropriate chemicals if populations are high. Destroy infested plants. Dispose of legume plants after harvest. Rotate legumes.

Disease	Symptoms	Control
'Bean common mosaic virus' (BCMV) or 'Sincama mosaic virus' (SMV)	SMV is probably identical to the BCMV. Irregular yellowing on leaves, brittle young shoots, downward cupping, and bluish-green mosaic patterns on leaves.	Control aphids. Destroy infected plants.
Root Rot	Fungal diseases that cause decay and root rot. Causes sunken lesions on young and mature roots. Plants often stunt and wilt.	Rotate planting areas and do not over-water.

Source: digitalcommons.usu.edu.

Two bruchids ~ *Acanthoscelides sanfordi* and *A. taboga* were isolated from two seed samples collected in Yucatan in 1985. Another disease spreading by insect vectors is the 'witch's broom', first diagnosed by Thung and Hadiwidjaja (1957) which is probably caused by mycoplasma-like organisms. The most likely vectors of these diseases are a number of species of leaf-piercing insect-pests. These are aphids (*Aphis rumicis, Brevicoryne brassicae, Oregma lanigera*), mealy bug (*Ferrisia virgata*) and white fly (*Orusius argentatus*). The BCMV or SMV may become a serious problem locally, particularly in fields bordering on wild vegetation with frequent *P. erosus* escapes. A bacterial leaf spot caused by *Pseudomonas syringae* has also been reported. Several fungi have been reported to cause severe damage in *P. erosus*. A high mortality rate recorded in young plants as a result of 'root attacks' by *Pythium* spp., *Corticium* spp. and *Macrophomina* spp. in multilocational field trials in Senegal. Mohanty and Behera (1961) reported a severe leaf spot disease in Bhubaneswar and succeeded in identifying the fungus as *Cercospora canescens* (Root Crops, 2023; Sorensen *et al.*, 1996).

The larvae of *Thecla jebus* and *Ferrisia virgata* have been reported to attack jicama crops in Central America. The seeds are often attacked by weevils in Mexico. In the Philippines, a mosaic disease is caused by a systemic virus which is transmitted through seed or root. A slight mottling or chlorosis and blistering of the leaves, together with the production of very small roots are characteristic symptoms (Root Crops, 2023).

Potato bean or Amazonian yam bean [*P. tuberosus* (Lam.) Spreng.]

Vernacular names: Ashipa, Chuin, Jiquima (Spanish - different cultivar groups found in Ecuador, Peru and Bolivia); Amazonische Jamsbohne or Amazonische knollenbohne (German); Pois cochon (French); Jacatupe (Brazil); Mbacucu (Guarani, Paraguay) [Sorensen *et al.* (1996)].

Introduction and Origin

Pachyrhizus tuberosus is native to Colombia, Ecuador and Venezuela and introduced to Bolivia, N,SE&WC Brazil, Haiti, India, Jamaica, New Caledonia, Puerto Rico, Sri Lanka and Trinidad-Tobago. It is a climbing tuberous geophyte which grows primarily in the seasonally dry tropical biome. It is used as a poison, a medicine and for food (POWO4, 2023). The species is distributed from the southern lowlands of Colombia and Venezuela (the Andean mountain ranges of Sierra de Perija and Cordillera de Merida) where it is known as 'nupe' (Munos, 1945) through the Amazonian parts of Ecuador as far west as the province of Manabi and in much of Brazil.

P. tuberosus is presently known to be cultivated by numerous migrating indigenous peoples in the Amazonian region. It is also found occasionally in the vicinity of remote villages situated within the premontane rainforest or 'selva alta' in Peru (Noda *et al.,* 1991; Sorensen, 1990; Sorensen *et al.,* 1996).

Botany

Amazonian yam bean is a climbing herbaceous plant with autogamous sexual reproduction (2–4% cross-pollination) (Sorensen, 1990; Leon, 1987). It is the largest species in its genus having stems that are more than 7m long along with terminal leaflets of 28×26cm. While the trailing or climbing genotypes can attain up to 10m length with moderately branched, trailing or climbing stems, the bushy erect genotypes reach 1–1.5m. The trifoliate leaves are abundant, large, dark and ternate with entire to slightly dentate margins and the terminal leaflet vary from being subentire to deeply trilobate. The petioles are 4–20.8cm long with pubescent pulvini and linear-lanceolate stipules (Sorensen *et al.,* 2007).

Only one inflorescence develops per leaf axil bearing thick, cylindrical and 10-25cm long peduncles having lateral branches with 7–33 flowers. The petals are usually pubescent but glabrous specimens occur. The colour of the flower is white or light to dark bluish-violet. The pods being largest in the genus have reddish irritant hairs and are 13–19cm long & 1.4–2.3cm wide. The seeds are either rounded kidney-shaped (1-1.5cm) or flat and almost square (1.5cm). Both kinds of seeds may have orange, reddish or black coats though kidney-shaped ones, in addition, these have white to cream *cum* black dots while square-seeded ones are claret- or cream-coloured with brown dots (Sorensen *et al.,* 2007).

The precise time and length of the flowering season of Amazonian yam bean cannot be determined due to the highly heterogeneous origin of both the wild and the cultivated plants, and the uncertainty over its exact status. Specimens in full bloom have been recorded throughout the year, except in February and July though most of the materials flowered from October to June and mature fruits from March to December. Locally, the flowering season can be well defined according to the cultivation period. This is the case with crops whose cultivation is linked to seasonal flooding, *e.g* the Chuins and the Jiquima cultivar groups (Sorensen *et al.,* 1996).

The tuberous roots are either mono-tuberous extending vertically into the soil or multi-tubered, spreading laterally from the main stem. They are carrot- or sausage-shaped, long, round or irregular and up to 30cm long and 25 cm wide. Their flesh is crunchy, succulent, white or yellow and sometimes with a touch of purple (Sorensen *et al.,* 2007).

Nutritional Value

Root dry matter content of *P. tuberosus* ranges from 21-34% compared to 16-20% in *P. ahipa*. Through interspecific hybridization (*P. ahipa* × *P. tuberosus* 'Chuin cultivar group'), the progenies can produce starch content up to 25%. Its starch is of high quality due to low amylose content (less than 16%) when compared to traditional starch from monocots. In commercial pilot production of *P. tuberosus* sundried meal, the estimated composition can increase up to 61% starch, 24% protein and 15% fibre (Sorensen *et al.,* 2023).

Phytochemicals

Investigations into the chemical constituents of the seeds of *Pachyrhizus tuberosus* resulted in isolating of seven components: five rotenoids (12a-hydroxyerosone, 12a-hydroxydolineone, erosone, 12a-hydroxyrotenone and rotenone), a phenylfuranocoumarin (pachyrrhizine) and an isoflavanone (neotenone). The compounds were isolated using several chromatography techniques and characterized and verified by NMR and HPLC/MS (Leunera *et al.,* 2013). It possesses the same toxicity as to other cultivated species.

Uses

Its uses are almost similar to *P. erosus* or *P. ahipa*. Storage roots can be eaten raw or cooked. The storage roots of the Jicuima cultivar group having white-cream flesh and also the Ashipa cultivar group are eaten raw or cooked as a vegetable. Extremely succulent, crispy and tasty storage roots can also be used to make a refreshing 'fruit' juice. In case of jicuima storage roots, the fruit juice may be mixed with milk. Chuin storage roots are cooked and eaten often in the same way as the common staple food manioc.

Cultivar groups

P. tuberosus forms a complex of four cultivar groups ~ the Jiquima (coastal Ecuador), the Ashipa (in the western Amazon basin), the Chuin (along Peruvian rivers), and the Yushpe (in Peru). The three main cultivar groups Ashipa, Chuin and Jiquima are found in Ecuador, Peru and Bolivia while Yushpe is very local in Peru. In Peru, the cultivar groups Ashipa, Yushpe, and Chuin occur, and are represented by three, one and five individual cultivars or landraces, respectively (Sorensen *et al.,* 2007).

Ashipa Cultivar Groups

Ashipa cultivars have much larger leaves and pods than Chuin and Yushpe cultivars, and also differ by having entire leaflets, straight pods, kidney-shaped seeds (also found in Yushpe) where most plants have multi-tuberous sausage-shaped roots extending sideways and outwards. However, an Ashipa cultivar—with mono-tuberous, turnip-shaped and vertical roots—is known from Ecuador (Sorensen *et al.,* 1997). Ashipa develops mature tubers in 10–12 months having low dry matter content therefore the tubers can be eaten fresh without preparation. Peruvian Ashipas can be divided into three subgroups (Sorensen *et al.,* 2007).

- **White Ashipa cultivar group** (*Ashipa blanca*): It is the most common subgroup, distributed along the Huallaga, Maranon and Ucayali rivers and their tributaries. The individual plant produces relatively few tuberous roots with sweet white flesh when fresh. Its seeds are black and kidney-shaped, and flowers white or violet.
- **Yellow Ashipa cultivar group** (*Ashipa amarilla*): It is found in many communities in the Tigre River basin, both Mestizo and native, and occasionally in Mestizo villages elsewhere along the Nanay, Maranon and Ucayali rivers. The fresh root has yellow and very sweet, fruit- or melon-flavoured flesh due to the high sugar content (Nielsen *et al.,* 1998). Kidney-shaped seeds are reddish with an orange to brown taint and the flowers are white or violet.
- **Ashipa cultivar group** (*San Martin*): It is known from the provinces of El Dorado, Lamas and Mariscal Caceres in the San Martin department. The fresh root has white and sweet flesh, seeds are white *cum* black and kidney-shaped and the flowers are white.

Yushpe Cultivar Group

Yushpe still occurs elsewhere along the upper reaches of the Ucayali River. It differs from Chuin by having much deeper lobed terminal leaflets and kidney-shaped rather than flat seeds. It differs from Ashipa by having smaller leaves, smaller and straight pods, black-dotted seeds and vertical mono-tuberous but irregular-shaped roots. Yushpe develops mature tubers in 6–7 months which are eaten fresh and uncooked like a fruit.

Chuin Cultivar Group

It is distributed in the Peruvian Amazon along the Ucayali and Maranon rivers with much variation in root shape and other qualities. Chuin is a less robust climber than Ashipa bearing much smaller leaves, more narrow leaflets, smaller pods, flat and almost square seeds and mono-tuberous vertical roots penetrating deep into the ground. Chuin develops mature tubers in 4–6 months and these can, depending on their dry matter content, be eaten fresh, cooked or processed into products like flour and soft drinks.

Two principal cultivar subgroups with the morphological and agronomic characteristics of Chuin exist, differing particularly in root dry matter content where plants with high dry matter content are eaten cooked and the ones with low content are consumed fresh. The former subgroup (re)discovered in 1993 (Sorensen *et al.*, 1997) is referred to as Pitichuin and the latter (re)discovered in 2000 as Cocotichuin. The first subgroup is more common and widespread than second and for this Shipibos therefore mostly refer to Pitichuin cultivars only as Chuin but apply Cocotichuin for the latter (Sorensen *et al.*, 2007).

- **Pitichuin**: Chuin cultivars with high root dry matter content:
- **White Pitichuin** (*Chuin blanco*) is recorded from Mestizo and native villages along the higher and the lower Ucayali river. It has irregular roots with white peel and flesh, and burgundy-coloured seeds.
- **Yellow Pitichuin** (*Chuin amarillo*) is recorded from the lower and middle Ucayali river. The irregular root has yellow flesh and peel, and the seed colour is maroon.
- **Purple Pitichuin** (*Chuin morado*) is the most common Chuin found along the entire Ucayali and the lower Maranon rivers. It has fairly regular oblong roots with a purple-dark violet peel and white flesh with a purple touch. The seeds are black.
- **Cocotichuin**: Chuin cultivars with low root dry matter content:
    - **Yellow Cocotichuin** (*Cocotichuin amarillo*) is known from three Shipibo communities along the upper Ucayali river. It has roots with almost globular to round shape having a very thin yellow peel and yellow flesh. Seeds are cream with brown dots.
    - **Purple Cocotichuin** (*Cocotichuin morado*) is recorded from a single Shipibo community along the upper Ucayali river. The tubers have very thin yellow peel with creamy-white flesh overlaid with purple touch. Seeds are cream to brown.

Jiquima Cultivar Group

The jiquima cultivar group is grown in the seasonally dry coastal province of Manabi (Ecuador). In contrast to the Mexican jicama (*P. erosus*) which is a popular food in Central America, Mexico, SE Asia and even in the United States, the jiquima is considered a very minor crop in Ecuador (Sorensen *et al.*, 1996).

The fleshy storage root of jicuima can weigh up to 3-4kg (Munos, 1945; Menezes and Oliveira Nunes, 1955, c.f. Sorensen *et al.*, 1996). The pruning practices give significantly varying yield of storage roots.

Cultivation

Plants are easily propagated by seed and, except for the bud and flower prunings to increase tuber growth, they require little agronomic attention. In case of the jiquima cultivars from Manabi, the seed rate used is 45-50kg/ha (Sorensen *et al.*, 1996). The ashipa and chuin cultivar groups are planted at significantly lower plant densities. Some of the tuberous roots produced by the multi-tuberous cultivars may occasionally be harvested individually leaving the plant with one remaining tuberous root in place to produce additional tubers and as a propagation material (Peckolt, 1880).

This species gives optimal performance in somewhat well-drained fertle sandy-loam soils having pH range of 4.3-6.8 (Munos, 1945; Sorensen, 1990; Sorensen *et al.*, 1996). It is cultivated in forest clearings or along rivers in the tropical regions of South America at altitudes from 550-2000m a.s.l. (Munos, 1945), precipitation range of 640-5000mm m.a.p.r. and temperatures varying between 21.3 and 27.4°C.

Fresh storage root weight of *P. tuberosus* is lower than the other two cultivated species. Yield of roots is 7-10t/ha but in fertile soils it may go up to 30t/ha, however, Munos (1945) reported 15kg/ha when planting at 60×30cm plant spacing or 35-50kg/ha when sown through broadcasting the seeds.

No special irrigation is applied to the crop. In general the farmers use the water remaining in the soil from the previous rainy season. During the rainy season the crops usually grown are maize or rice as they constitute better agricultural alternatives for the market. Harvesting can take place after approximately 5 months, owing to the short growth cycle of the jiquima. Fertilization with N, P, K fertilizer (10-30-10) may be used in some communities. Furthermore, a pesticide application of Ambush following recommendations for other crops is often used (Sorensen *et al.*, 1996).

The cultivation practices used in the three main cultivar groups are based on an agronomic study conducted in Venezuela (Munos, 1945; Sorensen *et al.*,

1996). The plants in the ashipa and chuin cultivar groups are generally grown in shifting cultivation by the Indians of the Amazon region (Sorensen *et al.*, 1996).

Principal Practices of Traditional Cultivation of *Pachyrhizus tuberosus*

Characteristics	*Ashipa*	*Chuin*	*Yushpe*
Planting site	Tierra firma, high-level levees	Sandy and clay beaches, low-level levees	Low-level levee
Cultivation period	Approximately one year	Between floods	Between floods
Cultivation system	With associated species	Monoculture, but in case of mosaic with other crops	With associate species
Planting period	November–February	May, August	May
Propagating material	Seed	Seed and root	Seed and root
Sowing depth	3 cm	3 cm	3 cm
Distance between plants	1 to 3m	0.6 to 1.5m	Not recorded
Field labour required	Weeding, removal of soil for sowing, earthing up	Weeding	Weeding
Staking/trellising	Always	For seed production	For seed production
Period to develop harvestableroots	8 months	4 months	6 months
Period to develop mature roots	11 months	6 months	6 months
Period to develop mature seeds	12 months	6 months	7 months
Pests	*Orthopthera* sp. on stems, *Caryedes icamae* in seeds, *Diabrotica* sp. on leaves	*Orthopthera* sp. on stems, *Caryedes icamae* in seeds, *Diabrotica* sp. on leaves	*Orthopthera* sp. on stems
Root storing period	23 days	34 days	(not recorded)

Source: Sorensen et al. (2007).

Ashipa Cultivar Group

Ashipa is found growing at 300-2000m a.s.1. where precipitation is 1500-5700mm a.p.r., average temperature 20.7-25.5°C (minimum and maximum tolerable temperatures 11.0-13.2°C and 29.7-35.4 °C, respectively) and relative humidity of 84-92% (Sorensen *et al.*, 1996).

The ashipa cultivar group plants are grown in areas with a permanently humid climate and mostly on land that is never inundated, *i.e.* largely infertile, acid and aluminium-loaded uplands (Salick, 1989; Sorensen *et al.*, 1996). On such soils, it takes 8-12 months for swollen roots to develop. In quite fertile land, *e.g.* on uneven flooded high levees along rivers, crops grow faster and produce larger fleshy roots. Reproductive pruning is not practiced in this cultivar group.

Alvarenga and Valio (1989) evaluated the effect of different temperature and photoperiodic regimes on the initiation of flowering and tuberous root formation in genotypes belonging to the ashipa cultivar group. They observed that flowering was only induced in this species at intermediate daylengths, *i.e.* 9-16 hours. The crop may be considered a short-day plant with respect to tuberization as this process only occurs at photoperiods of less than 16 hours. The day/night temperatures of 30/25°C delays and reduces flowering and inhibits completely the formation of the tubers ~ the processes where both are increased at day/night temperature regimes of 25/20°C and 20/15°C (Alvarenga and Valio, 1989).

Ashipas appear to be always intercropped in slash-and-burn fields and are typically mixed with crops such as plantain (*Musa* × *paradisiaca*), maniac and pineapple (*Ananas comosus*), *etc.* Fields are rotated and left for fallowing after a few years of cultivation. It is common practice to sow few ashipa seeds together at about 5cm depth on hills 3m apart. The plants are often trellised on 2m poles or alternatively allowed to climb the intercropped manioc plants (Sorensen *et al.*, 1996).

After attaining the harvestable fleshy root size, some producers harvest 1-2 of the laterally produced roots and leave the one closest to the vegetative top. This method allows subsequent production of new tuberous roots after about 8 months, and the production period may thus be changed from a strictly annual to an almost perennial one. Naturally, ashipas are seed-propagated (Sorensen *et al.*, 1996).

Chuin Cultivar Group

Chuins are found at 100-300m a.s.1. The limited altitudinal range is due to the restricted area in which landraces belonging to this cultivar group have so far been identified. The annual precipitation along the Ucayali is ±3000 mm (Sorensen *et al.*, 1996). Chuin cultivar group is also cultivated in a permanent humid climate, on floodplains as well as predominantly at the higher levees which are irregularly or only briefly inundated. The chuins are sown in May-June and harvested in February-March. At the time of the tuber harvest, chuins also produce seeds which are gathered and sown after the inundation. The top of the tuber may also be replanted which after 4 months will produce many seeds but no fleshy roots. These are known as high 'restingas' in Peru where ashipa is sown in May-June and harvested in February-March. The crop is also grown on lower levees which are flooded for up to 6 months each year, and on river banks (locally known as 'playas') that have not yet been stabilized by vegetation. These are flooded for up to 8 months each year. Production of the crop is limited by the short growth period available here (Sorensen *et al.*, 1996).

Chuins are intercropped with other floodplain crops. At the higher levees the intercropping may be done typically with plantain and maniac and at the lower levees the maize, beans and vegetables. As with ashipas, a number of seeds are sown in holes located about 3m apart and the plants are often supported by poles. They need a fertile but not too humid soil where these can be harvested just after 4 months. However, a longer growth period improves the size and quality of fleshy roots. When fleshy roots are ready for harvesting the chuins also produce seeds which are gathered and kept for sowing in the season following inundation.

Cultivation of Jiquima Cultivar Group

Jiquima is recorded at 30-350m a.s.1. with annual precipitation of 450-500mm m.a.p.r., maximum temperature of 31.l-31.6°C and relative humidity of 90% during rainy season (Sorensen *et al.,* 1996).

Jiquima cultivar group is grown in the seasonally dry coastal province of Manabi (Ecuador). In contrast to the Mexican jicama (*P. erosus*), which has become a very popular food in Central America, Mexico, SE Asia and even in the United States but even it is considered a very minor crop in Ecuador. The common planting months for the jiquima cultivar groups in Manabi are June to July, *i.e.* at the end of the rainy season. The short growth cycle of the jiquima allows harvest of both tuberous roots and seed to take place after approximately 5 months. Seed availability is, however, limited, as seeds are only produced on-farm (Sorensen *et al.,* 1996).

According to local sources, monocropping was and still is a common practice but intercropping with chilli, sesame, groundnut and tomato is also practiced. The jiquima is cultivated in small backyards, each containing some 60 to 100 plants in plots of 250-400m^2, *i.e.* a plant density of 1500-2400 plants/ha when intercropped. Approximately 70-90% of the plants from these plots are harvested when the tubers have reached a marketable size. The tubers are then sold locally, especially in the villages of Portoviejo, Montecristi and Jipijapa. Remaining 10-30% of the crop is left for seed production. A small number of the plants are used for home consumption and occasionally as presents to neighbours. The price per unit weight of jiquima is approximately three times that of manioc (Sorensen *et al.,* 1996). Insect-pests and diseases are the same as described under Jicama.

Ahipa or Andean yam bean [*P. ahipa* (Wedd.) Farodi]

Other common name: Ajipa.

Synonyms: *Dolichos ahipa* Wedd., *Pachyrhizus ahipa* var. *albifora* Parodi and *P.a.* var. *violacea* Parodi (POWO2, 2023).

Introduction and Origin

Ahipa, native to W&S Bolivia to NW Argentina (POWO2, 2023), is grown primarily by small communities located in the east of subtropical Andean valleys of Bolivia and N. Argentina (Sorensen *et al.*, 1996, 1997). Ahipa is one of the few leguminous plants cultivated for its edible fleshy roots. It is closely related to jicama (*Pachyrhizus erosus*) and goitenyo (*P. tuberosus*). While jicama is a plant of warm climate, ahipa is a plant of slightly less warm climates, preferring temperatures that are neither too warm nor too cold (Cultivariable, 2023). It grows up to 2000m above sea level in the highest Bolivian mountains and is little known outside of the Andes where it is mostly grown for personal or local consumption. The root is smaller and more elongated as compared to that of jicama (en.wikipedia.org1).

Ahipa roots

Yet its wild form is not traceable and the species exists only in cultivation (Cultivariable, 2023). The origin of the ahipa is most likely in the *Ceja de montanas* Andean region (Orting *et al.*, 1996). According to Sorensen *et al.* (1996), the greatest degree of diversity is found in Bolivia which probably is its region of origin. A few varieties are also found in Argentina, although they may be critically endangered and perhaps even absent from cultivation by now (Sorensen *et al.*, 1996; Cultivariable, 2023). Today, it is still in use in small native communities in Bolivia and N. Argentina. This crop at any time did not have wide distribution anywhere probably due to the very specific climatic adaptions it exhibits or may be due to the acquisition of Latin America by the Spanish and Portuguese conquest which had the general policy to destroy the traditional Andean agricultural systems. Today, the production is restricted to a few local villages and farmers (Sorensen *et al.*, 1997).

Botany

Ahipa is a short day perennial plant with no lateral axis. Its plants may be herbaceous and lignified depending on the genotype. The leaves are trifoliate with stipules or pinnately arranged leaflets with caduceus stipels (Sorensen *et al.*, 1997). Ahipa can grow in erect bush forms, sprawling bush (semi-erect) forms or in twining forms but only bush forms are widely available outside the Andes (Cultivariable, 2023). The erect bush type species can grow to about 15–40cm in height, the semi-erect forms up to about 30–60cm in length and the twining forms up to about 60–200cm (Orting *et al.*, 1996).

Ahipa is a hermaphrodite and self-pollinating plant which is pollinated by insects (PFAF, 2023). The flowers are white or pale-lavender which appear on short stalks and show a tubular calyx and a papilionaceous corolla. Generally, the flowers exhibit an internally curved stigma in close contact with the anthers. This habit is very unfavourable for insects to effectively pollinate the flowers. Moreover, the pollen fertility is often not very high ~ varying between 45 and 100%.The flowering pattern is not consistent and each season 100–800 flowers per plant can be produced. *P. ahipa* being a short-day plant so flowering takes place under decreasing day lengths. The pods are 13–17cm long and up to 1.6cm wide, each containing multiple seeds that are lilac, maroon, black or mottled with black and white. These are round *cum* kidney-shaped and about 0.8–1cm long. Seed production differs from plant to plant and lies between 20 and 100 per plant. The thousand grain weight is around 300g (Sorensen *et al.*, 1997).

Most varieties produce a single swollen storage root which is tapered to both the ends, though there are a few that produce multiple roots but multi-root varieties are not available outside the Andes. The roots are about 15cm long with a weight of 500-800g. Yellow skin of the root encloses a white flesh which is interwoven with a soft fibre (Orting *et al.*, 1996; Cultivariable, 2023).

Nutritional Value

Little is published about the nutritional values of ahipa, however, these are likely to be similar to its close relative jicama. The roots are rich in carbohydrate and provide calories, vitamins K, C and potassium (Leidi *et al.*, 2003; Doporto, 2014). Its dry matter ranges from 15-30%. Further, the tuber contains 48-54% starch which consists of 96-99.9% amylopectin. This is a very high value and especially interesting for food processing where low solubilisation and retrogradation are important. Ahipa contains 19-28% sugars in its roots. The protein content lies between 8 and 18%. About 80% of the protein is water-soluble and not extractable within a pH range of 2-10. The lipid content is below 1%. Altogether, the tuber shows a very good nutritional composition. However, because the tuber contains a lot of water, the protein and energy concentration may be rated as being medium (Sorensen *et al.,* 1997).

Toxicity

Most plant parts are toxic except the enlarged edible fleshy root. Foliage and especially the seedpods and their seeds contain the pesticide rotenone so are inedible (Cultivariable, 2023). As an adaption to insect predators, ahipa plants accumulate canavanine in its seeds. This is highly toxic to some insects and can pose a big problem in livestock foraging, especially if the ahipa plants escaped into the wild vegetation (Leidi *et al.,* 2003).

Uses

Ahipa root is usually eaten raw or can be prepared as a juice but it can also be baked, boiled or fried. The roots are an attractive addition to green salads, taste sweet and are crisp like an apple. Even after boiling the roots retain their crunchy texture. Ahipa roots are even considered to have a cleansing effect on the body. It is supposed to cure infections of the throat and the air passage (Sorensen *et al.,* 1997).

Because of the high starch and amylopectin content, ahipa is a good material for the starch industry (Sorensen *et al.,* 1997). After nine months, the starch granules are very homogenously distributed in the root. There is a high percentage of granules having 12-18μm diameter (en.wikipedia.org2).

The oil of the seeds has special properties such as high palmitic acid and γ-tocopherol concentrations as well as little of few linolenic acids. These aspects are of interest in the food industry (Leidi *et al.,* 2003).

Seed Production and Storage

Like most legumes, ahipa is primarily self pollinating and the pods are formed without human or insect intervention. The seed pods require either a warm summer or a long growing season in order to mature. Though reproductive pruning is usually practiced in order to produce good sized roots but few plants are left as such to produce seeds for the next sowing. Alternatively, little seed from all or most of the healthy and most vigorous plants are left to mature for seed production. Moreover, at harvest also the roots are visually evaluated and the best ones are retained for seed production (Cultivariable, 2023; Orting *et al.,* 1996).

Since ahipa is a perennial plant where grower can carry over roots from year to year for having a big head-start in seed production. In cool summer climates, some growers follow biennial seed production by employing total reproductive pruning in the first year and then allowing the plants to fully flower in the second year. Pollen fertility is relatively low in some varieties, but almost always greater than 50%, which does not create a significant impediment to breeding work. Pollinators are probably not required for ahipa since the flowers self pollinate but plenty of insects may get attracted by the flowers particularly bumble bees (Cultivariable, 2023).

As with most beans, the seeds are ready for storage when they cannot be dented with a fingernail. Growers should also store them with a desiccant to ensure that they are sufficiently dry. Seeds of ahipa appear to have good storage life, retaining high germination even after five years (Cultivariable, 2023). In tropical areas, seed is reported to retain good germination for 3-4 years (NRC, 1989) so a cool and dry temperature (10°C) is important for long term storage.

Cultivation

Propagation

Unlike most of the Andean root crops, ahipa is grown almost always from seeds similar to most other legumes. For better germination, seeds are pre-soaked in warm water for 12 hours and then sown in early spring in a warm greenhouse. Germination takes place within 2 weeks. As soon as the seedlings become large enough to handle, these are pricked out into individual pots of rich soil to be planted out after the last expected frosts. The plants should be given some cloche protection until these start growing well (PFAF, 2023).

Though vegetative propagation is not a traditional method of its cultivation but in need ahipa can be propagated from stem cuttings or side fleshy roots for specialized needs such as for breeding work or to multiply varieties for which little seed is available. When propagating through stem cuttings, growers should take about 15cm portions from the end of a stem and put the bottom 5cm in wet soil. It should then be covered with a humidity tent. Rooting takes in 6-8 weeks. Best practice is to take cuttings in late summer and then rooted for overwintering in a greenhouse (Cultivariable, 2023). Varieties that form multiple fleshy roots can be propagated from the smaller side roots and then potted for winter storage. Division of fleshy roots can be done during autumn. These roots should be stored in a cool but frost-free place over the winter, planting them into pots in the greenhouse in early spring and planting them out after the last expected frosts with cloche protection when in active growth (PFAF, 2023).

Factors Influencing the Growth

Climate

Ahipa grows to an elevation of 1800 – 3000m above sea level in the cool tropical and subtropical valleys on sun-facing slopes or on the borders between the warm and cold tropics. The precipitation in its native region is between 400 and 700mm, occurring within 4–6 months, with the rest being the dry season. Therefore, the climate is semiarid. The average temperature in this region is 16-18°C (Sorensen *et al.*, 1997) or 15 to 21°C (Cultivariable, 2023) though this depends entirely on time of the day (Sorensen *et al.*, 1997). Temperatures may range from near freezing to 32-35°C during the growing season. It does not like too hot or too cold temperatures as above 32°C causes wilting while below 10°C results in extremely slow growth (Cultivariable, 2023). Ahipa plant is not frost-hardy but could possibly be grown as a summer crop in cool temperate zones (PFAF, 2023). Ahipa has neither photoperiod dependency for root formation nor probably for flowering.

Soil

Ahipa cultivation is mainly carried out along loamy riverbanks, *vis-a-vis* on loamy hillsides. It prefers well-drained soils with pH levels of 6-8 (Sorensen *et al.*, 1997) though it tolerates quite acidic to neutral soil. Like other legumes, ahipa too fixes nitrogen in the soil with the help of certain nodulating bacteria but in its absence ahipa benefits from a balanced organic fertilizer (Cultivariable, 2023).

Season

The crop is usually sown from August to October. Reproductive pruning starts from November to March, *i.e.* 4-7 months after sowing and this matures from April to June (Sorensen *et al.*, 1997). However, Ahipa can be started at any time of the year when there is enough frost-free growing season remaining, thus giving the growers an option of planting at any time of year. In cool summer climates as in the Pacific Northwest, the best strategy seems to be starting ahipa indoors in '2 litre-mesh growbags' and then planting those growbags out once the soil temperature reaches about 15°C. This allows the plants to develop more quickly with minimum of transplant injury to which ahipa seems very susceptible (Cultivariable, 2023).

Field Preparation and Planting

Before sowing, the soil must be loosened to a depth of 15–25cm and thoroughly cleaned from weeds and stones. It is planted on ridges when it is flood irrigated, which is mostly the case in the Andean region. In Bolivia, *P. ahipa* is normally sown between August and October, depending upon the rainy season. The seeding rate is between 40 and 65 kg/ha. In the Andes, planting distances are usually kept 20–60cm between rows and 6–25cm between the plants in the same row (Sorensen *et al.*, 1997). Closer planting results in smaller roots but with greater yield.

Growers can directly sow ahipa when the soil temperature rises to at least 12°C though 18°C is better if it doesn't require too long a wait. These transplant well and can be started quickly indoors at a temperature of 21°C (Cultivariable, 2023).

Fertilizers

P. ahipa plants are able to form an efficient symbiosis with nitrogen-fixing bacteria such as *Rhizobium* and *Bradyrhizobium*, and is able to fix 58–80kg nitrogen per hectare so no further nitrogen is required (Sorensen *et al.*, 1997) for the crop. If the plants as such are left in the field, a substantial amount of the fixed nitrogen, *i.e.* 12-80kg/ha is returned to the soil. The enrichment is very important if a sustainable land-use system is desired. During nodulation, an increased phosphorus supply is advantageous as it promotes the root and seed yield in ahipa plants. The final recipients of phosphorus are the pod shells as well as the seeds. In total, about 7-9.6 kg P/ha were extracted from the soil. At harvest, the highest potassium contents can be found in pod shells and the roots. The removal of K at harvest was reported to be about 20.2 kg/ha (Leidi *et al.*, 2004a).

Irrigation

Pachyrhizus ahipa plants can tolerate long dry spells. However, an additional water supply may increase tuber yield (Orting *et al.*, 1996). Watering should be stopped in the last month before harvest as the root has a tendency to crack if watered heavily. In the regions, such as Pacific Northwest, where late summer rains tend to arrive at just the wrong time, plants can be grown in containers since they can be moved under cover when weather is likely to become wet (Cultivariable, 2023).

Weeding and Mulching

The crop is kept free from weeds and is often mulched to help conserve soil moisture and prevent weed growth.

Pruning

Cultivation period is 5-10 months and first flowers appearing roughly in 3-4$^1/_2$ months. Reproductive pruning is usually done by removing flowers and any young pods after which plant utilizes its full energy in root development which results into drastic improvement of the yield (Sorensen *et al.*, 1997). Reproductive pruning is though quite expensive but must be done once or twice a season (Leidi *et al.*, 2004b).

Intercropping and Crop Rotation

Ahipa doesn't seem particularly promising for companion planting and is normally sown in a pure culture though can be intercropped with maize. It really needs full sun and it is a fairly small plant that can be easily overpowered by stronger or larger companions. Vining types might be more suitable for mixed planting. In a crop rotation, it is sown prior to maize/potato, maize/tomato, maize/oca, groundnut or manioc (Sorensen *et al.,* 1997; Cultivariable, 2023).

Harvest

Ahipa will grow to full size in about six months in warmer climate whereas under cool summer climates it produces worthwhile roots in nine month's duration. In the Andes, the fleshy root may reach about one kg though in the Pacific Northwest 227g even though it is considered to be good (Cultivariable, 2023).

Mature pods normally appear from April through June. At about nine months the roots are completely tuberized and aerial parts become completely dry. Fleshy roots are harvested as soon as they show a marketable weight depending on consumer's preferences. Usually the preferred weight is about 0.4kg. The vegetative top is often left in the field for decaying with soil. The roots can be left in the soil until needed after cutting off their tops (Sorensen *et al.,* 1997).

Postharvest

Ahipa roots can be stored well up to two to three months at a temperature of about 10°C and 90% humidity. Storage life gets dramatically reduced when temperatures are much warmer or much cooler than the optimum. After 2-3 months of storage, the texture begins to change, quality declines and sprouting may begin. At this stage, any root that is meant for seed production should be potted. Ahipa can also be stored in the ground (NRC, 1989) and this will probably prove to be the best practice in climates where soil freezing is not a problem.

Storage alters the starch sugar balance in ahipa roots (Sorenson *et al.,* 1996, 1997). This is usually desirable because it produces a sweeter root. Ahipa also can be sweetened significantly by exposure to sunlight (Cultivariable, 2023).

It is important to avoid damage to roots while harvesting. If damaged, these may become susceptible to attack of certain pathogens such as *Rhizopus stolonifer*, *Cladosporium* sp., and *Penicillium* sp. An attack of one of those fungi leads to textural changes, decay and internal browning when stored at low temperatures and high relative humidity (>80%) (Sorenson *et al.,* 1996, 1997).

Yield

Though yield variance is expected from place to place and climate to climate but usually per hectare fleshy root production on an average comes to 28-53 tonnes though of dry seeds 1.3-2.7t/ha (Leidi *et al.,* 2003).

Pests and Diseases

Pachyrhizus ahipa plants are not very susceptible to pests. Nematodes such as *Meloidogyne* sp. can pose a serious problem which can be kept under control by applying 1kg/ha of Carbofuran which should be mixed in the soil at the time of field preparation. Further, some of the bean weevil species can be harmful which can be controlled by spraying the crop with 0.2% permethrin or bifenthrin. The most serious viral disease is the bean common mosaic virus (BCMV). Yields can be reduced up to 30% if infected by this virus. If a general irrigation management is lacking, rot is also likely to appear (Sorenson *et al.*, 1997). Such plants should immediately be removed and destroyed, aphids should be controlled by applying nicotinic acid and field sanitation should be done after the harvest.

References

Alvarenga, A.A. and I.F.M. Valio, 1989. Influence of temperature and photoperiod on flowering and tuberous root formation of *Pachyrrhizus tuberosus*. *Ann. Bot.*, **63**(44):411-4.

Chung, I.M., N. Paudel, S-H Kim, C.Y. Yu and B.K. Ghimire, 2019. The Influence of Light Wavelength on Growth and Antioxidant Capacity in *Pachyrhizus erosus* (L.) Urban. *J. Plant Growth Regul.*, **39**:296–312. doi: 10.1007/s00344-019-09982-1.

Cultivariable: Retrieved on 21 May, 2023. www.cultivariable.com/instructions/andean-roots-tubers/how-to-grow-ahipa/

Digitalcommons.usu.edu

Doporto, Maria Cecilia, 2014. Aprovechamiento integral de raices de ahipa (*Pachyrhizus ahipa*) y sus productos derivados con fines alimentarios (Tesis), 369p. (in Spanish). Universidad Nacional de La Plata. doi:10.35537/10915/34924. Retrieved 25 April 2014.

Duke, James A. 1992. Handbook of phytochemical constituents of GRAS herbs and other economic plants. Dr. Duke's Phytochemical and Ethnobotanical Databases. CRC Press. Archived from the original on September 23, 2015. Retrieved June 25, 2010.

en.wikipedia.org1: en.wikipedia.org/wiki/Pachyrhizus.

en.wikipedia.org2: en.wikipedia.org/wiki/Pachyrhizus_ahipa.

en.wikipedia.org3: en.wikipedia.org/wiki/Pachyrhizus_erosus.

Estrella-Parra, E.A., J.C. Gomez-Verjan, I. Gonzalez-Sanchez, E.R. Vazquez-Martinez, E. Vergara-Castaneda, M.A. Cerbon, D. Alavez-Solano and R.

Reyes-Chilpa, 2014. Rotenone isolated from *Pachyrhizus erosus* displays cytotoxicity and genotoxicity in K562 cells. *Nat. Prod. Res.,* **28**:1780–85. doi: 10.1080/14786419.2014.939081.

fdc.nal.usda.gov/fdc-app.html.

Global Bhasin, 2012. http://globalbhasin.blogspot.com/2012/07/globalization-of-foods-jicama.html.

Heredia, Z.A. and E.H. Garcia, 1994. San Miguelito, San Juan y Vega de San Juan: nuevas variedades de jicama (*Pachyrhizus erosus* (L) Urban) para el Bajio, Guanajuato, Mexico, pp. 257-72. In: Proceedings of the First International Symposium on Tuberous Legumes (ed. Sorensen, M.), Guadeloupe, FWI, April 21-24, 1992, Jordbrugsforlaget, Copenhagen, Denmark.

Jaiswal, V., S. Chauhan and H.J. Lee, 2021. The Bioactivity and Phytochemicals of *Pachyrhizus erosus* (L.) Urb.: A Multifunctional Underutilized Crop Plant. *Antioxidants* (*Basel, Switzerland*), **11**(1): 58. https://doi.org/10.3390/antiox11010058.

Kardono, L.B., S. Tsauri, J.M. Pezzuto, A.D. Kinghorn and K. Padmawinata, 1994. Cytotoxic isoflavonoids of *Pachyrrhius erosus seeds. J. Kim. Terap. Indones. (Indones. J. Appl.Chem.)* , **4**:2–5.

Leidi, E.O., R. Sarmiento and D.N. Rodriguez-Navarro, 2003. Ahipa [*Pachyrhizus ahipa* (Wedd.) Parodi]: an alternative legume crop for sustainable production of starch, oil and protein. *Industrial crops and products, an international journal,* **17**: 27-37.

Leidi, E.O., J.M. Vidueira, J. Cobo, de Castro and D.N. Rodriguez-Navarro, 2004a. Nutrient equirements of ahipa, tuberous-root crop. *J. Pl. Nutr.*, **27**: 931-45.

Leidi, E.O., D.N. Rodriguez-Navarro, M. Fernandez, R. Sarmiento, J. Semedo, N. Marques, A. Matos, A.P. Machado, B. Orting, M. Sorensen and M.C. Matos, 2004b. Factors affecting root and seed yield in ahipa (*Pachyrhizus ahipa* (Wedd.) Parodi), a multipurpose legume crop. *Eur. J. Agron.*, **20**:395-403.

Leon, J. 1987. Botanica de los cultivos tropicales. 1 edicion. IICA. San Jose, Costa Rica, 445 p.

Leunera, Olga; Jaroslav Havlikb; Milos Budesinskyc; Vladimir Vrkoslavc; Jessica Chud; Tracey Bradshawd; Jana Hummelovaa; Petra Miksatkovae; Oldrich Lapcike; Irena Valterovac; and Ladislav Kokoska, 2013. Cytotoxic constituents of *Pachyrhizus tuberosus* from Peruvian Amazon. *Nat. Prod. Commun.*, **8**(10): 1423-6.

Lukitaningsih E, 2014. Bioactive compounds in bengkoang (*Pachyrhizus erosus*) as antioxidant and tyrosinase inhibiting agents. *Indones. J. Pharm.*, **25**:68. doi: 10.14499/indonesianjpharm25iss2pp68.

Lukitaningsih, E., M. Bahi and U. Holzgrabe, 2013. Tyrosinase inhibition type of isolated compounds obtained from *Pachyrhizus erosus. Aceh Int. J. Sci. Technol.*, **2**:98–102. doi: 10.13170/AIJST.0203.05..

Menezes de, O.B. and de W. Oliveira Nunes, 1955. Esterilidade em Jacatupé (*Pachyrrhizus bulbosus* L.). *Revista Ceres* (Brazil), **10**:52-7.

Minh, T.T., H.K. Toan, H.T.L. Anh, T.T. Huong, T.T. Do and V.D. Hoang, 2021. Chemical constituents from the leaves of *Pachyrhizus erosus* collected in Vietnam. *Chem. Pharm. Bull.*, **69**:1136–9. doi: 10.1248/cpb.c21-00614.

Mohanty, N.N. and B.C. Behera, 1961. Leaf spot of *Pachyrhizus angulatus* Rich. *Sci. Nat.*, **27**(5): 54.

Morera, J.A. 1994. Reporte Anual 1993, investigación de la Jicama (*Pachyrhizus* spp.) en CATIE, Costa Rica, pp. 9-17 in First Bi-Annual Progress Report, STD3 Contract no. ERBTS3*CT920115 (ed. Sorensen, M.).

Munos, O.S. 1945. Nupe, a promising crop for Venezuela. *3rd Conf. Interamer. Agric., Caracas*, **38**:5-34 [in Spanish, no English summary].

Necha, L.L.B., S.B. Banos, L.B. Luna, F.J.L.G. Suarez, D.A. Solano and R.R. Chilpa, 2004. Antifungal activity of seed powders, extracts and secondary metabolites of *Pachyrhizus erosus* (L.) Urban (Fabaceae) against three postharvest fungi. *Rev. Mex. Fitopatol.*, **22**:356–61.

news.abs-cbn.com/lifestyle/03/09/10/heaviest-singkamas-found-ilocos.

Nielsen, P.E. 1995. Summary of progress report from Tonga (Oct 1993 – Sept. 1994; ed. Sorensen, O.), pp. 153-84 in Second Annual Progress Report, STD3 Contract no. ERBTS3*CT920115.

Nielsen, P.E., M. Halafihi and M. Sorensen, 1998. Germplasm evaluation of *Pachyrhizus tuberosus* (Lam.) Spreng. Proc. of 2nd Intern. Symp. on Tuberous Legumes (eds M Sorensen, M., J.E. Estrella-E, O.J. Hamann and S.A. Rios Ruiz), held at Celaya, Gto., Mexico, on 5–8 August 1996, pp. 253–60.

Noda, H., C.R. Bueno and D.F. Silva, 1991. Genetic erosion threatens native Amazonian vegetable crops. *Diversity*, 7(1/2):62-3.

NRC: National Research Council Staff, 1989. Lost crops of the Incas: Little-known plants of the Andes with promise for worldwide cultivation. National Academic Press, USA. https://www.cultivariable.com/instructions/andean-roots-tubers/how-to-grow-ahipa/.

Orting, B., J. Gruneberg and M. Sorensen, 1996. Ahipa [Pachyrhizus ahipa (Wedd.) Parodi] in Bolivia. *Genet. Res. Crop Evol.,* **43**: 435-46.

Peckolt, T. 1880. Jacutupe. Z. allg. ost. Apoth.-Ver., 13:193-7 and 14:209-213.

PFAF, 2023 – Plants for a future. Retrieved 21 May, 2023. pfaf.org/user/Plant.aspx?LatinName=Pachyrhizus+ahipa.

Phrutivorapongkul, A., V. Lipipun, N. Ruangrungsi, T. Watanabe and T. Ishikawa, 2002. Studies on the constituents of seeds of *Pachyrrhizus erosus* and their anti-herpes simplex virus (HSV) activities. *Chem. Pharm. Bull.*, **50**:534–7. doi: 10.1248/cpb.50.534.

POWO1: Plants of the World Online, 2023. *Pachyrhizus* Rich. ex DC. Royal Botanical Gardens, Kew. Retrieved on 20 May, 2023. powo.science.kew.org/taxon/urn:lsid:ipni.org:names:23121-1.

POWO2: Plants of the World Online, 2023. *Pachyrhizus ahipa* (Wedd.) Parodi. Royal Botanical Gardens, Kew. Retrieved on 20 May, 2023. powo.science.kew.org/taxon/urn:lsid:ipni.org:names:179929-2.

POWO3: Plants of the World Online, 2023. *Pachyrhizus erosus* (L.) Urb. Royal Botanical Gardens, Kew. Retrieved on 20 May, 2023. powo.science.kew.org/taxon/urn:lsid:ipni.org:names:179936-2.

POWO4: Plants of the World Online, 2023. *Pachyrhizus tuberosus* (Lam.) Spreng. Royal Botanical Gardens, Kew. Retrieved on 20 May, 2023. powo.science.kew.org/taxon/urn:lsid:ipni.org:names:512027-1.

Primiani, C.N., U. Lestari, M. Amin and S.B. Sumitro, 2013. The dynamics of daidzein chemical compound from yam tuber (*Pachyrhizus erosus*) and its potential in Myometrium. *Berk. Penelit. Hayati.*, **18**:122–8. doi: 10.23869/bphjbr.18.2.20139.

Ramaswamy, N., C.R. Muthukrishnan and K.G. Shanmugavelu, 1980. Varietal performance of Mishrikand [*Pachyrhizus erosus* (L.) Urban], pp. 199-200. In: National Seminar on Tuber Crops Production Technology, 21-22 November 1980, Faculty of Horticulture, Tamil Nadu Agricultural University, Coimbatore.

Root Crops, 2023. Humanity Development Library 2.0. Retrieved on 16 May, 2023 via https://www.nzdl.org/cgi-bin/library.cgi.

Salick, J. 1989. Ecological basis of Amuesha agriculture, Peruvian Upper Amazon. in Resource management in Amazonia: Indigenous and folk strategies (eds Posey, D.A. and W. Balee). *Adv. in Econ. Bot.*, **7**:189-212.

Sanderson, Helen; 2005. The Cultural History of Plants (eds Ghillean Prance and Mark Nesbitt). Routledge, 67p. ISBN 0415927463.

Sorensen, M. 1990. Observations on distribution, ecology and cultivation of the tuber-bearing legume genus *Pachyrhizus Rich.* ex DC. (Fabaceae: Phaseoleae). *Wageningen Papers*, **90**(3):1-38.

Sorensen, M., J. Heller and J. Engels, 1996. Yam bean (*Pachyrhizus* DC.). Promoting the conservation and use of underutilized and neglected crops. Vol. 2. Institute of Plant Genetics and Crop Plant Research, Gatersleben/ International Plant Genetic Resources Institute, Rome, p. 7-70. Publisher IPK and IPGRI, Rome, Italy. ISBN: 92-9043-282-9..

Sorensen, M., K.A. Vecht, E.O. Leidi Montes, S. Simonsen, 2023. Yam beans (*Pachyrhizus tuberosus* (Lam.) Spreng. and *Pachyrhizus erosus* (L.) Urb.—Fabaceae)—lowland South American and Meso-American cultivars and landraces with starch and protein market potential—their botany, agronomy, ethnobotany, and present uses (eds Marney Pascoli Cereda and Olivier François Vilpoux). In: Underground Starchy Crops of South American Origin, Varieties and Landraces. Academic Press, Vol. 2 (chapter 8), pp. 135-78. ISBN 9780323900577, https://doi.org/10.1016/B978-0-323-90057-7.00003-6. (https://www.sciencedirect.com/science/article/pii/B9780323900577000036).

Sorensen, M; Ore Balbin; K. Sorensen; Kuist Lars and Octavio Vasquez, 1979. Review of the *Pachyrhizus tuberosus* (Lam.) Spreng. Cultivar groups in Peru. *Pl. Gen. Res. NL,* (Rome, Italy:), **151**: 2-13.

Sorensen, M., S. Doygaard, J. Estrella, L.P. Kvist and P.E. Nielsen, 1997. Status of the South American tuberous legume *Pachyrhizus tuberosus* (Lam.) Spreng. *Biodiv. & Conserv.*, **6**:1581–1625.

Sorensen, M, W.J. Gruneberg and B. Orting, 1997. Ahipa [*Pachyrhizus ahipa* (Wedd.) Parodi]. In: Andean roots and tubers: ahipa, arracacha, maca and yacon. Promoting the conservation and use of underutilized and neglected crops (eds Herman, M. and J. Heller). Intern. Pl. Gen. Res. Inst., Rome, pp. 13-73.

Thung, T.H. and T. Hadiwidjaja, 1957. De Heksenbezemziekte bij Leguminosen. *T. Pl.Ziekten,* **63**:58-63.

United States 2009. The Complete Guide to Edible Wild Plants. USD Army, New York. Skyhorse Publishing. New York, 125p.. ISBN 978-1-60239-692-0. OCLC 277203364.

www.britannica.com/plant/jicama.

Yahara, S., N. Irino, T. Takaoka and T. Nohara, 1994. Studies on the constituents of *Pachyrhizus erosus. Nat. Med.*, **48**:312–3.

Zanklan, A.S., H.C. Becker, M. Sorensen, E. Pawelzik and W.J. Gruneberg, 2018. Genetic diversity in cultivated yam bean (*Pachyrhizus* spp.) evaluated through multivariate analysis of morphological and agronomic traits. *Genet. Res. Crop Evol.,* **65**: 811–43. https://doi.org/10.1007/s10722-017-0582-5.

Winged bean [*Psophocarpus tetragonolobus* (L.) DC.]

Not to be confused with Winged pea (*Lotus purpureus* Webb).

Family: Fabaceae.

Subfamily: Faboideae.

Other common names: Asparagus bean, Cigarillas, Dragon bean, Four-angled bean, Four-cornered bean, Goa bean, Manila bean, Mauntius bean and Princess bean (en.wikipedia.org/wiki/Winged_bean).

Introduction and Origin

The genus name of *Psophocarpus* means 'noisy fruit' as its pods make the popping noise when dehiscing and releasing the seeds. *Tetragonolobus* refers to the four sides of the fruit (Rajan, 2018). The Winged Bean's historical origin is uncertain. Many authors place the origin in the Papua New Guinea and Indonesian island regions where many genetic strains of Winged Bean exist (Echo community, 2023). The species is not found growing wild although it has been noted growing as an escape in Myanmar and the Philippines. The native range of this crop, as found in different literatures over internet, is Tanzania (POWO,

2023), New Guinea (Hymowitz and Boyd, 1977) and/or Africa (Madagascar or Mauritius), and may have spread to Asia. It is now cultivated usually as a market garden crop in southern India, Myanmar, Malaysia, New Guinea, Indonesia, the Philippines, China and Thailand, and to a lesser extent in Africa, mainly in Ghana and Nigeria, as well as in the West Indies (Root Crops, 2023).

Winged bean is widely recognized by farmers and consumers in South Asia for its variety of uses and disease resistance. Though it is cultivated mainly for root tubers (Root Crops, 2023), whole plant of winged bean is edible and has the potential to become a major multi-use food crop in the tropics of Asia, Africa and Latin America (Khan, 1982).

Botany

Winged bean possesses 2n=2x=18 chromosome numbers. It is an herbaceous perennial vine that climbs by twining to the left (Martin, 1978) and grows to a height of 3–4m, usually larger than the common bean (Etats-Unis, 2009). It has moderately thick stem that is slightly ridged and grooved, bearing trifoliate leaves on long and stiff petioles. The leaflets are ovate, 7.5-15cm long with terminal one usually longer than the laterals and attached to the petiole by a marked pulvinus (Root Crops, 2023). Leaf shape ranges from ovate to deltoid, ovate-lanceolate, lanceolate and long lanceolate (Khan, 1982). The green tone of the leaves also varies. The stem is most commonly green though sometimes boasts purple (en.wikipedia.org/wiki/Winged_bean)

The inflorescence is borne on about 15cm long axillary raceme with 2-10 blue, white or lilac flowers. It is usually bee-pollinated. The rectangular (sometimes appearing flat) bean pods with characteristic serrated wings running down the four corners contain 5-20 seeds which can vary in colour from white through varying shades of yellow and brown to black and may also be mottled (Root Crops, 2023). Depending on the genotype, the bean pod may be smooth or rough and cream, green, pink or purple. It is typically 15–22cm long having waxy skin and partially translucent flesh (in the young pods). A fully ripe pod turns ash-brown and splits open to release bean seeds that are often round though oval and rectangular seeds also occur (Khan, 1982). The beans are similar to soybeans in both use and nutritional content (29.8-39% protein) (en.wikipedia.org/wiki/Winged_bean).

Although the plant is perennial and produces new growth annually from shallow and persistent roots, it is usually grown as an annual for optimum results. Winged bean has numerous fibrous roots with the main lateral ones that run horizontally near the soil surface and usually become thickened and tuberous near the base of the plant after a few months though this does not occur in all the strains (Root Crops, 2023). Roughly cylindrical, starchy, fibrous and generally small tuberous roots are 8-12cm long and 2-4cm in diameter with weight of single root being seldom more than 50g (Martin, 1978; Root Crops, 2023; Bassal *et al.,* 2020). Tubers have brown and easy to peal skin with white or cream-coloured flesh that is firm and moist having distinctive nutty and earthy flavour (Growables, 2023).

Psophocarpus tetragonolobus	**Winged bean flower**	**Winged bean pods**

KENPEI. "*Psophocarpus tetragonolobus*, Osaka-fu Japan". Wikimedia Commons, 11 Aug. 2007, (CC BY-SA 2.5), commons.wikimedia.org/wiki/File:Psophocarpus_tetragonolobus4.jpg. Accessed 19 May 2023.	**Vinayaraj. "*Psophocarpus tetragonolobus*". Wikimedia Commons, 13 Oct. 2013, (CC BY-SA 4.0), commons.wikimedia.org/wiki/Category:Psophocarpus_tetragonolobus#/media/File:Psophocarpus_tetragonolobus_10.JPG. Accessed 19 May 2023.**	**By STRONGlk7 - Own work, CC BY-SA 3.0, https://commons.wikimedia.org/w/index.php?curid=21474925**

Varieties

Winged bean is a self-pollinating plant but mutations and occasional outcrossing may produce variations in the species (Khan, 1982). There are many varieties of winged bean varying in length, width and pod colour. Motis (2019) mentions these varieties 'Bogor', 'Chimbu', 'Day Neutral', 'Flat', 'Ribbon', 'Siempre' and 'Square' in their 2006 research in finding a niche to determine which part(s) and attributes of these plant parts will be of greatest benefit to farmers. Varieties differ greatly in production of various parts (Growables, 2023; Motis, 2019).

Nutritional Value

All parts of the plant including leaves, flowers, roots, young bean pods and seeds have high nutrient composition, especially proteins, vitamins and minerals. The starchy tubers have high percentage of proteins ~ by weight up to 20% compared to other vegetables, leaves and flowers from 5-15% and the seeds 32-37% which is similar to the amount of proteins found in soybeans though higher than that found in other beans (Sri Kantha and Erdman, 1984).

Tuberous Roots

The water content of the fresh tubers is 52-68%. Dry matter composition has revealed that fleshy roots contain protein 13-20%, fat 0.6-1.4%, carbohydrate 63-77%, fibre 1.5-21%, ash 1.7-3.9%, calcium 40mg/100g, iron 3mg/100g and phosphorus 64mg/100g. The wide range of values appears to be associated with variety and possibly with maturity of the root tubers. The carbohydrate contains about 80% starch and 20% sugars. The non-protein nitrogen was found to be 7.4-15% of the crude protein but in some other reports the values are as high as 50%. Sulphur-containing amino acids are low though aspartic acid content is exceptionally high (Root Crops, 2023).

Seeds

Nutritional value per 100g of winged bean seeds	
Energy	**1,711 kJ (409 kcal)**
Carbohydrates	41.7 g
Dietary fibre	25.9 g
Fat	16.3 g
Saturated	2.3 g
Monounsaturated	6 g
Polyunsaturated	4.3 g
Protein	29.65 g
Vitamins	**%DV† (Quantity)**
Vitamin A	0 IU
Thiamine (B1)	90% (1.03 mg)
Riboflavin (B2)	38% (0.45 mg)
Niacin (B3)	21% (3.09 mg)
Pantothenic acid (B5)	16% (0.795 mg)
Vitamin B6	13% (0.175 mg)
Folate (B9)	11% (45 μg)
Vitamin C	0% (0 mg)
Minerals	**%DV† (quantity)**

Nutritional value per 100g of winged bean seeds	
Calcium	44% (440 mg)
Iron	103% (13.44 mg)
Magnesium	50% (179 mg)
Manganese	177% (3.721 mg)
Phosphorus	64% (451 mg)
Potassium	21% (977 mg)
Sodium	3% (38 mg)
Zinc	47% (4.48 mg)
Units: µg = micrograms; mg = milligrams; IU = International units †Percentages are roughly approximated using US recommendations for adults. ***Source***: USDA FoodData Central	

Source: en.wikipedia.org/wiki/Winged_bean.

The seeds also contain 23-40% carbohydrates (National Research Council, 1981; Misra *et al.,* 1987; Mohanty *et al.,* 2015) and vitamins such as B1, B2, B3, B6, B9, C, A and E. Mature seeds contain 14-25% fats by weight out of which 94% is in free form whereas the rest in complex form, and with carbohydrates and proteins (Mohanty *et al.,* 2015).

The seed of winged bean is rich in minerals including calcium, iron, phosphorous, potassium, sulphur, sodium, magnesium, zinc, manganese, boron, barium, copper, chromium and strontium (Misra *et et al.,* 1987; King and Puwastien, 1987; Jaffe and Korte, 1976). Seeds contain 54-75% of unsaturated fatty acids of which 38.6% is mono-unsaturated and 36.9% polyunsaturated (Mohanty *et al.,* 2015; Varriano-Marston *et al.,* 1983) but no trans fatty acids. Winged bean oil contains 30-40% saturated fatty acids which represents more than the content observed in soybean (Khor *et al.,* 1980). Oleic and linoleic acids represent nearly 50% of unsaturated fatty acids. Unsaponifiable lipids are mainly represented by β-sitosterol (66.4%) and stigmasterol (25.1%) (Mohanty *et al.,* 2015). Moreover, winged bean oil is better than soybean oil due to its high oxidative and high thermal stabilities (Makeri *et al.,* 2016).

Phytochemicals

Phytochemicals present in different parts of the winged bean are given below in the table.

Plant part	Phytochemical name	Reference
Fruit	Phaseollidin	ISBN:9788185042084
Fruit	Isomedicarpin	ISBN:9788185042084
Leaf	beta-Carotene	ISBN:9788172362461
Seed	Myristic acid	ISBN:9788172362461
Seed	Stearic acid	ISBN:9788172362461
Seed	Maritimetin	ISBN:9788185042114
Seed	N6-Methylagmatine	ISBN:9788172362461, ISBN:9788185042145
Seed	Docosanoic acid	ISBN:9788172362461
Seed	Palmitic acid	ISBN:9788172362461
Seed	Isovitexin	ISBN:9788185042114
Seed	Arachidic acid	ISBN:9788172362461
Seed	Oleic acid	ISBN:9788172362461
Seed	Vitexin	ISBN:9788185042114
Seed	Linolenic acid	ISBN:9788172362461
Seed	Linoleic acid	ISBN:9788172362461
Seedling	N6-Methylagmatine	ISBN:9788185042145
	Ascorbic acid	DOI:10.15482/usda.adc/1239279
	4-Carvomenthenol	ISBN:9788185042114
	beta-Carotene	DOI:10.15482/usda.adc/1239279
	Betulinic acid	DOI:10.15482/usda.adc/1239279

Source: IMPPAT (2023).

Analysis of the mutual genotypic relationships among twenty four genotypes of *P. tetragonolobus* through Mantel test found a significant correlation (r = 0.839) between similarity matrices of the results obtained from the use of the RAPD and ISSR molecular markers. Quantification of total polyphenols, flavonoids and tannin revealed the highest percentage of occurrence of kaempferol (1.07-790.5 μg/g) and the lowest percentage of gallic acid (0.09-3.49 μg/g) in the seeds. (Rajan, 2018) .

A study was done to evaluate the physical properties and proximate composition of the seeds and the physico-chemical properties of fatty oil from fully mature seeds. The fatty oil compositions of immature, mature and fully mature seeds were evaluated by GC-FID, GC/MS and 1 H-NMR. The result revealed that the fatty oil from fully mature seeds contained high proportion of unsaturated fatty acids (75.5 %), compared to immature seeds which showed higher percentage of saturated fatty acid (61.3 %). Unsaponification matter (0.25 %) of fatty oil was identified as stigmasterol (66.4 %) and β- sitosterol (25.1 %). Total lipids of fully mature seeds were extracted and isolated as neutral, glyco- and phospholipids. Thus the fatty oil of fully mature seeds is enriched with mono-unsaturated fatty acids (38.6 %) and poly-unsaturated fatty acids (36.9 %) without trans-fatty acids thus meeting the edible oil standard (Rajan, 2018).

These beans are also rich in tocopherols as well as antioxidants that facilitate vitamin-A utilisation in the body (en.wikipedia.org/wiki/Winged_bean).

Anti-nutritional Factors

Winged bean contains many anti-nutritional factors such as trypsin inhibitors, chymotrypsin inhibitors (WCI), hemagglutinins, amylase inhibitors, phytates, phytic acid, flatulence factors, hydrogen cyanide, saponins, tannins and other phenolic compounds (Kortt, 1984; Sri Kantha *et al.,* 1986). The seeds and fleshy roots have high levels of trypsin inhibitory activity (13,500 – 30,100 IU/mg fresh weight) and this is inactivated through cooking (Root Crops, 2023).

The main phytochemicals of *Psophocarpus tetragonolobus* responsible of its anti-oxidant and anti-inflammatory activities are:

Metabolites	***Activity***	***Plant Part***	***Country***	***References***
Phytate	Anti-nutritional factors, affinity for specific blood cell antigens	Seed	Malaysia	Sri Kantha *et al.,* 1986; Tan *et al.,* 1983
Tannin	Nonspecific enzymes inhibitors, hemagglutinin activity	Seed	Malaysia	Tan *et al.,* 1983
Psophocarpin	Chemotrypsin inhibitory activity	Seed, pods	India	Mit and Singh, 1986, 1988
Lectin	Hemagglutinin activity	Mainly in seed, roots	India	Kortt, 1984, 1985; Manjunath and Madaiah, 1987
Albumin 1 (WBA-1)	Kunitz-type trypsin inhibitors	Seed	Malaysia	Kortt *et al.,* 1980
Phaseolin	α-amylase inhibitor	Seed	Malaysia	Yao *et al.,* 2016

Source: www.mdpi.com/2223-7747/9/12/1730.

Culinary Uses

All parts of the plant including leaves, flowers, roots, young unripe seed pods and seeds have high nutrient composition and each part except the ripe seeds can be eaten raw or cooked though mature seeds are consumed only after cooking and mature seed pods become too fibrous to eat. Smoked pods, dried seeds, fleshy roots (cooked and uncooked) and leaves have been sold in domestic markets in South East and E&S Asia (Khan, 1982).

The nutrient-rich fleshy roots have a distinctive nutty *cum* earthy flavour. Highland tribesmen in Papua New Guinea esteem it so highly that they hold winged bean sing-sings (feasts) at harvest time. Researchers in Thailand also had made tasty snacks from winged bean fleshy roots sliced thin, fried and salted or softened in sugar syrup (Growables, 2023).

Boiled winged bean roots served as a snack in Myanmar

By Wagaung, CC BY-SA 3.0, https://commons.wikimedia.org/w/index.php?curid=18506943

Tender pods which are most widely eaten part of the plant are best when eaten before they exceed 2.5cm in length. They are ready for harvest within three months of planting. The flowers are used to colour rice and pastry. The young leaves can be picked and prepared as a leaf vegetable similar to spinach (en.wikipedia.org/wiki/Winged_bean).

Bean seeds are rich not only in protein but also in tocopherols ~ antioxidants facilitating vitamin-A utilisation in the body (en.wikipedia.org/wiki/Winged_bean). They can be made into milk when blended with water and an emulsifier. Winged bean milk is similar to soy milk but without the bean-rich flavour (Yang and Tan, 2011).

Other Uses

Similar to soybean oil, its seeds can also be processed to produce oil which may be used for culinary purposes, illumination and for soap making (Growables, 2023). Expressed oil cakes are fed to livestock. The vegetative parts or whole plant is used as animal feed (Root Crops, 2023).

Growth

Winged bean grows quite fast. Plants start flowering in 40-140 days of sowing. Pods reach full-length about two weeks after pollination, after 3 weeks these become fibrous and after six weeks the mature seeds are ready to harvest (en.wikipedia.org/wiki/Winged_bean). Fleshy root development and flower production vary according to genotype and environmental factors. Some winged bean varieties do not produce tuberous roots. The winged bean is a tropical plant and flower only when day length is shorter than 12 hours though some varieties have been reported to be day-length neutral. Since winged bean is a vine so must grow over a support. Some examples of support systems include: growing against exterior walls of houses, huts, buildings, supporting against larger perennial trees, through stakes placed in the ground vertically and through structures made from posts and wires (Khan, 1982).

Cultivation

Propagation

Propagation of winged bean is usually done by seeds that remain viable for approximately one year (Root Crops, 2023) and can be transplanted or direct-seeded (Growables, 2023). Seeds do not require pre-treatment in tropical climate but scarification or soaking in water overnight before planting enhances the germination rate of seedlings. Seed soaking may be used in conjunction with scarification to increase the germination speed. Seedlings under natural field conditions have been reported to emerge between 5-7 days (Khan, 1982).

In certain areas of Myanmar and the Philippines the crop is treated as a perennial and the tubers are left in the ground to produce fresh plants. A great deal of work is currently underway breeding new cultivars to suit specific conditions and for tuber production (Root Crops, 2023).

Climate

Winged bean is a tropical crop resistant to high temperature and grown between 20°N and 15°S latitude, at elevations up to about 2,000m in the tropics (Root Crops, 2023). Although it can be grown at temperatures between 14 to 40°C, the optimum growing temperature is 18-30°C (Growables, 2023), most ideal being 25°C. Lower temperatures suppress germination and extremely high temperatures inhibit yield. Growing winged bean at lower temperatures can increase tuber production though leaf expansion rate is higher in a warmer climate (Khan, 1982).

For highest yield, the soil should remain moist throughout the plant's life, however, it is highly susceptible to moistre stress and waterlogging. If the plant matures during the drier part of the growing season, yields are higher (Khan, 1982). A well-distributed rainfall in excess of 150cm per year is required and the plant thrives in areas with an annual rainfall of 250cm or more. With adquate irrigation facilities but no waterlogging, it can be grown even as a dry season crop provided there is adequate irrigation facilities. Despite its perennial nature and extensive root-system, it does not survive prolonged drought though drought-resistant cultivars are being developed (Root Crops, 2023).

The winged bean requires short days for normal flower induction, since when grown under a long photoperiod there is excessive vegetative growth at the expense of flowers. Recent experiments have shown that there is an interaction between day-length and day/night temperatures. Flowering and fleshy root growth are inhibited in 16 hour days but with 8 hour days fleshy roots develop regardless of temperature though flowering is dependent upon temperature regime used (Root Crops, 2023).

Soil, Its Preparation and Manuring

If the soil is well-drained, winged bean is not very demanding in its soil requirements. Neither it tolerates waterlogging nor saline soils. Well-cultivated fertile sandy-loam soils are best for optimum pod *cum* fleshy root yield, however, clay soils are not favourable as there fleshy root development is poor, *via-a-vis* lack in flavour. While preparing the soil for sowing, some 10q/ha well-decomposed dung manure should be added in the soil thoroughly at the time of third ploughing. Almost successfully it can be grown in nitrogen deficient soils due to its exceptional ability to rhizobium strain nodulation which fixes nitrogen in the soil provided effective rhizobium strains are available though in Nigeria recently the plants were observed with chlorotic light green foliage due to non-availability of most effective cowpea group rhizobial strains. The manurial requirements of the winged bean have not been studied in detail but the crop responds favourably to nitrogen fertilization. When grown as a vegetable the routine application of standard NPK, *i.e.* 8:8:8 NPK fertilizer at intervals of 14-21 days has been found quite useful (Root Crops, 2023).

Seed Sowing

Winged bean pre-soaked seeds are sown usually at the beginning of the rainy season. For pod and seed production, planting is usually done on the flat bed and the seeds are dibbed in holes about 2.5cm deep. When a little grown, the plants are generally provided with supports such as bamboo poles arranged singly or in tripods and trellis or wire fence. When poles are used the plants may grow so tall that picking is difficult but use of a trellis or wire fence 1-1.5m high is most convenient (Root Crops, 2023).

If the crop is being grown on a field scale for fleshy root production as in Myanmar, the seeds are normally planted 5-7.5cm deep on ridges and earthed up, stakes are frequently dispensed with and there the plants are left to ramble over the ground reaching a height of up to 30cm. Use of stakes may slightly increase the yield of roots but stake cost may well offset any increase in yield (Root Crops, 2023).

For root production in Myanmar, they sow the seeds on 20cm high ridges with plant density of 200000-1000000/ha, *i.e.* plant to plant distance being 7.5-15cm and line to line 60cm, however a plant density of 150000/ha has been found to be optimum and for sowing one hectare of land some 5q/ha seeds are required when 2-3 seeds are sown per hole (Root Crops, 2023).

Weeding, Interplanting and Plant Shearing

Seedlings make slow growth for the first 3-5 weeks, therefore at this time monthly weeding is important, however, this practice can be stopped when the plants are well established (Root Crops, 2023). The crop may be interplanted with sweet potato, taro, sugarcane, bananas and certain other vegetable crops. If flowers are not allowed to come up, the root yield increase is obtained roughly four times (Root Crops, 2023) but then there would not be pods or seeds.

Harvest and Postharvest

Depending upon season, variety used and cultivation practices adopted, harvesting of fleshy roots starts from 5 to 12 months of sowing (Growables, 2023). Fleshy roots are usually harvested when these attain 7.5-12cm length and 2.5-5cm diameter. Lifting is usually through shovel or fork but in no case the roots should be injured. Lifting is quite convenient when the crop has been raised on ridges. Digging in case of flat-grown crop is of little inconvenience and there chances of root injury also increase. However, crop raised on flat beds certainly becomes easier in lifting if two days before digging the field is flooded with water. There is little information on storage of the root tubers as they are normally eaten immediately after harvest, but it has been found that under normal tropical conditions deterioration is rapid (loss of moisture, loss of vitamin C and long cooking time) but if stored at lower temperatures with 90% humidity, these can remain fresh for 2-3 weeks, however, during this period there should be sufficient ventilation to avoid fungal infection (Root Crops, 2023).

Young leaves of large maturing plants can be harvested at any time but especially during the off-season when vegetative growth is abundant and the plants are not in bloom. Periodic harvest of up to 10% of the foliage is not likely to reduce flowering and seed yield. During the period of production of pods only light harvesting of leaves should be done (Growables, 2023). Pods are usually harvested 2-3 weeks after fruit set while they are still tender and not very fibrous. They grow rapidly after the flowers are produced and reach their mature size in about 21 days. As the pods begin to mature, these lose their shiny appearance and develop fibre. On Guam, the most popular local winged bean variety is cultivated seasonally or left in the field to grow year-round. Immediately after harvest, fresh immature winged bean pods should be cooled and stored at 10°C at 90 per cent relative humidity to slow down water loss and decay (Growables, 2023). Dried bean seed can be stored for long periods under cool and dry conditions.

Yield

Fleshy root yield is greatly influenced by growing conditions, altitude of growing, variety used and the growing conditions. Virtually all the reported yields are either from crops grown primarily for pod or seed. Malaysia has reported 2.5-6t/ha and the Philippines 2.3t/ha of roots, though root yield in Indonesia when grown at higher altitudes were rercorded 6-16t/ha, however, in lowland areas it was much less. In Papua New Guinea also, higher yields were reported from the highlands though recently in Malaysia it was 14t/ha from staked plants of selected strains. In none of these cases reproductive pruning was followed (Root Crops, 2023).

Pests and Diseases

If winged bean is grown in mixed-market garden culture or shifting agriculture it is generally free from major insect-pests and diseases, however, there may be casual visitors such as caterpillars, leaf miners, grasshoppers, spider mites and nematodes (Root Crops, 2023).

Probably the most serious disease which occurs in Papua New Guinea, Java, the Philippines and Malaysia is false rust caused by *Synchytrium psophocarpi* which may be controlled through use of copper fungicides or through 0.2% spraying of mancozeb, myclobutanil, penconazole or bupirimate with triforine. In plantations *Cercospora* spp. causes leaf spot though not to the extent of causing defoliation. Its infection produces light to dark brown and zonate spots on leaves and pods. It is controlled through 0.2% spraying of Kavach or 0.05% Topsin M. *Podalia* spp. caterpillars have been reported from Brazil as serious on winged bean feeding voraciously on every aerial part of the plant. These also bore the flowers and pods.Their infestation is brought down through spraying the crop with 0.1% Dichlorvos 76EC or Lambda cyhalothrin 5EC at 0.05%. The mites (*Polyphagotarsonemus laetus* and *Tetranychus virticae*) start their feeding from underside of leaves and cause webbing as may be obserbed on old leaves. These are controlled through spraying with 0.04% of Spiromesifin 240SC or 0.2% Dicofol 18.5EC. Viral symptoms have been observed in Nigeria and the Ivory Coast. This disease spreads through aphids ~ the virus vectors so infected plants should be gently uprooted and destroyed as well as the crop is sprayed with 0.2% Dimethoate 30EC to kill the aphids and other insects if any. It is probable that, as plantation-type plantings increases, disease problems will become more severe and many disease-causing organisms have been listed in recent literature. In particular, root-knot nematodes (*Meloidogyne incognita, M. javanica* and *M. hapla*) have been reported to cause up to 70 per cent losses in roots (Root Crops, 2023). In case of infested soil, pre-plant bed treatment with dazomet at 40g/m^2 followed by application of 1kg/m^2 of neem cake a fortnight later of sowing.

References

Bassal, H., O. Merah, A.M. Ali, A. Hijazi and F. El Omar, 2020. *Psophocarpus tetragonolobus*: an underused species with multiple potential uses. *Plants*, **9**(12):1730. https://doi.org/10.3390/plants9121730.

Echo community, 2023. Retrieved on 17 May, 2023. https://www.echocommunity.org/en/resources/3e995069-d3e7-4a2b-8188-a50f2d168193.

en.wikipedia.org/wiki/Winged_bean.

Etats-Unis, 2009. The Complete Guide to Edible Wild Plants, 57p. United States Department of the Army. New York: Skyhorse Publishing, New York. ISBN 978-1-60239-692-0. OCLC 277203364.

Growables: Grow Florida Edibles, 2023.. Retrieved on 19 May, 2023 from www.growables.org/informationVeg/BeanWinged.htm.

Hymowitz, T. and J. Boyd, 1977. Ethnobotany and Agriculture Potential of the Winged Bean. *Econ. Bot.,* **31** (2): 180–88. doi:10.1007/bf02866589. S2CID 5944034.

IMPPAT: Indian Medicinal Plants, Phytochemistry and Therapeutics. Retrieved on 18 May, 2023. cb.imsc.res.in/imppat/phytochemical/Psophocarpus%20 tetragonolobus.

Jaffe, W.G. and R. Korte, 1976. Nutritional characteristics of the winged bean in rats. *Nutr. Rep. Int.*, **4**: 449–55. www.mdpi.com/2223-7747/9/12/1730.

Khan, T. 1982. Winged Bean Production in the Tropics, 1p. FAO of the United Nations.

Khor, H.T., N.H. Tan and K.C. Wong, 1980. Winged bean seed: Potential food source of the tropics. In Proc. 6th Malaysian Biochemical Society Conference, Kuala Lumpur, Malaya, held on 22–23 August 1980, pp. 157–62. www.mdpi.com/2223-7747/9/12/1730.

King, R.D. and P. Puwastien, 1987. Effects of germination on the proximate composition and nutritional quality of winged bean (*Psophocarpus tetragonolobus*) seeds. *Food Sci.,* **52**:106–8. www.mdpi.com/2223-7747/9/12/1730.

Kortt, A.A. 1984. Purification and properties of the basic lectins from winged bean seed [*Psophocarpus tetragonolobus* (L.) DC]. *Eur. J. Biochem.*, **138**: 519–25.

Kortt, A.A. 1985. Characterization of the acidic lectins from winged bean seed (*Psophocarpus tetragonolobus* (L.) DC). *Arch. Biochem. Biophys.*, **236**: 544–54.

Kortt, A.A., P.M. Strike and J.D. Jersey, 1989. Amino acid sequence of a crystalline seed albumin (winged bean albumin-1) from *Psophocarpus tetragonolobus* (L.) DC. Sequence similarity with Kunitz-type seed inhibitors and 7s storage *globulins. Eur. J. Biochem.,* **181**: 403–8.

Makeri, M.U., R. Karim, M.S. Adbulkarim, H.M. Ghazali, M.S. Miskandar and K. Muhammad, 2016. Comparative analysis of the physiochemical, thermal and oxidative properties of winged bean and soybean oils. *Int. J. Food Prop.*, **19**: 2769–87. www.mdpi.com/2223-7747/9/12/1730.

Manjunath, S. and S. Madaiah, 1987. Distribution of lectin activity at different stages in the tissues of winged bean (*Psophocarpus tetragonolobus* (L.) DC). *Plant Sci.,* **53**: 161–5.

Martin, F.W. 1978. Vegetables for the Hot, Humid Tropics Part 1. The Winged Bean, *Psophocarpus tetragonolobus*. Agric. Res. Service, U.S. Dept. of Agriculture, Southern Region. ia801207.us.archive.org/9/items/CAT79711173/CAT79711173.pdf.

Misra, P.S., G. Misra, D. Prakash, R.D. Tripathi, A.R. Chaudhary and P.N. Mishra, 1987. Assay of some nutritional and anti-nutritional factors in different cultivars of winged bean *Psophocarpus tetragonolobus* (L.) DC seeds. *Plant Foods Hum. Nutr.*, **36**: 367–71. www.mdpi.com/2223-7747/9/12/1730.

Mit, R. and Singh, 1986. Purification of a storage protein of *Psophocarpus tetragonolobus. Phytochem.*, **25**: 595–600.

Mit, R. and M. Singh, 1988. Psophocarpin B1, a storage protein of Psophocarpus tetragonolobus, has chymotrypsin inhibitory activity. Phytochem., **27**: 31–4.

Mohanty, C.S., R.C. Pradhan, V. Singh, N. Singh, R. Pattanayak, O. Prakash, C.S. Chanotiya and P.K. Rout, 2015. Physicochemical analysis of *Psophocarpus tetragonolobus* (L.) DC seeds with fatty acids and total lipids compositions. *J. Food Sci. Technol.*, **52**: 3660–70. www.mdpi.com/2223-7747/9/12/1730.

Motis, T. 2019. Winged Bean Revisited. Echo Research, Echo Community, edn.link/hwtqkj. Accessed 25 May 2019.

National Research Council, 1981. Winged Bean: A High-Protein Crop for the Tropics. The National Academies Press: Washington DC, USA. www.mdpi.com/2223-7747/9/12/1730.

POWO, 2023. Plants of the World Online. Kew Royal Botanical Gardens. Retrieved: 2023-05-15. https://powo.science.kew.org/taxon/urn:lsid:ipni.org:names:515970-1.

Rajan, D. 2018. A Review on Pharmacognostical, Phytochemical and Pharmacological aspects of *Psophocarpus tetragonalobus. Res. J. Pharmacognosy and Phytochem.*, **10**(4): 331-5. doi: 10.5958/0975-4385.2018.00053.5.

Root Crops, 2023. Humanity Development Library 2.0. Retrieved on 16 May, 2023 via https://www.nzdl.org/cgi-bin/library.cgi.

Sri Kantha, S. and J.W. Erdman, 1984. The winged bean as an oil and protein source: A review. *J. Am. Oil Chem. Soc.*, **61**: 515–25. www.mdpi.com/2223-7747/9/12/1730.

Sri Kantha, S., N.S. Hettiarachchy and J.J.W. Erdman, 1986.. Nutrient, antinutrient contents and solubility profiles of nitrogen, phytic acid and selected minerals in winged bean flour. *Cereal Chem.*, **63**: 9–13.

Tan, N.H., Z.H.A. Rahim, H.T. Khor and K.C. Wong, 1983. Winged bean (*Psophocarpus tetragonolobus*) tannin level, phytate content and hemagglutinating activity. *J. Agric. Food Chem.*, **31**: 916–7.

Varriano-Marston, E., A. Beleia and C.C. Lai, 1983. Structural characteristics and fatty acid composition of *Psophocarpus tetragonolobus* seeds. *Ann. Bot.*, **51**: 631–40. www.mdpi.com/2223-7747/9/12/1730.

www.mdpi.com/2223-7747/9/12/1730.

Yang, J. and H. Tan, 2011. Winged Bean Milk. International Conference on New Technology of Agricultural Engineering, held at Zibo, pp. 814–7. doi:10.1109/ICAE.2011.5943916. ISBN 978-1-4244-9574-0. S2CID 22046459.

Yao, Y., Y. Hu, Y. Zhu, Y. Gao and G. Ren, 2016. Comparisons of phaseolin type and α-amylase inhibitor in common bean (*Phaseolus vulgaris* L.) in China. *Crop J.*, **4**: 68–72.

Zombi pea (*Vigna vexillata* (L.) A.Rich)

Family: Fabaceae.

Subfamily: Faboideae.

Other names: Wild cowpea and Narrow-leaved wild sweetpea (Eng.) (pza. sanbi.org)

Synonyms: *Dolichos cylindricus* Desv.; *D. vexillatus* (L.) Kunth; *Phaseolus capensis* sensu Thunb.; *P. glycinaeformus* Weinm.; *P. humifusus* Savi; *P. pulniensis* Wight; *P. quadriflorus* A. Rich.; *P. sepiarius* Dalzell; *P. vexillatus* L.; *Plectrotropis angustifolia* Schumach & Thonn.; *P. hirsuta* Schum. & Thonn.; *Strophostyles capensis* E. Mey.; *Vigna capensis* (Thunb.) Burtt Davy; *V. carinalis* Benth.; *V. crinita* A. Rich.; *V. davyi* Bolus; *V. dinteri* Harms; *V. dolichoneura* Harms; *V. golungensis* Baker; *V. hirta* Hook.; *V. lobatifolia* Baker; *V. phaseoloides* Baker; *V. scabra* Sond.; *V. senegalensis* A. Chev.; *V. thonningii* Hook. f; *V. tuberosa* A. Richard (Quattrocchi, 2016)

Introduction and Origin

Genus *Vigna* is named in honour of Italian botanist Domenico Vigna (*d.* 1647), who was a professor of botany and Director of the Botanical Garden of Pisa. The specific epithet *vexillata* means 'having a standard petal', referring to the widely spreading standard petals (pza.sanbi.org). *Vigna vexillata* is an underutilized legume crop that can be found in diverse areas of Africa, America, Australia and Asia (Tomooka *et al.*, 2011). It is cultivated for edible storage roots and seeds. Zombi pea is a highly heterogeneous legume species (Dachapak *et al.*, 2017). Former works revealed that some varieties of zombi pea adapt well to environmental stresses including infertile soil (Karuniawan *et al.*, 2006), alkaline soil (Lawn and Watkinson, 2002), drought (Roecklein and Leung, 1987) and waterlogging (MIller and Williams, 1981).

Flower of *Vigna vexillata*

By Jayesh Patil from Mumbai, India - Zombi Pea, CC BY 2.0, https://commons.wikimedia.org/w/index.php?curid=32507018

Vigna vexillata probably originated in the Old World tropics with centres of diversity in SE Africa and SE Asia (tropical.theferns.info). It is a pan-tropical herbaceous legume distributed widely throughout tropical and sub-tropical regions of Africa, Asia, Australia and America in its wild form (Duke, 1991). The species often develops storage roots that are edible and consumed by people in Africa, India, Australia and SE Asia (Lawn and Cottrell, 1988; Sasikumar and Sardana, 1988; Karuniawan *et al.*, 2006). It is an important food crop in several areas such as Namibia where it is commonly harvested from the wild for local use (tropical.theferns.info). Its use has spread beyond its native range and the plant is now sometimes cultivated for its edible fleshy roots. The cultivated form of *V. vexillata* is found in a limited area of Africa and Asia (Bali, Indonesia) where seeds and fleshy roots both are consumed (Amkul *et al.*, 2019).

Botany

Zombi pea possesses 2n=2x=22 chromosome numbers. It is a strong perennial twiner with fusiform fleshy roots producing stems that are usually covered with brownish silky hairs (trichomes). Its dark green, oval and tip-pointed leaflets have appressed trichomes on both the surfaces. Leaflets appear in threes with terminal leaflet being 7.5–15cm long. Sweetly scented 2.5cm long pink or purplish to yellow flowers are borne on 7.5–30cm long 2-4-flowered peduncles with the keel prolonged into an uncurved beak (Quattrocchi, 2016). Linear *cum* recurved silky pod is 7.5–9cm long (Palaniswami and Peter, 2008). Plant produces a number of fleshy roots along with thread-like feeding roots quite deep into the soil (tropical.theferns.info). *V. vexillata* seeds due to presence of a dipeptide para-aminophenylalaline have developed a significant defense mechanism which has brought inherent resistance to the attack of predators such as parasitic beetles (pza.sanbi.org).

Wild and Cultivated Varieties

Though zombie pea has wild distribution to all the five continents (Lawn and James, 1991; Pienaar and Kok, 1991; Sasikumar and Sardana, 1988) but only its cross-incompatible varieties are being cultivated only in Sudan and Indonesia (Karuniawan *et al.*, 2006) which suggests that wild forms have substantial in-built genetic variation within a species complex. The Sudanese cultivated form is generally cultivated for seeds which is moderately tolerant to drought, salt and certain biotic stresses. However, in Bali where it is cultivated for its fleshy roots, the varieties are more susceptible to many kinds of stresses (Wettberg *et al.*, 2020).

V. vexillata* var. *macrosperma

Source: cropped from Wettberg et al. (2010)

Infraspecific Botanical Varieties

Source: (POWO, 2013).

- *Vigna vexillata* var. *angustifolia* (Schumach. & Thonn.) Baker: It is native to tropical and subtropical Old World where it is found growing primarily in the seasonally dry tropical biome.
- *Vigna vexillata* var. *davyi* (Bolus) B.J.Pienaar: It is native to S. Africa where it has been found growing primarily in the subtropical biome.
- *Vigna vexillata* var. *macrosperma* Marechal, Mascherpa & Stainier: This cultivated variety is native to Indo-China where it is found growing primarily in the wet tropical biome.
- *Vigna vexillata* var. *ovata* (E.Mey.) B.J.Pienaar: It is native to South Africa which has been recorded growing primarily in the subtropical biome.
- *Vigna vexillata* var. *sepiaria* (Dalzell) Babu & S.K.Sharma: A native to India which has been found growing primarily in the seasonally dry tropical biome.
- *Vigna vexillata* var. *tsusimensis* Matsum.: It is native to South Korea and Tsushima (Japan) where it is found growing primarily in the temperate biome.
- *Vigna vexillata* var. *vexillata*: It is native to tropical and subtropical areas growing primarily in the seasonally dry tropical biome.
- *Vigna vexillata* var. *wightii* (Benth. ex Bedd.) Babu & S.K.Sharma: It is native to India where it grows primarily in the seasonally dry tropical biome.
- *Vigna vexillata* var. *youngiana* F.M.Bailey: It is native to Australia where it is found growing primarily in the seasonally dry tropical biome.

Nutritional Value

Fleshy roots as well as the seeds of *V. vexillata* contain 15% protein content which is much greater than found in potato, yam and cassava (Chandel *et al.*, 1972; tropical.theferns.info). Across all the regions where it is consumed, it often serves an important food security role, particularly in times of famine or emergency so it is a very potent crop for food security across the world though is quite under-exploited than should be.

Uses

Its soft and easy to peel fleshy roots having cream-coloured tasty flesh can be relished raw or cooked. They are boiled or roasted similar to sweet potato or cassava fleshy roots. Young leaves as well as young shoots are cut into small pieces and cooked with salt *cum* chilli and then garnished with mustard seeds, curry leaves and onions in oil and can be relished as vegetable. Young pods are also cooked and eaten as a vegetable. Seeds are cooked and consumed as a pulse. Its seeds contain L-Dopa which is a medicine for treating Parkinson's disease (Gangarao *et al.*, 2011). In poor as well as other Indian soils it is also grown as a green manure and ground cover crop (tropical.theferns.info).

Habitats

Zombi pea is naturally found in grasslands with dispersed trees, thickets, bushlands, in deserted cultivated land and even in forests at altitudes ranging from 0–2500m elevations and prefers southern slopes (pza.sanbi.org). Plants can be found in the deep shade of trees in the wild (tropical.theferns.info).

Cultivation

In India, zombi pea flourishes at an altitude of 1200–1500m in the foothills of the Himalayas and in the hills of eastern and NE India. In Australia, it grows in the north where monsoonal rains bring about 1250–1500mm of rainfall with a common long dry season (eol.org; PROSEA, 2023). The plant can survive in drought-prone areas and can thus perform well when cultivated in drier areas (pza.sanbi.org).

The optimal mean annual temperature for zombi pea is 20–28 °C but is tolerant of 15-32°C. It prefers a mean annual rainfall of 1300–2100mm, but can survive in areas with lower or higher rainfall (650-4200mm). Zombi pea prefers full exposure to sun with well-drained light to medium soils having pH range of 5.5-6.5 though tolerates even a pH of 8 (tropical.theferns.info; pza.sanbi.org). It can also survive poor soils or in areas prone to flooding but it takes more time to grow in too much water.

This can be grown directly in the beds or in a spacious container filled with a commercial potting mixture. It is propagated through seeds that take about 10 days to germinate. Though not necessary but sowing of scarified seeds is a sure success. Scarification in this case is elimination of testa with scalpel without injuring the endosperm or through rubbing these with sandpaper, freeze-thawing, heat and through acid treatment (pza.sanbi.org).

A small amount of water should be sprinkled to moisten the soil before planting in the garden bed. Seeds are sown 1cm deep with the seed eye facing downwards. After sowing, soil and water should be sprinkled to cover the seeds. The stems of mature plants tend to twine and find support on surrounding vegetation or a 1-1.5m long stick per plant can be used as a support on which the crop can twine. Irrigation twice per week is sufficient (pza.sanbi.org). Flowering occurs 3-4 months after planting and the seedpods ripen about one month later and by this time the fleshy roots are also ready for harvesting. Bhattacharyya (1984) reported the yield of fleshy roots up to 7t/ha.

Since this crop is nowhere in the country being grown as mandated crop so very little is known about its insect-pest and disease problems.

References

Amkul, K., L. Wang, P. Somta, S. Wang and X. Cheng, 2019. Construction of a high density linkage map and genome dissection of bruchid resistance in zombi pea (*Vigna vexillata* (L.) A. Rich). *Sci. Rep.*, **9**: 11719. https://doi.org/10.1038/s41598-019-48239-5.

Bhattacharyya, P.K. 1984. Grow Vigna vexillata for protein-rich tuber-cum-pulse crop in North-eastern hill region. *Seeds & Farms*, **10**: 33–6.

Chandel, K.P.S., R.K. Arora and B.S. Joshi, 1972. *Vigna capensis* Walp. (*V. vexillata)* an edible root legume. *Curr. Sci.*, **41**: 537.

Dachapak, S., P. Somta, S. Poonchaivilaisak, T. Yimram and P. Srinives, 2017. Genetic diversity and structure of the zombi pea [*Vigna vexillata* (L.) A. Rich] gene pool based on SSR marker analysis. *Genet.*, **145**: 189–200.

Duke, J.A. 1981. *Vigna unguiculata* (L.) Walp. spp. unguiculata. In: Legumes of World Economic Importance (ed. Okeson, O. N.), 303–5. Plenum Press.

eol.org/pages/655241/articles.

Gangarao, B., Ch..K.V.L.S.N..Anjana Male, T. Hari priya, Venna Naga Malleswari and S.K..Reeshma, 2011. A phytophasrmalogical review on *Vigna species*. *Pharmanest* - an international journal of advances nn pharmaceutical sciences, **2** (1). www.pharmanest.net. ISSN 0976 - 3090 (Print) 2231 - 0541 (Online). https://pharmanest.net/journal_pharmanest/uploads/1/414_pdf.pdf.

Karuniawan, A., A. Iswandi, P.R. Kale, J. Heinzemann and W.J. Gruneberg, 2006. *Vigna vexillata* (L.) A. Rich. cultivated as a root crop in Bali and Timor. *Genet. Resour. Crop Evol.*, **53**: 213–7.

Lawn R.J. and A. Cottrell, 1988. Wild mungbean and its relatives in Australia. *Biologist*, **35**: 267–73.

Lawn, R.J. and A.T. James, 1991. Inheritance of selected traits in accessions of *Vigna vexillata* (L.) A. Rich of Australian and African origin. *Aust. J. Bot.*, **39**: 415-29.

Lawn, R.J. and A.R. Watkinson, 2002. Habitats, morphological diversity and distribution of the genus *Vigna savi* in Australia. *Aust.. J. agric. Res.*, **53**: 1305–16.

Miller, I.L. and W.T. Williams, 1981. Tolerance of some tropical legumes to six months of stimulated waterlogging. *Trop. Glassland,* **15**: 39–41.

Palaniswami, M.S. and K.V. Peter, 2008. Tuber & Root Crops, 67p. Horticulture Science Series. New India Publishing House, New Delhi. ISBN 9788189422530.

Pienaar, B.J. and P.D.F. Kok, 1991. The Vigna vexillata complex (Fabaceae) in southern Africa. *S. Afr. Tydskr. Plantk.*, **57**(5):236-45.

POWO, 2013. Plants of the World Online – Kew Royal Botanical Garden: Retrieved 12 May, 2013 via https://powo.science.kew.org/taxon/urn:lsid:ipni.org:names:525561-1.

PROSEA, 2023. Plant Resources of South-East Asia, vol. 11. Auxiliary Plants, 261p. Yayasan Obor Indonesia. ISBN 9789797990930. Retrieved on 11 May, 2023.

pza.sanbi.org/vigna-vexillata-var-vexillata.

Quattrocchi, U. 2016. CRC World Dictionary of Medicinal and Poisonous Plants: Common Names, Scientific Names, Eponyms, Synonyms, and Etymology (reprint ed.), 3909p. CRC Press, USA. ISBN 9781482250640.

Roecklein, J.C. and P.S. Leung, 1987. A Profile of Economic Plants. Transaction Publishers: Piscataway/New Brunswick, NJ, USA.

Sasikumar, B. and S. Sardana, 1988. *Vigna vexillata* (Fabaceae), a pulse *cum* tuber crop of northeastern hill region of India. *Econ. Bot.,* **42**: 292.

Tomooka, N., A. Kaga, T. Isemura and D. Vaughan, 2011. *Vigna.* In Wild Crop Relatives: Genomic and Breeding Resources: Legume Crops and Forages (ed. Kole, C.). Springer: Berlin/Heidelberg, Germany, pp. 291–311.

tropical.theferns.info/viewtropical.php?id=Vigna+vexillata.

Wettberg, Eric B. von; Ken Naito, Andi Kur and Ndiko Ludidi, 2020. The unusual biogeography of zombi pea, *Vigna vexillata*. Legume Perspectives. *J. intern. Leg. Soc.*, No. 19, pp. 8-9. www.researchgate.net/profile/Taofeek-Adegboyega/publication/348430504_Winged_bean_Psophocarpus_tetragonolobus/links/5ffea468a6fdccdcb84d8bf1/Winged-bean-Psophocarpus-tetragonolobus.pdf.

15

Skirret & Water parsnip

Skirret or Skirwort (*Sium sisarum* L.)

Family: Apiaceae

Other names: in Scotland: Crummock; in Irish: Cearrachan; in Danish: Sukkerrod; in Dutch: Suikerwortel (Smith, 1998) and in German: Zuckerwurzel (translates as 'sugar root') (en.wikipedia.org, 2023a).

Homotypic Synonyms: *Apium sisarum* (L.) Calest.; *Carum sisarum* (L.) Baill.; *Pimpinella sisarum* (L.) Jess.; *Selinum sisarum* (L.) E.H.L.Krause; *Seseli sisarum* (L.) Crantz (POWO, 2023a)

Heterotypic Synonyms: *Berula lancifolia* Besser; *Pimpinella sisaroidea* (DC.) Manden; *Sisarum sisaroideum* (DC.) Schischk; *Sium brevifolium* DC.; *S. lancifolium* M.Bieb.; *S. podolicum* Besser ex Rchb.; *S. podolicum* G.Don; *S. sisaroideum* DC.; *S. sisarum* subsp. *sisaroideum* (DC.) Soo; *S. sisarum* var. *lancifolium* Thell. and *S. sisarum* var. *vulgare* Alef. (POWO, 2023a).

Introduction and Origin

Common name 'skirret' has been derived from the Dutch 'suikerwortel', meaning 'sugar root' (Stephens, 2023). The native range of the species is EC&SE Europe to SW Siberia and N. Afghanistan (POWO, 2023a). This minor European root crop with a wild distribution across Asia can be easily grown in the Pacific

Northwest and throughout the northern US. It is usually cultivated for its sweet roots. The cultivated skirret doesn't appear to be much different from wild forms scattered about Europe and Asia (Cultivariable, 2023).

Flowers of *Sium sisarum*	Roots of *Sium sisarum*

By Cephas - Own work, CC BY-SA 4.0, https://commons.wikimedia.org/w/index.php?curid=122870746	Cultivariable, 2023

Two possible wild ancestors of skirret are *S. sisaroideum*, a plant found primarily in the Middle East, and *S. ninsi* which is found in eastern Asia. Hybridization might be a source of new traits. It is also possible that hybridization could be accomplished with other members of the genus though there may be danger of toxicity in such progenies. Not every plant in the genus *Sium* is safe for human consumption and most are untested (Cultivariable, 2023).

Botany

Skirret is a true perennial that will come back year after year. It grows to a height of 90-120cm with large, shining, dark green, compound and pinnate leaves. The plant can produce small umbellate white flowers in its very first season. Its grayish-white, sweetish, cylindrical, 15-20cm long, 1.25cm wide, somewhat jointed and little aromatic roots come up in clusters from the stem base similar to those of sweet potatoes. If grown well, its fleshy roots appear sweet though some of the roots may have non-edible woody core (Stephens, 2023; en.wikipedia.org, 2023a; Cultivariable, 2023). Being perennial, it can be left in the ground for many years. The yield gets a lot better in the second year and then increases a little more in following years. Because it can survive freezing soil, it can be left in the ground year round for harvesting whenever the soil is workable (Cultivariable, 2023).

Nutritional Value

A study of the nutritional composition of skirret showed that the root has a greater content of dry matter (average being 16%). It is rich in sugars especially sucrose (65% of dry matter). The sprouts have a level of dry matter near 8% and are rich in reducing sugars and proteins (12 and 25% of dry matter, respectively).

The sprouts have higher vitamin contents than the root. The sprouts growing under greenhouse are higher in vitamins than those cultivated in the open field (Leclerc and Peron, 1989).

Uses

Skirret roots are used as a vegetable in the same manner as the common salsify, black salsify and the parsnip (Grieve, 1931). Its sweet flavour is variously described as resembling parsnip, carrot and parsley. Cooked roots are very soft *cum* sweet and have a combined flavour of potato and carrot (Cultivariable, 2023). Skirret roots can be stewed, baked, roasted, fried in batter as fritter, or creamed and also be grated and used raw in salads (en.wikipedia.org, 2023a). The fibrous central core must be removed before cooking as after cooking it is not easy to remove. Seedlings differ in the amount of core. The root texture is firm, the taste is sweet and the colour is very white (Stephens, 2023). Its spring shoots are also used as vegetable (Cultivariable, 2023).

Propagation

Seeds of unimproved varieties may be sown in seedbeds for transplanting or sown directly in the garden. Plants may also be started from root division (Stephens, 2023). Improved varieties are propagated from offsets. Skirret is a diploid and so it should be possible to produce true breeding seed lines but all of the superior varieties produced so far are clones (Cultivariable, 2023).

Skirret is easily propagated by the offsets that are basically rooted cuttings. Offsets with large and straight shoots with a fringe of rootlets forming around the base should be selected for planting. These should be cut free with a small knife. Although not necessary, it would be better if the offsets are cut free with one or more storage roots attached to it (Cultivariable, 2023). Though offsets can be planted directly but these are very small and vulnerable to pests and rough weather. Hence, it would be better to pot and when have attained 15cm length and growing well, these are transplanted. These are generally planted in a 10-cm pot filled with wet soil with the tip of the sprout just below the surface of the soil, perhaps 3mm (Cultivariable, 2023).

Being entomophilous, it produces a lot of seeds. In case of controlled crosses its inflorescence may be bagged and manually the pollen of some other desired variety may be placed over the stigmas of bagged flowers. For the pollen receiver, all but the central part of the umbel are cut off, preventing self pollination. Male parts develop before female so growers should watch for the anthers to wither and then begin introducing pollen from the chosen donor with a brush. In most cases, it would be easier to plant parent varieties side by side and net or cage them to prevent pollination from other undesirable plants. The grower will get both self pollinated and crossed seed, but the process involves a lot less labour. If the weather is wet when skirret seeds are maturing, it may be preferable to cut the seedheads and take them indoors to finish (Cultivariable, 2023). Per gramme there are some 500 seeds of skirret. Seeds retain good germination for at least three years if stored at a temperature of 21°C or lower and five years if stored at 10°C or lower (Cultivariable, 2023).

Climate

Skirret is a fairly adaptable crop and grows wild in Europe and Asia. In the wild, it is largely a streamside plant but grows well when grown in a garden as well. Although it prefers full sun but tolerates shade but on little cost of root production. If irrigated properly, it can handle a fairly wide range of temperatures. If temperatures rise above 32°C for long, it will benefit from afternoon shade. It can survive fairly cold winters and can probably be grown as a perennial down to USDA zone 5. Skirret has no photoperiod dependencies (Cultivariable, 2023). Like parsnip, the skirret roots become sweeter after first frost if left in the ground.

Soil

Skirrets grow well in a well-drained acidic to neutral soil rich in organic matter. However, it grows well even in heavy and swampy soils. Skirret is very responsive to nitrogen but excess of nitrogen produces lush green top growth at the expense of root yield (Cultivariable, 2023).

Planting

Skirrets can directly be sown any time from the mid-fall to late spring. In spring planting, it is best to wait until after the last frost, since the new sprouts may die back in a hard freeze. These will usually recover but too many freezes will kill the plant. In order to produce full size roots the first year, the seeds should be started indoors in moist climates. It will be better to sow seeds eight weeks before the last frost. Skirret seed germinates slowly and erratically over a long period of time and may take three weeks or even more to germination and up to six weeks before 50% have emerged. The soil must be kept moist. If possible, 21°C bottom heat can be given until 50% of seed germination and thus seedlings would become some 15cm tall at the time of transplanting (Cultivariable, 2023).

The inter-plant spacing of 20 cm in rows spaced 45cm apart is ideal.

Management

Skirret is basically a wild plant and does not need much attention during the growing season. However, the soil should not get drying out at any time (Cultivariable, 2023) as lack of moisture can also make the root more fibrous (en.wikipedia.org, 2023a). If the crop is grown from seeds, thinning is required to maintain proper planting distance, however, thinned out plants can be eaten.

Companion Planting

In the wild, skirret grows in shaded areas and forest edges. It grows fairly well in shade though with slow root development. It makes a fine understorey plant and can be used around the edges of orchards or in a transition zone between a cultivated area and native trees and shrubs (Cultivariable, 2023).

Harvest

Skirret is ready for harvest once the top of the plant has died back at the end of its first year. It can be left in the ground for another two years or even more. In case there are lots of plants, growers can harvest some and leave the others to grow larger. Once a plant has passed its first year, it can be harvested any time, so this can give a year round harvest (Cultivariable, 2023).

Storage

Skirret stores perfectly well in the ground as long as the soil doesn't get too wet in the winter or voles are not a problem. If they are not left in the ground, any offset that will be used for propagation should be removed and then the intact and unwashed root ball should be stored at a temperature and humidity as close as possible to 4°C and 95%, respectively. Under these conditions, skirret lasts three to four months in storage. The intact root balls can be kept in good condition for about a week at room temperature (Cultivariable, 2023).

Pests and Diseases

Pests

Small rodents, called voles, love to eat the roots of skirret. Tiny seed feeding beetles can be a real problem if the growers are trying to save seeds from particular plants for breeding. Root- knot nematodes can apparently be a problem for skirret being grown in high soil populations of root-knot nematodes. In case the population of nematode is low then frequent crop rotation with non-host crops is recommended (Cultivariable, 2023).

Diseases

Very little is known about skirret diseases. Major skirret diseases are black blight (*Septoria apiicola*). This disease can wipe out plants, starting as pinprick spots on the leaves. In its attack, leaves yellow quickly and turn necrotic *cum* papery and then spreads to the stems, developing lesions and wilt. Plants that survive to overwinter show worst infection in early spring and such plants usually do not survive (Cultivariable, 2023).

Another condition is observed in England which causes partial wilting and blackening of the stems but that doesn't kill the plants. The symptoms are consistent with organisms like *Fusarium* and *Verticillium* though real cause is still unknown (Cultivariable, 2023).

Water parsnip or Hemlock water parsnip [*Sium suave* Walter (1788)]

Introduction and Origin

The common name 'water parsnip' is due to its similarity to parsnip (*Pastinaca sativa*) *cum* its wetland habitat. The alternate common name, 'hemlock waterparsnip', is due to its similarity to the highly poisonous 'spotted water hemlock' (*Cicuta maculata*). The native range of this species is Temperate Asia, subarctic America to U.S.A. (POWO, 2023b). The species grows in sandy and non-sandy wetland habitats such as wet prairies, bottoms of seeps, low areas along springs, soggy thickets, swamps, borders and shallow water of ponds, marshes and ditches (en.wikipedia.org, 2023b; IllinoisWildFlowers, 2023).

Water parsnip (*Sium suave*)	**Flowers of water parsnip** 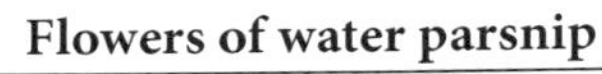

By I, KENPEI, CC BY-SA 3.0, https://commons.wikimedia.org/w/index.php?curid=4456732	**Source: IllinoisWildFlowers, 2023**

Water parsnip is very similar to skirret. It grows under conditions very similar to skirret but can tolerate even wetter soils. The aerial parts and roots of both the plants are quite similar in structure and appearance and would be easy to confuse though the roots of water parsnip are not as large as that of skirrets. Its flavour is admiring but not as sweet as skirret. It was used as a staple food by the Okanagan people of Washington and British Columbia. There are many warnings not to eat it due to its resemblance to the toxic water hemlock (*Cicuta douglasii*). Mistaken identity isn't likely to be a problem in cultivation though (Cultivariable, 2023).

Botany

Water parsnip is a stout semiaquatic perennial herb with fibrous roots originating from the lower stem nodes. The plant grows up to 3m tall with marginal branching in upper part. The light green, glabrous and longitudinally veined stems are 5cm in diameter and terete or angular-terete in cross-section (IllinoisWildFlowers, 2023). Leaves along the stems are alternate, stem-sheathing, 5-12.5cm long, odd-pinnate and compound with 7 to 17 lance to linear leaflets, and becoming smaller in the upper plant, typically with tiny, sharp teeth around the edges but occasionally smooth (Minnesota Wildflowers, 2023). Depending on its level of exposure to standing water, it may produce terrestrial, semi- to fully aquatic leaves during early stages of its development. On moist ground it initially forms a rosette of basal leaves. In shallow water, it initially forms a cluster of aquatic or semi-aquatic leaves of variable length. The terrestrial basal leaves are very similar to the alternate leaves except the leaflets which in former are wider. Fully aquatic leaves are double-pinnate or double-pinnatifid with filiform leaflets or lobes that are pale green and glabrous. Semi-aquatic leaves are primarily odd-pinnate with linear to linear-lanceolate leaflets and these leaflets usually have narrow pointed lobes along their sides. In addition, semi-aquatic leaves may have whorls of secondary leaflets at the bases of the primary leaflets. These secondary leaflets are similar to the primary leaflets except that former are smaller in size (IllinoisWildFlowers, 2023).

Roots of water parsnip

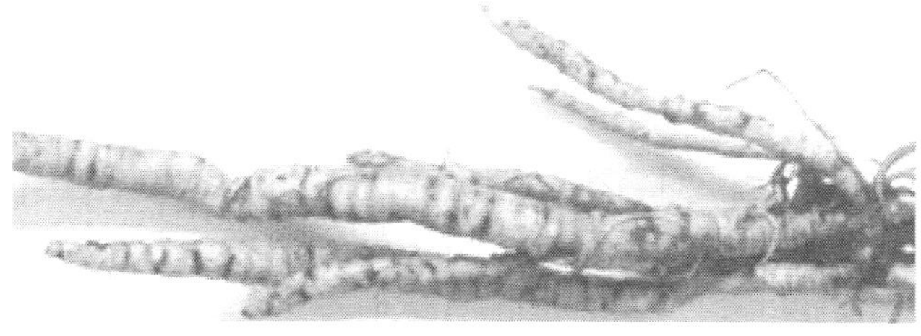

***Source*: https://cdsc-wsu.org/nwnativeplants/digital-heritage/sium-suave**

The upper stems terminate in compound umbels of flowers that are up to 12.5cm across, each consisting of groups (umbellets) of 10-20 flowers. Flowers are about 3mm across with 5 white petals that are somewhat heart-shaped and may be of unequal size (Minnesota Wildflowers, 2023). Its blooming period occurs from mid-summer to early fall, lasting for about 1½ months. Each flower is replaced by an oblongoid-ellipsoid fruit about 3mm long that is slightly compressed and strongly ribbed; each fruit consisting of a pair of seeds that are joined together. The root system consists of shallow fibrous grayish-white and fleshy roots (IllinoisWildFlowers, 2023) that may resemble small parsnips in colour and shape. About two years old plants have an average root diameter of 1cm. The epidermis puckers at rootlet nodes giving it an almost banded appearance.

Uses

Water parsnip resembles the very toxic spotted water hemlock (*Cicuta douglasii*) and caution should be taken when using this plant for food. Edible parts of *Sium suave* include the root in the spring and fall either raw or cooked as it has a nutty flavour. The leaves are also sometimes used for condiments. Crushed water parsnip roots have also been used as an analgesic (pain reliever) in cases of broken limbs.

Cultivation

Water parsnip should be grown in wet mucky soil/sand or in standing water up to 2.5cm deep. It also prefers partial to full sun (IllinoisWildFlowers, 2023). The seeds should be sown in late winter to early spring during a cold frame. The plant flowers from July to August while the seeds ripen from September to October. It has been shown to accumulate arsenic and heavy metals near abandoned mine tailings in South Korea. In the presence of added N03-N to the ground, water parsnip shows increased production (en.wikipedia.org, 2023b).

References

Cultivariable, 2023. Skirret. Retrieved on 5 June, 2023. www.cultivariable.com/instructions/root-crops/how-to-grow-skirret/.

en.wikipedia.org, 2023a: en.wikipedia.org/wiki/Sium_sisarum. Retrieved on 5 June, 2023.

en.wikipedia.org, 2023b: en.wikipedia.org/wiki/Sium_suave. Retrieved on 5 June, 2023.

Grieve, Maud, 1931. *A Modern Herbal*. Peregrine Books,742p.

IllinoisWildFlowers, 2023. Water Parsnip (*Sium suave*). www.illinoiswildflowers.info/wetland/plants/wt_parsnip.html. Retrieved on 6 June, 2023.

Leclerc, J. and J.Y. Peron, 1989. Mineral, Sugar and Vitamin Contents of Skirret (*Sium sisarum* L.). *Acta Hortic.,* No. 242, pp. 325-8. DOI: 10.17660/*ActaHortic*.1989.242.42.

Minnesota Wildflowers, 2023. *Sium suave* (Water Parsnip). Minnesota Wildflowers. www.minnesotawildflowers.info. Retrieved 6 June, 2023..

POWO, 2023a. Plants of the World Online. *Sium sisarum* L. Royal Botanical Gardens, Kew. Retrd. on 5 June, 2023. https://powo.science.kew.org/taxon/urn:lsid:ipni.org:names:849059-1.

POWO, 2023b: Plants of the World Online. *Sium suave* Walter. Royal Botanical Gardens, Kew. Retrieved on 5 June, 2023. https://powo.science.kew.org/taxon/urn:lsid:ipni.org:names:849061-1.

Smith, K. 1998. Growing Uncommon Fruits & Vegetables in Australia, 131p. New Holland, Australia. ISBN 1864363541.

Stephens, James M. 2023. Skirret – *Sium sisarum* L. IFAS Extension. Univ. of Florida. Retrieved 5 June, 2023. https://edis.ifas.ufl.edu/publication/MV134.

Tartarican Colewort/Tartar Bread Plant (*Crambe tatarica* Sebeok)

Family: Brassicaceae.

Other names: In German - Krambe tatarenkohl (Magic Garden Seeds, 2023) or Tatarischer meerkohl and in Russian - Katran tatarskij (Mansfeld, 2023).

Synonyms: *Crambe pinnatifida* R. Br., *Crambe buschii* (O.E. Schulz) Stankev., *Crambe litwinowii* Gross, *Crambe tatarica* Pall., *Crambe aspera* M. Bieb. and *Crambe gibberosa* Rupr (Mansfeld, 2023).

Introduction and Origin

The generic name *Crambe* has been derived from the Latin *crambe* which comes from Ancient Greek *krambe*, a kind of cabbage (Shorter Oxford English dictionary, 2007). *Crambe tatarica* is native to Central Europe to SW Siberia and Caucasusis (POWO, 2023). It grows in steppes, on stony-gravelly slopes, and on chalky outcrops and is harvested from the wild for local use as a food. In Kazakhstan, it is found in the northern Pre-Caspian, Aktobe, Atyrau and Kostanay regions where the eastern border of its range is located. Outside Kazakhstan, the species is found in the south of the European part of the former USSR, in Crimea, in Caucasus, in western Siberia and in southern Europe (Pavlov, 1961; Red Data Book, 2014; Gouz, 2016).

Before 19th century, *Crambe tatarica* was often collected as wild vegetable for its starchy rhizomes and roots as well as its leaves as in Hungary but in 19th and early 20th century it was cultivated for some time as a vegetable in France where it was used like *Crambe maritima*. Recently it was recommended once more for cultivation in the former Soviet Union (Mansfeld, 2023). There is some confusion over the correct spelling of the name of this species it is written as *C. tatarica* in most literatures though *C. tataria* in many others.

Botany

Tartar bread plant is a perennial, single or multi-stemmed and up to 90cm tall polycarpic plant growing from 60–120cm long, fusiform and juicy roots that can be as thick as a person's arm. The stems are erect with sparse, short and stiff hairs though naked later on. Leaves are fleshy and grey-green, basal leaves are large and with petioles up to 30cm long and 20cm wide, bipinnate with oblong-linear, toothed or incised lobes and stiff hairs especially on the abaxial side but later on naked while the uppermost leaves are linear and entire. The honey-smelled flowers are white, 4.5–5 mm in length and arranged in dense paniculate inflorescence. Pods are spherical where upper pod segment is 4–5mm in diameter, tetragonous, mesh-rugose and with four sharply prominent ribs on the stalks up to 10mm in length (Bauyrzhan *et al.*, 2022).

Crambe tatarica

Source: Magic Garden Seeds, 2023

Chemical Composition

The aboveground parts of *C. tatarica* are richer in alkaloids, carotenoids, monosaccharides and phospholipids than the root system though in roots larger amounts of tannins, coumarins, polysaccharides, triterpenoids, phenolic and hydroxycinnamic acids and flavonoids have been recorded (Bauyrzhan *et al.*, 2022).

The chemical composition of *C. tatarica*, as reported by Bauyrzhan *et al.* (2022) are kaempferol-3-(2-hydroxypropionyl) glucoside-4'- glucoside, kaempferol-3-(p-coumaroyl) glucoside-4'- glucoside, kaempferol-3-feruloylglucoside-7-glucoside, kaempferol-3-(p-coumaroyl)-glucoside-7,4'-diglucoside, kaempferol-3- (p-coumaroyl) feruloyl-diglucoside-7- rhamnoside, quercetin-3-malonylglucoside-4'-glucoside, quercetin-3-feruloylglucoside-4'-glucoside, quercetin-3- feruloylglucoside-7-glucoside, quercetin-3- feruloylglucoside-7,4'-diglucoside, vitamin C, carotene (Amirkhanov *et al.*, 1974; Sokolov, 1986; Budantsev and Lesiovskaya, 2001; Grudzinskaya *et al.*, 2014), up to 43.4% of fatty oil acids (Grashchenkov, 1959; Miller and Earle, 1965; Sokolov, 1986; Budantsev and Lesiovskaya, 2001; Grudzinskaya *et al.*, 2014): oleic-28.69%, linoleic-22.17%, erucic-27.00%, palmitoleic-0.32%, eicosene-21.00%, linolenic-11.00%, palmitic-2.00%, eicosadienic-1.29%, stearic-0.80%, arachidic-0.50% and myristic-0.05% (Miller and Earle, 1965; Dolia *et al.*, 1977; Sokolov, 1986; Pushkarova *et al.*, 2016).

Aqueous infusions of *C. tatarica* roots have been used as a general tonic and vitamin supplementation for children (Sokolov, 1986; Budantsev and Lesiovskaya, 2001). In Indian medicine, a decoction of the roots of *Crambe cordifolia* is used to treat scabies (Sokolov, 1986; Budantsev and Lesiovskaya, 2001; Chopra and Nayar, 2006; Grudzinskaya *et al.*, 2014); and a decoction of leaves of *Crambe orientalis* and roots of *Crambe pinnatifida* is used to treat scurvy (Gammerman, 1982; Budantsev and Lesiovskaya, 2001; Grudzinskaya *et al.*, 2014).

Uses

The leaves, young stems and roots are consumed as to that of radish raw or cooked. It can be used raw in salads or cooked as a vegetable. It can be dried and ground into a powder and then mixed with cereal flours when making bread. Root is rich in starch and sugars (PFAF, 2023). Its roots can also be used as a natural calcium accumulator (Bauyrzhan *et al.*, 2022).

Cultivation

Propagation

Seeds

Seeds can be sown early to mid-spring in a seedbed outdoors and then either the plants are thinned out or moved to their permanent positions when about 10cm tall. The young plants attract slugs so these require to be protected from them. Germination can be slow and usually takes place in 3-5 weeks around 15°C so it is best to sow the seeds in pots in a cold frame. The seedlings should be pricked out into individual pots as soon as they are large enough to handle and then can be planted out into their permanent positions when these attain at least 10cm length (Temperate Plants Database, 2023).

Division

These are propagated through division in spring or autumn. The root clumps are dug up and cut off into the desired section making sure that each section has at least one growing point. The larger of these divisions can be planted out straight into their permanent positions though smaller ones are best potted up and grown on in a cold frame until they are established (Temperate Plants Database, 2023).

Root Cuttings

In spring. root cuttings of 3-10cm long can be planted straight into the open ground or can be potted up in the greenhouse and then planted out once they are growing well (Temperate Plants Database, 2023).

Other Requirements

Crambe tatarica is quite hardy and easy to grow plant that can tolerate different conditions fairly. It is a deep rooted plant and dislikes root disturbance.

Sunlight	Full sun; in a position sheltered from strong winds
Soil	Slightly alkaline good loamy soil; dislikes acid soils
Tolerance	Tolerates some shade and poor soil
Indoor pre-culture (seedling stage)	January-February
Sowing time outdoors	November-December
Sowing depth	2 -2.5cm
Optimum temperature for germination	10–20°C
Germination time	Usually 7–14 days but sometimes 21-35 days, depending on growing conditions
Planting distance	50×50cm
Main harvest	April to June

Source: Temperate Plants Database (2023) and Magic Garden Seeds (2023).

References

Amirkhanov, N.A., K.R. Mukumov and S.S. Khamrakulov, 1974. The content of vitamins and trace elements in some species of *Crambe* L. *Grows. Resour.*, **10**(3): 422- 6.

Bauyrzhan, T., K. Natalya, I. Zarina, K. Andrey, K. Meruert, A. Karime and B. Aliya, 2022. Phytochemical study of *Crambe tatarica* Sebeok, a promising source of medicinal plant extracts. *Pak.. J. Bot.*., **54**(2): 653-66.

Budantsev, A.L. and E.E. Lesiovskaya, 2001. Wild useful plants of Russia. SPHFA publishing house. St. Petersburg. https://www.pakbs.org/pjbot/papers/1643958545.pdf.

Chopra, R.N. and S.L. Nayar, 2006. Glossary of Indian Medicinal Plants, New Delhi, 330p.

Dolia, V.S., E.N. Shkurupiy, N.A. Kaminskiy and E.D. Megerya, 1977. Seed oils of nine species of the genus *Crambe. Chem. Nature. Con.*, **1**: 18-20.

Gammerman, A.F. 1982. Overview of medicinal plants in Turkmenistan. *Turkm. Phil. Acad. Sci. USSR,* **2**: 55-92.

Gouz, G.V. 2016. The spatial distribution of Crambe tatarica (Brassicaceae) across Striltsivsky steppe. In: *Proc. 4rth intern. Conf. 'Rare Plants and Fungi of Ukraine and Adjacent Areas: Implementing Conservation Strategies'*, held at. Palyvoda (Kyiv), pp. 67-69.

Grashchenkov, A.E. 1959. Katrans, their biology and introduction to culture near Leningrad. *Tr. Bot. Inst. USSR Acad. Sci. Introduction and Green Building,* **7**: 124-7.

Grudzinskaya, L.M., N.G. Gemedzhieva, N.V. Nelina and Zh. Karzhaubekova, 2014. Annotated list of medicinal plants in Kazakhstan, Almaty. https://www.pakbs.org/pjbot/papers/1643958545.pdf.

Magic Garden Seeds, 2023. Tatarican Colewort (*Crambe tatarica*). Retrieved on 7 June, 2023. https://www.magicgardenseeds.com/The-Good-To-Know/Tatarican-Colewort-(Crambe-tataria)-A.CRA02-

Mansfeld, 2023. Mansfeld's World Database of Agricultural and Horticultural Crops. Retrd. Jyne 7, 2023. https://mansfeld.ipk-gatersleben.de/apex/f?p=185:46:5995064301271::NO::module,source,akzanz,rehm,akzname,taxid:mf,botnam,0,,Crambe%20litwinowii,24215.

Miller, R.W. and F.R. Earle, 1965. Search for new industrial oils: Oils from 102 species of Cruciferae. *J. Amer. Oil Chem. Soc.*, **42**(10): 817-21.

Pavlov, N.V. (ed.), 1961. Flora of Kazakhstan. Vol: 4. Academy of Sciences Kaz SSR Publishing House, Alma-Ata. https://www.pakbs.org/pjbot/papers/1643958545.pdf.

PFAF, 2023. *Crambe tatarica* Sebeok. Plants For A Future. Retrieved on 6 June, 2023. https://pfaf.org/user/Plant.aspx?LatinName=Crambe+tatarica.

POWO, 2023. Plants of the World Online. *Crambe tatarica* Sebeok. Royal Botanical Gardens, Kew, England. https://powo.science.kew.org/taxon/urn:lsid:ipni.org:names:281682-1. Retrieved on 6 June, 2023.

Pushkarova, N., M. Kalista, M. Kharkhota, D. Rakhmetov and M. Kuchuk, 2016. *Crambe tatarica* Sebeok seeds and plants grown *in vitro* and *in vivo* for fatty acid composition comparison. *Potravinarstvo*, **10**(1): 494-8.

Red Data Book of Kazakhstan, 2014. Plants 2nd ed. (vol. 2. part 1)., Astana. https://www.pakbs.org/pjbot/papers/1643958545.pdf.

Shorter Oxford English Dictionary (6th ed.), 3804p, 2007. Oxford University Press, United Kingdom.

Sokolov, P.D. 1986. Plant resources of the USSR, Leningrad. Flowering plants, their chemical composition, use. Families Paeoniaceae - Thymelaeaceae. *Science*, 336p.

Temperate Plants Database, Ken Fern. temperate.theferns.info. 2023-06-07. <temperate.theferns.info/plant/Crambe+tatarica>17.

17

Bitterroot/Sand Rose (*Lewisia rediviva* Pursh)

Family: Montiaceae.

Introduction and Origin

The specific epithet *rediviva* means 'revived' or 'reborn', referring to its ability to regenerate from dry and seemingly dead roots (William Curtis, 1801). To honour Meriwether Lewis (1774-1809), an American explorer, the genus was named *Lewisia*, since he encountered the species in 1806. In 2009, the genus was moved from the purslane family (Portulacaceae) with adoption of the APG III system, which established its family as Montiaceae (en.wikipedia.org/wiki/Bitterroot). The native habitat of *Lewisia* species is rocky ground and cliffs in western North America (en.wikipedia.org/wiki/Lewisia). The gray-crowned rosy finch (*Leucosticte tephrocotis*) feeds on the seeds of *Lewisia* spp. (Anderson and Roderick, 2003).

Bitterroot ~ the state flower of Montana (Anon., 1929) is a perennial herb found at wide range of elevations ranging from 60 to 3000m. It is found growing in open woodlands and sagebrush shrublands with pine, oak or juniper in many soil types such as shale, sand, clay, granite, serpentine or talus. It is found in the San Jacinto Mountains and the Transverse Ranges in southern California, the Coast Ranges of California from Mt. Pinos to Mt. Diablo and Lake County, the

Klamath Mountains, in the eastern Cascade Ranges of Oregon and Washington, north to British Columbia and east to Montana, Colorado and Arizona (Anderson and Roderick, 2003).

Bitter-root	Flower in University of California Botanical Garden

By Walter Siegmund (talk) - Own work, CC BY-SA 3.0, https://commons.wikimedia.org/w/index.php?curid=8728744	**By Stan Shebs, CC BY-SA 3.0, https://commons.wikimedia.org/w/index.php?curid=3234403**

Botany

By Andrew Kratz, 2023. Bitterroot (*Lewisia rediviva*). USDA Forest Service.

Bitterroot is a perennial low-growing herb with thick fleshy branched taproot and a short, simple or branched stem-base. The 1–3cm tall leafless flower stem bears at the tip a whorl of 5–6 linear bracts which are 5–10mm long. The plant has numerous succulent basal leaves that are 1.5-5cm long, linear to club-shaped and round in cross section. The leaves usually wither by flowering time which occur from April through July. A single whitish to deep pink or lavender flower appears on each flower stem with 10-19 oblong petals (1.8–3.5cm long), multiple stamens (40-47) and 5–9 oval-shaped sepals. At maturity, the bitterroot produces egg-shaped capsules with 6–20 nearly round seeds (https://web.archive.org; en.wikipedia.org/wiki/Bitterroot; Anderson and Roderick, 2003).

Uses

Almost all the species of *Lewisia* are edible. Local Native Americans eat the roots, which have also been used to treat sore throats (en.wikipedia.org/wiki/Lewisia). The large edible root of *Lewisia rediviva* became a food source for local Native Americans (Vizgirdas and Rey-Vizgirdas, 2009). The roots are prepared for eating by peeling and then boiling, steaming or pit-roasting. Cooking reduces the bitterness of roots (Anderson and Roderick, 2003). Native people chew the bitterroot to relieve a sore throat and also to promote milk flow during lactation.

Cultivation

Lewisias are usually planted in rockeries as it mimics its natural habitat and also provides free drainage that these plants require to prevent rotting of their roots. These may also be planted in pots though these need to be well drained and protected from sustained wet weather (Clayton and Drury, 2012).

Bitterroot is not an easy plant to grow. Digging the plant in the wild kills the plant, therefore, it should be started from seed in early October. Seeds are placed on top of the soil in 10-15cm pots and a light layer of soil is sprinkled over the seeds. To hold the seeds in place, about 7.5mm gravel is placed on top of the soil. The container should be placed outside in full sun. The pots should be watered only if there is no rain. The seeds will germinate in late winter or early spring and thereafter light occasional watering may be required. In May when the leaves turn yellow, watering should be stopped and the pots should be placed in the hottest spot until fall when the rains come. Bitterroot requires hot and dry summer baking. This cycle may be performed for over two years. During the summer or fall of the third year when the plants are dormant, the plants can be planted outdoor in full sun and in well-drained soil. It is recommended that growers should crown the plants at the soil surface. Interplant distance should be 10-15cm. Growers must mark the spot where the bitterroot is planted as it does not require watering and should wait for the rains to occur. Protection from deer and other wildlife is required. Periodic weeding is required as this plant needs full sun (Anderson and Roderick, 2000).

References

Anderson, M. Kat and Wayne Roderick, 2000. Plant Guide. Article Edited: 05 dec 2000 and 21 may 2003. USDA NRCS National Plant Data Center & East Bay Regional Parks Botanic Garden https://www.fs.usda.gov/wildflowers/plant-of-the-week/lewisia_rediviva.shtml.

Anon..1929. Dept. of Public Instruction. Montana Educational Directory, Montana.. 30p. The Montana State Flower is the Bitter Root (*Lewisia rediviva*).

Clayton, P. and Anita Drury, 2012..Succeed with Lewisia. *The Garden*, **137** (6): 36–7.

en.wikipedia.org/wiki/Lewisia.

en.wikipedia.org/wiki/Bitterroot.

https://web.archive.org/web/20160304113715/http://linnet.geog.ubc.ca/Atlas/Atlas.aspx?sciname=Lewisia%20rediviva.

Vizgirdas, R. and Edna Rey-Vizgirdas, 2009. Bitter root, pp. 153-4. In: Wild Plants of the Sierra Nevada, 384p. University of Nevada Press. ISBN 9780874177893.

William Curtis, 1801. The specific name rediviva is given by Pursh in consequence of the root, long preserved in the herbarium, and apparently dead, having been planted, revived in a garden in Philadelphia. *The Curtis's bot. Mag.*, USA, 123p.

Tuberous-rooted Chervil/Chervil Root (*Chaerophyllum bulbosum* L.)

Family: Apiaceae.

Other names: Turnip-rooted chervil, Tuberous-rooted chervil, Bulbous chervil, Parsnip chervil (USDA, 2023).

Synonyms: *Chaerophyllum bulbosum* f. *glabrum* Cheshm. & Delip.; *C. bulbosum* var. *caucasicum* Fisch. ex Hoffm.; *C. caucasicum* (Fisch. ex Hoffm.) Schischk.; *C. laevigatum* Vis.; *C. neglectum* W.J.Zinger; *C. rapaceum* Alef.; *C. verticillatum* Pers.; *Myrrhis tuberosa* Jundz.; *Polgidon bulbosum* (L.) Raf.; *Scandix bulbosa* Roth; and *Selinum bulbosum* E.H.L.Krause (WFO, 2023).

Introduction and Origin

Chaerophyllum is a combination of the Greek words *charein* ('be glad') and *fyllos* (leaf), the reasons for being glad may probably be due to the appealing colour of the leaves and the pleasant fragrance; however, some of the poisonous species of the family have intoxicating characteristics (NatureGate, 2023).

The native range of tuberous-rooted chervil is Europe to N. Turkiye and Central Asia (POWO, 2023) and is cultivated on a small scale in parts of Europe for the edible roots which look like a dark gray carrot with yellowish-white flesh (en.wikipedia.org, 2023). The native wild populations of *Chaerophyllum bulbosum* are distributed in Europe from Ukraine to France and from Sweden to Italy. In France and Germany, wild populations can be found in the Rhine and the Weser River basins. It can also be found in W&C Asia as in Turkey and Caucasus. In the wild, this is found growing in dense populations mainly along river banks. Seeds are dispersed by hydrochory (Le Clerc, 2014).

Tuberous-rooted chervil is somewhat challenging to grow and is utilized similarly to other root vegetables in culinary applications. It is generally harvested from the wild for local use as a food. Cultivation for its edible taproot began on a small scale in the 1700-1800s in Europe, especially in France as then it was spread across United States. Today in the fresh markets of speciality grocers in Europe, especially in France and USA it can be found in the early fall through winter but only in limited quantity (Specialtyproduce, 2023).

Botany

It has been reported possessing 2n=22 chromosome numbers (Kamari *et al.*, 2004; Alexeeva *et al.*, 1994; Daushkevich *et al.*, 1995). At the first glance, the tuberous-rooted chervil with its small leaflets can easily be confused with wild chervil (*Anthriscus sylvestris*) but leaves in tuberous-rooted chervil are even more delicately lacelike, stem base with long hairs and the joints being distinctly bulbous and with little digging with the plants one can find its roundish-globose fleshy as final proof (NatureGate, 2023).

Chaerophyllum bulbosum is a biennial plant which in the first year develops leaf rosettes and a small, bulbous and stubby taproot having rounded ends and dimension of 5-8×2cm (length and width). Its root skin is thin, rough, brown to tan and covered with shallow eyes. The starchy flesh is dense and ivory to cream-coloured (Specialtyproduce, 2023; MagicGardenSeeds, 2023). In the second year this grows up to a height of 70cm or maximally 2m (Reuther and Claben-Bockhoff, 2013).

Early forms of *C. bulbosum* are said to have blackish or grey skinned roots so this colour difference in the roots of the wild plants may be a good way to distinguish between the two similar species in the field though none of the floral keys mention this. More recently released French cultivars of *C. bulbosum* do have golden skins but dark skinned forms were and may still be grown commercially (Dupey, 2010).

Its smooth and solid stems are mottled reddish-brown. These in the upper part are solid, branched and glabrous with bulbous joints but lower part is often blue-rimmed and sparsely stiff-haired (NatureGate, 2023). The alternate and spirally arranged stem-sheathing leaves have pod-like bases and are pinnately compound. As the plant grows, the bases of the seed leaves, near the taproot are pushed apart. The stem located just above the ground is compressed with indistinct internodes (en.wikipedia.org, 2023).

Chaerophyllum bulbosum

By Ouicoude - Own work, CC BY-SA 3.0,

Stem tip narrows and becomes pointed as the seed stalk elongates for flowering in July-August (en.wikipedia.org, 2023). The stem extends upward to become a highly branched inflorescence forming a compound umbel ~ primary umbel without bracts though narrowly ovate and long-tapered secondary umbels have 3–5(–7) bracteoles of different sizes. A plant produces 10-200 umbels and 1,000-36,000 protandrous flowers. Styles elongate only after shedding of pollen thus preventing selfing almost completely under wild. Tuberous-rooted chervil is an andromonoecious plant, *i.e.* hermaphrodite and male flowers occurring on the same plant. Sex ratio between functionally male and hermaphrodite flowers is about 80% to 20% (Reuther and Claben-Bockhoff, 2013; NatureGate, 2023).

Alternate, compound leaves

Corolla is white, regular, less than 6mm wide; outer corollas often slightly zygomorphic and bigger than central ones; petals 5, notched, with an incurved point; sepals missing; stamens 5; pistil of 2 fused carpels and styles are 2. Broadly oblong and 4–6mm long fruits are two-parted schizocarp that are brown, glabrous and slightly *cum* bluntly ridged; and the styles are bent outward (NatureGate, 2023).

Fruiting umbellets in the foreground and male flower umbellets in the background

Cultivars

Though the plant has been grown since the Middle Ages, it only survived as a garden crop in France without any named varieties. In modern times, the first named cultivar 'Altan' (1986) seems to have been developed in France, followed by new cultivars 'Vega' and 'M4.10', and these cultivars have found a niche market among home. The new varieties are mainly characterized by low seed embryo dormancy which normally is removed by the natural winter conditions occurring in the field. Through breeding achievements of French scientists, this alternative crop gained agricultural importance in regions like Loire Valley (France) and the north of Brittany (Peron and Briard, 2003; Peron and Dubost, 1992).

Nutritional Value

The root of tuberous-rooted chervil is high in fibre and has the same starchy quality as a potato. It contains favourable amounts of vitamins B and C and mineral salts. The rich carbohydrate reserves of the root increase after a period of cold storage, making it a highly desirable crop for diversity and use as a food source. Such high levels of sucrose and starch in one root vegetable are rare (Specialtyproduce, 2023).

Since the root starch content at harvest time is usually very high (37% dry matter, containing 76% starch) which is exceptional for a species cultivated under a temperate climate. *C. bulbosum* could be a good option for diversification of starch sources under this climate. A two-month period of cold storage (4°C) is necessary to reach the consumable stage which together results into a dramatic change of the carbohydrate composition. After this period, sucrose content reaches 25% of root dry matter (DM). Sucrose and starch (35% of root DM) are the two major components. The combination of high rates of these two in a single organ is quite uncommon and of great interest for a vegetable so this makes very interesting in human diet. Indeed it is a very tasty edible root combining a quick digestible sugar with a more slowly digestable one which results in long-term availability of energy. As it can be used for industrial starch production and food production both, *C. bulbosum* offers a good opportunity to diversify horticultural crops (Ayala Garay *et al.*, 2003).

Phytochemicals

Chemical composition of the hydrodistillated essential oils (EO) as well as the volatiles obtained by static headspace method (HS) of *C. bulbosum* as analyzed by GC and GC/MS and the results are presented in table given below. In general, monoterpene hydrocarbons were found to be the main class of all investigated samples with (E)-β-ocimene being the most abundant component (Stamenkovic *et al.*, 2020).

Headspace volatiles of the aerial parts of the plant consisted 100% of mono-terpene hydrocarbons with (E)-β-ocimene accounting 96.1% of the sample. In the root sample for which twenty compounds were identified (representing 99.9 % of the total), the main class was also represented by monoterpene hydrocarbons (98.7%) with (E)-β-ocimene (46.1%), followed by limonene (14.9%), γ-terpinene (12.5%), p-cymene (11.1%) and α-pinene (10.9%) as main components. Sesquiterpene hydrocarbons were also present but only 0.4% (Stamenkovic *et al.*, 2020).

Table: Chemical composition (average of triplicates±SD) of the *Chaerophyllum bulbosum* essential oils and headspace volatiles; compounds listed in order of elution on a HP-5MS column (Stamenkovic *et al.*, 2020).

RI	RA	Compound	Content, %			
			HS1	HS2	EO1	EO2
928	924	α-Thujene	–	0.1±0.01	–	–
935	932	α-Pinene	0.3±0.02	10.9±1.05	0.4±0.02	2.0±0.2
950	946	Camphene	–	0.1±0.04	–	–
975	969	Sabinene	–	0.7±0.05	tr	0.9±0.05
978	974	β-Pinene	0.4±0.05	0.2±0.01	0.3±0.02	0.2±0.01
991	988	Myrcene	0.1±0.01	1.0±0.35	0.2±0.02	0.9±0.07
1004	998	n-Octanal	–	0.2±0.05	–	0.5±0.03
1006	1002	α-Phellandrene	–	Tr	–	–
1018	1014	α-Terpinene	–	0.1±0.02	–	0.1±0.02
1026	1020	p-Cymene	0.6±0.05	11.1±0.55	0.8±0.02	3.3±0.54
1030	1024	Limonene	0.3±0.02	14.9±1.13	0.3±0.02	15.2±1.25
1039	1032	(Z)-β-Ocimene	0.4±0.05	0.9±0.32	1.6±0.24	1.0±0.03
1051	1044	(E)-β-Ocimene	96.1±2.17	46.1±1.15	91.5±1.18	38.5±0.95
1061	1054	γ-Terpinene	1.9±0.35	12.5±0.95	4.1±0.08	11.7±0.55
1091	1089	p-Cymenene	–	0.7±0.11	0.1±0.01	0.7±0.05
1121	1118	cis-p-Menth-2-en-1-ol	–	–	–	0.6±0.02
1140	1137	(E)-Epoxy-ocimene	–	–	–	1.6±0.16
1158	1157	(E)-2-Nonen-1-al	–	–	–	0.4±0.02
1178	1174	Terpinen-4-ol	–	–	–	0.3±0.01
1186	1184	Dill ether	–	–	–	0.2±0.01
1187	1184	Cryptone	–	–	–	tr
1191	1186	α-Terpineol	–	–	–	tr
1196	1195	cis-Piperitol	–	–	–	0.1±0.02
1198	1195	Methyl chavicol	–	–	–	0.2±0.02
1208	1207	trans-Piperitol	–	–	–	0.5±0.01
1220	1215	trans-Carveol	–	–	–	tr
1229	1232	Thymol, methyl ether	–	–	–	0.2±0.02
1234	1232*	Isothymol, methyl ether	–	–	–	0.7±0.05
1256	1 2 4 9 Piperitone	Piperitone	–	–	–	tr
1260	1260	(E)-2-Decenal	–	–	–	0.2±0.04
1287	1287	Bornyl acetate	–	Tr	–	0.4±0.01
1293	1289	Thymol	–	–	–	tr
1315	1315	(E,E)-2,4-Decadienal	–	–	–	tr
1358	1356	Eugenol	–	–	–	tr
1379	1374	α-Copaene	–	–	–	tr
1382	1380	Daucene	–	–	–	0.2±0.02
1402	1396	α-Chamipinene	–	–	–	0.2±0.02
1415	1407	α-Barbatene	–	–	–	tr

RI	RA	Compound	Content, %			
			HS1	HS2	EO1	EO2
1422	1413	β-Funebrene	–	0.1±0.04	–	2.3±0.34
1426	1419	β-Cedrene	–	–	–	0.5±0.05
1436	1429	cis-Thujopsene	–	–	–	0.1±0.02
1441	1436	Isobazzanene	–	–	–	tr
1444	1440	(Z)-β-Farnesene	–	–	–	tr
1452	1450	β-Barbatene	–	0.1±0.01	–	1.2±0.08
1456	1454	(E)-β-Farnesene	–	–	–	0.5±0.01
1461	1456*	Amorpha-4,11-diene	–	0.1±0.01	–	1.5±0.05
1483	1476	β-Chamigrene	–	–	–	0.5±0.03
1486	1484	Germacrene D	–	–	0.1±0.01	–
1496	1493	α-Zingiberene	–	–	–	tr
1499	1498	β-Alaskene	–	–	–	0.4±0.05
1501	1500	Bicyclogermacrene	–	–	0.2±0.01	–
1502	1501	Epizonarene	–	–	–	0.1±0.01
1512	1505	β-Bisabolene	–	0.1±0.02	–	1.5±0.12
1516	1512	α-Alaskene	–	–	–	0.5±0.02
1524	1517	Myristicin	–	–	–	5.1±0.64
1534	1530	Dauca-4(11),8-diene	–	–	–	0.3±0.02
1537	1532	γ-Cuprenene	–	–	–	0.2±0.01
1556	1555	Elemicin	–	–	–	0.2±0.01
1583	1577	Spathulenol	–	–	–	tr
1596	1589	allo-Cedrol	–	–	–	0.2±0.03
1608	1600	Cedrol	–	–	–	0.6±0.05
2034	2035	(Z)-Falcarinol	–	–	–	1.9±0.12
Number of constituents			8	20	12	57
Total identified			100	99.9	99.6	98.4
Monoterpenes hydrocarbon (MH)			100	99.3	99.3	74.5
Monoterpenes oxygenated (MO)			–	–	–	4.8
Sesquiterpenes hydrocarbon (SH)			–	0.4	0.3	9.8
Sesquiterpenes oxygenated (SO)			–	–	–	0.8
Others (O)			–	0.2	–	3.2
Phenylpropanoids (PP)			–	–	–	5.3
RI: experimentally determined retention indices on the mentioned column by co-injection of a homologous series of n-alkanes C8–C32. RA: Adams retention indices. *: identified by NIST Chemistry WebBook Retention indices. tr: trace (<0.05 %). –: not detected.					Headspace samples: HS1 – aerial parts HS2 L– root; Essential oil samples: EO1 – aerial parts EO2 – root	

Similar situation was observed regarding essential oil samples. In the essential oil obtained from the aerial parts, monoterpene hydrocarbons make up over 99% of the sample with (E)-β-ocimene as the main component. Though both of the samples were dominated by (E)-β-ocimene, two main features distinguished essential oil obtained from the root ~ the number of identified constituents (57:12) and the presence of the oxygenated fraction which was not even detected in aerial parts. In the root sample (E)-β-ocimene was present with the contribution of 38.5%, followed by limonene (15.2%) and γ-terpinene (11.7%), unlike the aerial parts where (E)-β-ocimene was over 91.5% of the total. These differences could be attributed to the fact that the essential oils which have been isolated from root and aerial parts of the plant have significantly different chemical composition (Stamenkovic *et al.*, 2020).

Composition of essential oils of the aerial parts of *C. bulbosum* collected from different geographic regions was found to be quite different. Analysis of volatile fraction of *Chaerophyllum bulbosum* ssp. *bulbosum* growing wild in Greece demonstrated the presence of apiol (37%), 3,7,11-trimethyldodeca-1,6,10-trien-3-ol (8.5 %), linalool (7.7 %), myristicine (6.9%) and eugenol (5.8%) (Kokkalou and Stefanou, 1989). Essential oil of *C. bulbosum* from Iran had (E)-β-farnesene (22.3%), (Z)-β-ocimene (18.8%) and myristicin (17.1%) as the major components in this oil. Other notable compounds in the oil were caryophyllene oxide (6.6%), allo-ocimene (5.1%), and (E)-β-ocimene (4%) (Masoudi *et al.*, 2011). In the oil from the epigeal part of *C. bulbosum* grown in Azerbaijan (Mamedova and Akhmedova, 1991) showed linalool (18.3 %) and α-pinene (7.8 %) which were predominant (Stamenkovic *et al.*, 2020).

The differences in chemical composition can be explained by many factors including genetic variation, plant ecotype or variety, plant nutrition, application of fertilizers, geographic location of the plants, climatic conditions, seasonal variations, stress during growth or maturity and the post-harvest drying and storage (Barra, 2009). All these factors influence one another so it is difficult to say with certainty which of these factors have the greatest effect on the chemical composition of essential oils (Stamenkovic *et al.*, 2020).

Taste

The taste of raw tuberous-rooted chervil is somewhat sharp like radish but the cooked root gives a subtle taste of potatoes and chestnuts with a slightly pungent taste of celery and parsnip (MagicGardenSeeds, 2023). When cooked, the root develops a soft consistency similar to texture of a cooked potato and has a sweet flavour reminiscent of chestnuts, carrots and parsnips. Much of the flavour comes from skin of the roots which also contain much of the not insubstantial nutritive content of the vegetable. The aroma intensifies with longer plant storage. The leaves of the plant are also edible and have a taste similar to parsley (Specialtyproduce, 2023; growingtaste.com, 2023).

It is interesting that at harvest time the root is not edible so traditionally when these are stored at 10°C for 4-5 months the stage is reached at which it becomes edible. Storage enables changes in carbohydrate reserves responsible for taste (Polat *et al.*, 2013).

Uses

The chervil roots are eaten raw or cooked and utilized similarly to other root vegetables in culinary applications. In raw use, it can be grated into green salads or utilized as a crunchy garnish but it is most popularly used in cooked applications such as boiling, pureeing, steaming, and roasting. Much of the flavour and aroma is from the skin of the tuber, so it is advised not to peel the chervil root prior to preparing it. Roots should also be thoroughly washed before use to remove any soil or debris. Chervil root can be braised alongside meats or other vegetables, boiled and mashed or pureed. Chervil root pairs well with meats such as poultry, beef, lamb, and fish and other root vegetables such as carrots, parsnips and celeriac (Specialtyproduce, 2023).

C. bulbosum is also reported to be a traditional medicinal plant used by locals in E. Turkey as there fleshy roots are consumed raw to increase appetite, to treat diabetes and to lower high cholesterol levels (Polat *et al.*, 2013).

Cultivation

The cultivation of tuberous-rooted chervil is similar to carrots and parsnips. However, it is somewhat challenging to grow and it surely takes 9-10 months from the sowing to the finished dish.

Propagation

Mainly it is propagated through seeds after harvesting the roots. Seeds having poor viability so only fresh seeds should be used every year. Keeping the seeds in cool and slightly damp sand might help sustain their vitality a bit longer (growingtaste.com, 2023).

Seeds require 42-56 days of cold moist stratification before sprouting. This can be accomplished either under the snow with fall-sown seed or in the more controlled conditions of a sealed container of moist sand in the refrigerator. Newly sprouted seedlings are favourites of rodents so these require to be protected (Dupey, 2010).

Climate and Sowing Season

Tuberous-rooted chervil probably performs quite well in cool maritime climates with long cool and moist spring weather. It is cultivated only on a small scale in Europe for its roots, especially in France (Specialtyproduce, 2023). This requires good exposure to sun and like almost all other roots they are not frost-sensitive (growingtaste.com, 2023). It is sown during autumn. Under optimal conditions these sprout by end of February or early March and then by July the full-sized fleshy roots are formed, after which soil dries up and plants yellow and go dormant (Dupey, 2010).

The seeds require exposure to cold temperatures (vernalization) to break dormancy in spring, which ends with 6-8 weeks of continuous humidity and temperatures below 5°C. After stratification, germination is at optimum with a temperature range between 5 and 10°C. At temperatures above 25 to 30°C, significant germination inhibition occurs (Auge *et al.*, 1988). Direct sowing is done from September to November. Very early sowing in cold spring can also be effected but then the crop may face some losses (MagicGardenSeeds, 2023).

Soil

Tuberous-rooted chervil prefers deep, permeable and continually moist sandy-loam soils rich in humus (MagicGardenSeeds, 2023) with a pH range of 6.5-6.8 or even a little lower (growingtaste.com, 2023).

Sowing

The seeds of tuberous-rooted chervil can be sown in a deep container or in a deep-dug or raised bed. Its seeds are sown quite shallowly and then the soil is kept moist throughout especially till emergence. Chervil seeds can have a low germination rate even under optimum conditions, therefore, it is wise to plant 2 or 3 seeds at each point where a plant is wanted and after emergence these can be thinned if necessary (growingtaste.com, 2023).

Distances apart are akin to carrots as here plant to plant distance is maintaioned at 4-6cm in the rows spaced 20-25cm (MagicGardenSeeds, 2023; Becker-Dillingen, 1950).

Cultural Practices

Chervil seedlings are very weak in competition. Regular hoeing protects from weed and promotes regular ground humidity which this crop prefers. The cultivation should be performed well but carefully and not too deep. In summer heat the plants might require a little shade which can easily be managed through erecting a light framework with latticing to provide sun/shade strips that will move as the sun moves across the sky. If the plant is being grown in containers, they should be placed at the east or west side of the house to reduce sun hours (growingtaste.com, 2023).

Fertilizers and Irrigation

Chervil is light feeder so if sufficient compost is incorporated in the soil at the time of sowing no further nutrient is required except a little quantity of phosphorus afterwards (MagicGardenSeeds, 2023; growingtaste.com, 2023). Regular irrigation is required to keep the soil moist. Similar to other crops, chervil is also grown on the same land every four years to keep the ground healthy (MagicGardenSeeds, 2023).

Harvest

The roots become mature when the foliage becomes yellow in June or July. Harvesting of chervil starts in July and lasts until September. The fleshy roots are either harvested, stored and subjected to periods of cold exposure before being eaten or left in the ground (when there's no danger of voles) to be harvested and eaten in the late fall and winter. Root-crown buds formed in the fall of the first year are a good indication that the roots had undergone enough sweetening and flavour enhancement to be ready for consumption (Dupey, 2010).

The roots develop the best flavour when these are wrapped in moist sand at cool temperatures and stored for another 6-8 weeks. The edibility of the roots becomes effective only at the end of the autumn when root reserves have undergone a biochemical change favoured by cold temperature (MagicGardenSeeds, 2023).

Due to the requirement of a dormancy period, it is almost mandatory to sow the seeds directly after harvest for the upcoming season. Because of this and due to the low yield and poor germination of the seeds chervil has hardly spread in commercial cultivation (Becker-Dillingen, 1950).

Storage

After harvest, the roots will keep up to three months when stored in a cool, dry and at dark place (Specialtyproduce, 2023). During storage, sugar content increases through hydrolysis of starch by amylases (Geoffriau *et al.*, 2005).

Pests and Diseases

Voles are a common pest for this crop in the field. The newly sprouted seedlings are very attractive to rodents and most will be eaten away quickly if not protected timely (Dupey, 2010). Tuberous-rooted chervil may get infected by similar diseases as for carrots, parsnips and as other members of Apiaceae. These are mainly the carrot fly, storage decay and root aphids. Furthermore, it can be host for other aphids, *Erysiphe heracley* or celery mosaic virus (Vogel, 1996) and their control are the same as to those of carrot.

References

Alexeeva, T.V., M.G. Vasil'eva, J.V. Daushkevich and M.G. Pimenov, 1994. IOPB chromosome data 8. *Int. Organ. Pl. Biosyst. Newslett. (Zurich)*, **23**: 11–2.

Auge, R., P. Bourgeais and J.Y. Peron,1988. Germination conditions of bulbous-rooted chervil seeds (*Chaerophyllum bulbosum* L.). In: International Symposium on Diversification of Vegetable Crops, 242, pp. 239-48.

Ayala Garay, O..J., M. Briard, J.Y. Peron and V. Planchot, 2003. *Chaerophyllum bulbosum*: a new vegetable interesting for its root carbohydrate reserves. *Acta Hortic.*, **598**: 227-34.

Barra, A. 2009. *Nat. Prod. Commun.*, **4**: 1147. (https://doi.org/10.1177/1934578X0900400827).

Becker-Dillingen, Joseph, 1950. Handbuch des gesamten Gemüsebaues (ed. Paul Parey), pp. 689–92. https://en.wikipedia.org/wiki/Chaerophyllum_bulbosum. Accessed 10 June 2023.

Daushkevich, J.V., T.V. Alexeeva and M.G. Pimenov, 1995. IOPB chromosome data 10. *Int. Organ Pl. Biosyst. Newslett.* (Zurich), **25**: 7–8.

Dupey, S. 2010. Pursuing the giant root chervil. Accessed on 11 June, 2023. http://www.edimentals.com/blog/?tag=chaerophyllum-bulbosum.

en.wikipedia.org, 2023. https://en.wikipedia.org/wiki/Chaerophyllum_bulbosum. Accessed on 10 June, 2023.

Geoffriau E., A. Suel, M. Briard, J.Y. Peron and O.J.A. Garay, 2005. Evolution of amylase activity in tuberous-rooted chervil (*Chaerophyllum bulbosum* L.) roots during storage at various temperatures. *Acta Hortic.*, No. 682, pp. 1153-8. doi:10.17660/actahortic.2005.682.152.

growingtaste.com, 2023. Chervil Root. Accessed 10 June 2023 via https://growingtaste.com/vegetables/chervilroot.shtml.

Kamari, G., C. Blanche and F. Garbari (eds), 2004. Mediterranean chromosome number reports 14. *Fl. Medit.*, **14**: 423-53. ISSN 1120-4052.

Kokkalou, E. and E. Stefanou, 1989. *Pharm. Acta Helv.*, **64**:133. (https://pubmed.ncbi.nlm.nih.gov/2755958/).

Le Clerc, V., A. Suel, E. Geoffriau, S. Huet and M. Briard, 2014. Evolution temporelle de la diversite genetique de *Chaerophyllum bulbosum*: consequences sur la gestion des ressources genetiques. *Comptes Rendus Biologies*, **337**(5): 352–9. doi:10.1016/j.crvi.2014.03.008. PMID 24841963.

MagicGardenSeeds, 2023. Root Chervil (*Chaerophyllum bulbosum*). Accessed on 10 June, 2023. https://www.magicgardenseeds.com/The-Good-To-Know/Root-Chervil-%28Chaerophyllum-bulbosum%29-A.CHA01-

Mamedova, S.A. and E.R. Akhmedova, 1991. *Chem.. Nat. Compd.*, **27**: 248. (https://doi.org/10.1007/BF00629777).

Masoudi, S.H.., A. Faridchehr, S. Alizadehfard, N. Zabarjadshiraz, F. Chalabian, R. Taghizadfarid and A. Rustaiyan, 2011. *Chem. Nat.. Compd.*, **47**: 829. (https://doi.org/10.1007/s10600-011-0076-1).

NatureGate, 2023. Bulbous Chervil (*Chaerophyllum bulbosum*). Accessed 10 June 2023. https://luontoportti.com/en/t/1844/bulbous-chervil.

Peron, J. and M. Briard, 2003. Breeding Advances in Tuberous-Rooted Chervil (*Chaerophyllum bulbosum* L.). A New 'Old Vegetable' Among the Apiaceae. In: Intern. Symp. Sustain. *Use Pl. Biodiver. Promote New Opport. Hortic. Prod.*, held in 2001, pp. 235-42.

Peron, J.Y. and D. Dubost, 1992. Revalorization of Lost Vegetables : A Contribution to Preservation of Genetic Resources. *Acta Hort.*, No. 318, pp. 263–70. doi:10.17660/actahortic.1992.318.37.

Polat, R., U. Cakilcioglu and F. Satıl, 2013. *J. Ethnopharmacol.*, 148: 951.

POWO, 2023. *Chaerophyllum bulbosum* L. Plants of the World Online. R.B.G., Kew. https://powo.science.kew.org/taxon/urn:lsid:ipni.org:names:840155-1. Accessed on 10 June, 2023.

Reuther, K.. and R. Claben-Bockhoff, 2013. Andromonoecy and developmental plasticity in *Chaerophyllum bulbosum* (Apiaceae–Apioideae). *Ann. Bot.*, **112**(8): 1495-1503.

Specialtyproduce, 2023. Chervil Root. Accessed 10 June 2023. https://www.specialtyproduce.com/produce/Chervil_Root_12880.php.

Stamenkovic, J.G., A.S. Dordevic, G.S. Stojanovic, V.D. Mitic and G.M. Petrovic, 2020. Phytochemical analysis of volatiles and biological activities of *Chaerophyllum bulbosum* L. essential oils. *J. Serb. Chem. Soc.*, **86** (3): 257–67.

USDA, Agricultural Research Service, National Plant Germplasm System, 2023. Germplasm Resources Information Network (GRIN Taxonomy). National Germplasm Resources Laboratory, Beltsville, Maryland. https://npgsweb.ars-grin.gov/gringlobal/taxon/taxonomydetail?id=10020. Accessed 10 June 2023.

Vogel, Georg, 1996. Handbuch des Speziellen Gemüsebaues (Manual of Specialized Vegetable Production) (in German), pp. 1046-9. Hohenheim, Stuttgart, Ulmer. ISBN 978-3800152858. OCLC 247631055.

WFO, 2023: *Chaerophyllum bulbosum* L. Published on the Internet; http://www.worldfloraonline.org/taxon/wfo-0000598957. Accessed on: 10 Jun 2023.

19

Siberian or Russian Chervil (*Chaerophyllum prescottii* DC.)

Family: Apiaceae.

Other common names: Prescott chervil, Cerfeuil de Prescott (Dupey, 2010).

Synonyms: *Chaerophyllum bulbosum* subsp. *prescottii* (DC.) Nyman, *C. rapaceum* var. *prescottii* (DC.) Alef. (POWO, 2023).

Introduction and Origin

Siberian Chervil, often considered as a subspecies, *Chaerophyllum bulbosum* subsp. *prescottii* (DC.) Nyman, is a biennial or perennial plant native to N&E Europe to Mongolia and Caucasus, and introduced to Finland and Sweden (POWO, 2023). Siberian chervil is not to be confused with the better known tuberous-rooted chervil (*Chaerophyllum bulbosum*). While *C. bulbosum* has been improved through cultivation and selection to a larger commercially grown form, *C. prescottii* seems to have been abandoned and long forgotten after an initial period of enthusiastic interest and experimental cultivation in the mid to last half of the 19th century (Dupey, 2010).

Botany

Siberian chervil does not become usually as high as the ordinary tuberous-rooted chervil though its final height is above 1.15m. The lower stems are covered with stiff downward sloping hairs. Glabrous and twice winged leaves are finely divided in fewer divisions but more elongated and narrow, especially at the top of the stem. When compared with *C. bulbosum*, white flowers generally have styles that are less divergent, involucre bracts are longer, fruits larger, solitary (rarely 2 or more) golden yellow fleshy roots are longer with white and delicate flesh. In a good soil in the second year its fleshy roots may attain 2.8cm diameter (Dupey, 2010).

Phytochemicals

Distillation of the fresh fowering tops of *Chaerophyllum prescottii* collected from Altai region of Siberia yielded 0.22% mobile colourless oil. Analysis of the oil by GC, GC/MS for the frst time revealed the presence of 18 substances with the major components consisting of (E)-β-ocimene (35.6%), (Z)-β-ocimene (19.4%), γ-terpinene (18.8%), myrcene (10.6%) and terpinolene (4.6%) which represented 89% of the total oil while 11% of oil comprised of other minor components (Letchamo *et al.*, 2005). Its nutritional value is almost the same as tuberous-rooted chervil.

Cultivation

Though tuberous-rooted chervil is full-sized when fruit is completely ripe but Siberian chervil continues to grow in size for a long time yet. In August when fruits are well mature on the stem, its plants can be sheared off at 10cm height and the ground covered with old half-rotted manure which causes these often forming root-collar buds that can be used to propagate the plant of the first year. We then get tubers that are less bulky but one can use them for planting in the spring if desired. In case grower wants to harvest more and larger fleshy roots, the smallest fleshy roots should be left in the ground even when these are pea-size or even when these are uprooted then these should be replanted later in the autumn. However, one should be careful not to pull out fleshy roots too early as in such cases initially these show some growth but afterwards die. The best thing is to dig up these out of the soil not before the August end after thorough drying of the plants and then these are stored, stratified and then planted during spring (Dupey, 2010).

Frost-treatment is necessary requirement for getting vigorous plants when perenniating these for more than two years. It should be spaced out in rows wider than for the tuberous-rooted chervil. After the autumn sowing, it is recommended to cover the rows with a layer of 7.5cm of old manure. Growth can also be accelerated by occasional sprinkling of guano-water mixture on the plants (Dupey, 2010)

References

1. Dupey, S. 2010. Pursuing the giant root chervil. Accessed on 11 June, 2023. http://www.edimentals.com/blog/?tag=chaerophyllum-bulbosum.

2. Letchamo, Wudeneh, Elena Korolyk and Alexey Tkachev, 2005. Chemical screening of essential oil bearing flora of Siberia. V. Composition of the essential oil of *Chaerophyllum prescottii* DC tops from Altai region. *J. Essent. Oil Res.*, **17**: 560-62. 10.1080/10412905.2005.9698994.

3. POWO, 2023. *Chaerophyllum prescottii* DC. Plants of the World Online. RBG,Kew. https://powo.science.kew.org/taxon/urn:lsid:ipni.org:names:840249-1. Accessed on 10 June, 2023.

Ramkand (*Agave sisalana* Perrine)

Family: Agavaceae

Synonyms: *Agave rigida* var. *sisalana* (Perrine) Engelm. (1875); *Agave sisalana* var. *armata* Trel. (1913); *Agave sisalana* f. *armata* (Trel.) Trel. (1920); *Agave amanuensis* Trel. & Nowell (1933); *Agave segurae* D.Guillot & P.Van der Meer (2005).

Introduction

Common names: Century plant, sisal agave, "yacci", "babkij" (Maya); "sisal" (Italian, English, French, Portuguese, Spanish, German), Hemp sisal. Commonly called as Ramkand/ kandmool/ ketki in India.

Though the so-called kandmool/Ramkand tuber has been marketed in India at a number of places, particularly at holy sites as street snack, its origin is one of the best-kept secrets by its vendors. Due to lack of available research sources, the precise identity of this distinctive street food has not yet been made public. According to the narrative given by the merchants, it appears that Lord Rama consumed the tubers during his exile. During 2011, Nimbalkar and his co-workers used the slice sold in the market for anatomoical study which showed typical monocotyledonous vascular bundle arrangement. But this only added to the confusion, as monocots have adventitious roots and not a tap root system. From one of the vendors from Jyothiba hill temple at Wadi Ratnagiri, Kolhapur District, Maharashtra they purchased tuber slices for DNA barcoding studies. They selected plastid locus for maturase k (*matk*) for identification of the plant species. The plastid and mitochondrial DNA have been viewed as the most appropriate regions to sequence for species identification in plants and animals respectively (Chase *et al.*, 2005). The similarity search showed 89% identity with the partial sequence of the plastid locus maturase of *Agave sisalana* (Nimbalkar *et al.*, 2011). Further to confirm the identification, plants of Agave were visited and their leaves enclosing the rosette and juvenile inflorescence were excised, which exposed the core of the rosette which was found soft and of similar dimension to that of the Ramkand being sold (Nimbalkar *et al.*, 2011). However, DNA test is yet to confirm the identity of this plant. Though there is a possibility that other species of Agave being processed and used for the same purpose, but as per present records, this may be treated as *Agave sisalana*.

The name of the genus comes from the Greek "agaué, feminine of "agauos" = noble, splendid; the name of the species comes from Sisal, Yucatan seaport from where the fibre obtained from its leaves was exported. Sisal plants have a lifespan of 7–10 years, producing 200–250 usable leaves containing fibers used in various applications. Sisal farming initially led to environmental degradation, but it is now considered less damaging than other farming types. It is an invasive species in Hawaii and Florida.

Uses: Sisal is mainly cultivated for fibre production from its leaves. These fibres have high tensile intensity and tenacity, alkali and acid resistance, abrasion, corrosion and seawater resistance. It is also used in making low-cost and specialty paper, dartboards, buffing cloth, filters, geotextiles, mattresses, carpets, handicrafts and wire rope cores. The fibre element, which accounts for only about 4% of the plant by weight, is extracted by a process known as decortications and the remaining material is commonly discarded by sisal farms (Bandeira and Silva, 2006; Botura *et al.*, 2013). After fibre extraction, the leaf pulp may be used as manure in the field. Waste material is occasionally fermented to produce alcohol or biofuel. Sisal can also be grown as a fence plant, a field delineator, and a soil stabilizer. On the trash piles of sisal manufacturers in Tanzania, a number of palatable mushrooms, most notably *Coprinus cinereus*, proliferate profusely and offer a year-round source of food. Flowers, leaves, stalks, basal rosettes, sap and root of agave are edible (Davidson, 2006; Deane, 2012). It is also used as an ornamental. In India, large cylindrical roots are used as street snack seen in road sides of Karnataka, Kerala and Maharashtra (Kumari, 2021). Roots possess diaphoretic and diuretic, syphyllis medicinal properties (Chopra *et al.*, 1986,

Bown, 1995). In traditional medicine in East Africa, a root decoction is drunk as a diaphoretic (https://uses.plantnet-project.org/en/Agave_sisalana) while it was used for the treatment of blackleg by Zay people in Ethiopia (Giday, 2001).

Phytochemical Properties in Agave

The occurrence of steroidal saponins in genus *Agave* is well documented (Blunden *et al.*, 1980, 1986; Ding *et al.*, 1989, 1993). Sterols, steroidal sapogenins, steroidal alkaloids and alkaloidal amines derived from *Agave sisalana* provide the starting material for steroid production. In China, licogenin and tigogenin from *Agave* species are used as industrial precursors for the partial synthesis of steroidal drugs (Chen *et al.*, 1994). Hecogenin from sisal leaves has been used as a precursor in the partial synthesis of corticosteroids such as cortisone, hydrocortisone and prednisone. This has been found in the leaves of several other *Agave* spp. and is a possible source of the production of cortisone, but cultivars selected particularly for sisal fibre show reduced hecogenin. Hecogenin is most abundant (0.235%) in the leaves of the old plant.

Diversity and Distribution

There are over 275 species of *Agave*, most of which are found in dry and semi-arid regions stretching from the south-west of the United States through western Panama, the Caribbean, and Venezuela. The genus has been divided into numerous sections and subgenera due to its extremely complex taxonomy. Different opinions do, however, exist about the classification of *Agave*. Except for *Agave sisalana* which is placed in section Sisalanae, all of the long-fibre cultivar species belong to the subgenus *Euagave*.

Though the native origin of *Agave sisalana* is uncertain, it is known to be probably of southern Mexican origin, but wild forms are not known. Traditionally it was deemed to be a native of the Yucatán Peninsula, but there are no records of botanical collections from there. They were originally shipped from the Spanish colonial port of Sisal in Yucatán (thus the name). *Agave sisalana*, is probably of hybrid origin, with *Agave angustifolia* Haw. and *Agave kewensis* Jacobi as possible parents (https://uses.plantnet-project.org/en/Agave_sisalana_(PROTA). In the 1960s in Kenya a chance selection was made from *Agave sisalana* which was taken as cv. 'Hildana'. It is characterized by a large habit, many but shorter leaves and delayed flowering. It is grown in East Africa. Most other long-fibre *Agave* cultivars also derive at least partly from *Agave angustifolia*, a species classified in the section 'Rigidae'.

At present it is grown mainly in Florida, the Caribbean islands, Brazil and Africa (notably Tanzania and Kenya) though naturalized in other parts of Mexico, Spain, Libya, Morocco, the Canary Islands, Cape Verde, many parts of Africa, Madagascar, Réunion, Seychelles, China, the Ryukyu Islands, India, Pakistan, Nepal, Burma, Cambodia, Thailand, the Solomon Islands, Queensland, Polynesia, Micronesia, Fiji, Hawaii, Florida, Central America, Ecuador, and the West Indies. In India, sisal is found in many regions of India including Orissa, Andhra Pradesh, Madhya Pradesh, Bihar, Jharkhand, Chhattisgarh, Maharashtra, Tamilnadu and Karnataka.

Genetics: Cytological studies revealed that the basic chromosome number in *Agave* is 30 and the 2n chromosome numbers are 90 for *A. tequilina*, 120 for *A. americana*, 150 for *A. sisalana* and 180 for *A. decipiens* (Banerjee and Sharma, 1989).

Climate: Sisal is a tropical and subtropical plant, thriving in temperatures above 25°C and sunshine. *Agave sisalana* is a hardy plant that displays exceptional drought and temperature tolerance. It grows well all year round in hot and extremely dry climate. They have the inbuilt ability to survive more than one season without rainfall and can tolerate extremely hot and low temperatures ranging from −16.1°C to 61.4°C (Nobel *et al.,* 1983). The crassulacean acid metabolism (CAM) mechanism makes them possible to utilize water 4–2x more efficiently (Gross *et al.*, 2014).

Botany: Monocarpic perennial succulent, stem short, covered by leaf bases. Leaves 1-2m long, robust, monocarpic, 3–9m tall when flowering, with numerous leaves crowded in a dense rosette; roots fibrous, originating from the base of the leaf scars at the bottom of the stem, spreading horizontally up to 3(–5)m, and vertically up to 150cm, but concentrated in the upper 30–40cm of the soil; stem short and thick, 120cm × 20cm, with an apical meristem; rhizomes arising from buds in the axils of leaves below ground level, numbering 5–10 at one time and about 20 in the total life span, 1.5–3cm in diameter, growing to about 2m in length before surfacing and producing suckers. Leaves succulent, arranged in an ascending spiral; stipules absent; petiole absent; blade linear-lanceolate, 75–185cm × 10–15cm × 2–4.5cm, base fleshy, bulbous, triangular in cross section, blade gradually broadening to the middle and narrowing into a terminal sharp, lignified, dark brown spine up to 3cm long, concave above and convex below, margin usually spineless, surface dark green but covered with a white waxy layer. Inflorescence a panicle on a long peduncle, 2–8m tall, branches widely spreading, 30–100cm × 2cm, apically 5–6 times branched trichotomously, bearing about 40 flowers per branch. Flowers erect, protandrous; pedicel short; perianth tubular, 6-lobed, 5–6cm long, pale green, tube 1–2cm long, lobes oblong, on inner side of the top with a tuft of hairs; stamens 6, attached above the middle of the perianth tube, accrescent during anthesis, finally 6–8cm long;

ovary inferior, 3-celled, style much accrescent during anthesis and finally 6–8cm long, stigma 3-lobed. Fruit (rarely produced) an ellipsoid capsule, tapering at base, green and fleshy when young and black and dry when ripe, with about 150 seeds. Seeds rounded-triangular, thin, flat, papery, black. Bulbils copiously produced on the inflorescence branches, consisting of a meristem, 6–8 reduced leaves and a rudimentary stem with rudimentary adventitious roots. 8cm wide, widest about the middle, erect, neck not sharply constricted, greyish green, often glabrous, channelled to about 2/3 rd its length, the basal 1/3 rd flat above, marginal spines lacking. Inflorescence a large panicle. Flowers arranged on the terminal portion in dense clusters, sessile, 4-5cm long. Perianth segments six, spreading, equalling the ovary segment. Stamens 6, filaments longer than the perianth segment, 3-4cm long anthers versatile. Style exserted, stigma 3-lobed. The plant blooms when 8-10 years old, producing a terminal panicle inflorescence, tall up to about 5m, with greyish yellow flowers, tubular, about 6cm long and with a rather unpleasant smell. The inflorescences of different *Agave* species are one of the largest in the plant kingdom, which may reach upto 10 meters above ground (Abarca *et al.*, 2013). Plants are usually sterile and die after blooming once. Small plantlets (bulbils) form in inflorescence that fall to ground after flowering and form new plants.

Inflorescence of *Agave sisalalna* source: https://en.wikipedia.org/wiki/Sisal

Flower of *Agave sisalalna* source: https://en.wikipedia.org/wiki/Sisal

The mysterious street snack that has baffled botanists for decades, source: Kumari, 2011

a. Ramkand; b. excised rosette of *Agave sisalana*, source: Nimbalkar et al., 2011

Propagation and Planting

Sisal is vegetatively propagated through bulbils or suckers. Although suckers can be picked right from the field, bulbils are frequently preferred since they are generated in greater quantities, which allows for selection and results in a more consistent yield. After they have fallen, bulbils can be collected off the ground, or the pole can be cut and the bulbils shook into sacks. Bulbils that are at least 10cm long are planted in nurseries with a 50cm × 25cm spacing and a 1.3cm depth. The growth of plants is aided by the application of sisal waste in nurseries. The seedlings are ready to be planted in the field after 12 to 18 months. The fibrous roots that surround the base of plantlets are typically cut off during transplantation, and the lowest leaves may be plucked off. The soil is removed mechanically or manually, and it may be shallowly ploughed, prior to planting on the field. Depending on the soil and climate, 4000-6000 plants/ha is the ideal density. Sowing of legume cover crops such as *Calopogonium mucunoides*, *Centrosema pubescens* and *Pueraria phaseoloides* is recommended. On fertile soils or with proper fertilization, young sisal may be intercropped with maize, beans or cotton without any adverse effect unless the other crops are planted very close to the sisal rows. Rotation is not necessary if sisal processing waste is returned to the field though it is usually not practised in East Africa.

Crop Management

Fertilizer recommendations depend on soil characteristics and cropping history. Lime application is recommended in highly acidic soils. 'Purple leaf tip', in which the leaf tip becomes reddish-purple and the leaf margins curve upwards, is associated with exhausted acid soils and a shortage of calcium, but other factors may also be involved. Potassium deficiency causes 'banding disease', characterized by 10-15cm wide horizontal bands of purplish-brown necrotic tissue, especially at the transition between the leaf base and the leaf blade, resulting in wilting and bending over of the leaf blade. Nitrogen application tends to shorten the crop cycle, but the total number of leaves is not affected. Care should be taken in using ammonium sulphate which may increase soil acidity. The nutrient removal per tonne of fibre is about 27–33 kg N, 5–7 kg P, 59–80 kg K, 42–70 kg Ca and 34–40 kg Mg, but the majority of the removed nutrients can be returned to the field with the waste material after fibre extraction (https://uses.plantnet-project.org/en/). Weeds should be controlled in the first 2–3 years after transplanting either by hand, or through mechanical or chemical means. Important weeds in sisal plantations include: couch grass (*Cynodon dactylon*), nut grass (*Cyperus* spp.), African couch (*Digitaria abyssinica*), lalang (*Imperata cylindrica*), cow-itch (*Mucuna pruriens*) and Guinea grass (*Panicum maximum*). Suckers should be removed and used for propagation.

Diseases and Pests

The most serious disease of sisal is **bole rot** caused by *Aspergillus niger* which enters through the leaf bases when leaves are cut. It causes a wet rot which becomes yellowish-brown and soft, with a pinkish margin, and may lead to death of plant. The incidence can be reduced through removal of infected material and harvesting under dry conditions. **Zebra disease** caused by *Phytophthora nicotianae* makes striped lesions on the leaves and may also cause bole rot. It mainly occurs on poorly drained soils. **Korogwe leaf spot** is a viral disease occurring in Tanzania which gives very poor grade fibre and often renders the crop unusable.

Agave weevil or Mexican sisal weevil (*Scyphophorus acupunctatus*) is the only serious insect-pest of sisal. The larvae damage the subterranean parts of young plants and may cause substantial losses. They also feed on leaves in the central bud, giving a shot-hole effect, whereas the adult weevil damages the crop by feeding on the youngest leaves before and shortly after unfurling. To control the disease, planting should be done before or in the early rains and insecticides may be applied around young plants.

References

Abarca, N. A., E.A. D. Alvarado, J. A. Reyes, J.N.U. Soto and L.S.G.Valdez, 2013. The phenols of the Genus *Agave* (Agavaceae). *J. Biomater Nanobiotechnol.*, **4**:9-16

Bandeira D.A. and O.R.R.F. Silva, 2006. *Aproveitamento de resíduos.* In: Andrade, W. (Ed.), O sisal do Brasil. Sindifibras, Salvador, pp. 56–61.

Banerjee, S. and A. K. Sharma, 1989. Structure and behaviour of chromosomes in four different species of agave. *Cytologia,* **54** (4): 667-72.

Blunden, G., A. Carabot and K. Jewers, 1980. Steroidal sapogenins from leaves of some species of *Agave* and *Furcraea*. *Phytochem.,* **9**:2489-90.

Blunden G., A. V. Patel and T. A. Crabb, 1986. Barbourgenin, a new steroidal sapogenin from *Agave sisalana* leaves. *J. Nat. Prod.,* **49**: 687-689.

Botura, M. B., J.D.G. Santos, G.D. Silva, H.G. Lima, J.V.A. Oliveria, M.A.O. Almeida, M.J.M. Batatinha and A. Branco, 2013. *In vitro* ovicidal and larvicidal activity of *Agave sisalana* Perr. (sisal) on gastrointestinal nematodes of goats. *Vet. Parasitol.,* **192**: 211–7.

Bown D., 1995. Encyclopaedia of Herbs and their Uses. Dorling Kindersley, London.

Chase, M. W., N. Salamin, M. Wilkinson, J. M. Dunwell, R. P. Kesanakurthi, N. Haider and V. Savolainen, 2005. *Philos. Trans. R. Soc. London,* Ser. B, **360**, 1889–95.

Chen, Y. and Y. Wu. 1994. Progress in Research and Manufacturing of Steroidal Sapogenins in China. *J. Herbs. Spices & Med. Plants,* **2**(3): 59-70

Chopra, R. N., S. L. Nayar and I.C. Chopra, 1986. Glossary of Indian Medicinal Plants, Council of Scientific and Industrial Research, New Delhi.

Dan S. (April 21, 2009). "Hoping Mezcal Can Turn the Worm". The New York Times.

Davidson, A. 2006. The Oxford Companion to Food (2nd ed.). Oxford University Press, Oxford, UK, 907p.

Deane, G. Agave century plant. http://www.eattheweeds.com/century plant/ Accessed 10 Sep. 2023.

Ding, Y., R.H. Tian, C.R. Yang, Y.Y. Chen and T. Nohara, 1993. Two new steroidal saponin from dried fermented residues of leaf juice of *Agave sisalana. Chem. Pharm. Bull.,* **41**: 557-60.

Ding Y., Y.Y. Chen, D. Z. Wang and C.R. Yang, 1989. Steroidal saponin from a cultivated form of *Agave sisalana*. *Phytochem.,* **28**: 2787-91.

Fazli, F.R., 1968. Contraceptives and other steroid drugs: their production from steroidal sapogenins. *Pak. J. Sci.,* **20**(1, 2); 64-7.

Giday. M. 2001. An ethnobotanical study of medicinal plants used by Zay people in Ethiopia.- *CBM Skrifserie* **3**:81-99.

Gross, S. M., J.A. Martin, J. Simpson, M. J. Abraham-Juarez, Z.Wang and A. Visel, 2013. *De novo* transcriptome assembly of drought tolerant CAM plants, *Agave deserti* and *Agave tequilana*. *BMC Genomics,* **14**: 563.

https://en.wikipedia.org/wiki/Sisal (retrieved on Aug 10, 2023)

https://uses.plantnet-project.org/en/Agave_sisalana_(PROTA) (retrieved on Aug 10, 2023)

https://www.monaconatureencyclopedia.com/agave-sisalana (retrieved on Sep 10, 2023).

Kumari Barkha (Aug 5, 2021), https://www.atlasobscura.com/articles retrieved on Sep 1st , 2023

Nimbalkar, M. S., P. R. Kshirsagar, V. B. Shimpale, N. V. Pawar and G.B. Dixit, 2011. DNA barcoding for identification of the enigmatic plant Ramkand. *Curr. Sci.,* **100**(9): 1277.

Nobel, P. S. and S. D. Smith, 1983. High and low temperature tolerances and their relationships to distribution of agaves. *Plant, Cell & Environ.,* **6**: 711–19.

Spensley, J.C., C.M. Wathes, N. K. Waran and J.A. Lines, 1995. Behavioral and physiological responses of piglets to naturally occurring sounds. *Appl. Anim. Bahav. Sci.,* **44**: 277.

Colour Plates

pg. 1

pg. 24

pg. 24

pg. 46

pg. 46

pg. 47

pg. 48

pg. 50 pg. 51

pg. 66

pg. 67 pg. 68 pg. 88

pg. 95 pg. 112 pg. 112

pg. 112 pg. 115 pg. 117

pg. 119 pg. 119 pg. 121

pg. 122 pg. 151

pg. 164 pg. 166 pg. 189

pg. 189

pg. 189

pg. 192

pg. 192

pg. 194

pg. 204

pg. 222

pg. 223 pg. 224 pg. 240

pg. 241 pg. 260 pg. 260

pg. 261 pg. 264 pg. 269

pg. 270 pg. 275 pg. 276

pg. 277 pg. 284 pg. 285

pg. 294 pg. 295 pg. 295

pg. 300 pg. 310 pg. 311

pg. 329 pg. 343

pg. 348

pg. 358

pg. 359

pg. 366

pg. 366

pg. 370

pg. 370

pg. 371 pg. 376 pg. 382

pg. 382 pg. 387

pg. 388 pg. 407

pg. 408

Index – Common and Regional Names

D

M

Q

R

S

T

W

X

Y

Z